Dynamical Properties of Solids

Volume 3

Dynamical Properties of Solids

Volume 3

Metals, Superconductors, Magnetic Materials, Liquids

edited by

G. K. Horton
Rutgers University
New Brunswick, USA

A. A. Maradudin
University of California
Irvine, USA

1980

North-Holland Publishing Company
Amsterdam · New York · Oxford

ISBN 0 444 85314 6

Publishers: North-Holland Publishing Company
Amsterdam · New York · Oxford

Sole distributors for the U.S.A. and Canada:
Elsevier North-Holland, Inc.
52 Vanderbilt Avenue, New York, N.Y. 10017

Printed in The Netherlands

Preface

In the first two volumes of this series the foundations of the contemporary theory of lattice dynamics were presented and applications of the fundamentals to particular subareas of lattice dynamics were discussed. Both fundamental aspects of the subject, and applications are presented in this volume.

The microscopic theory of the lattice dynamics of non-transition metals, covalent crystals, and ionic solids was presented in the first volume of this series. In the present volume the theory of the lattice dynamics of transition metals is presented by Sinha, and his chapter completes the presentation of microscopic theories of the lattice dynamics of all the most important types of crystalline solids.

The phenomenon of superconductivity in most metals owes its existence to the electron–phonon interaction. The role of phonons in establishing the superconducting transition and in determining the superconducting transition temperature is reviewed by Allen.

The theory of anharmonic effects in crystals has been developed quite extensively in the first volume of this series. In this volume Fleury examines an important consequence of strong anharmonic interactions in solids and liquids: the formation of collective pair excitations. The richness of the types of excitations that can be formed is probably not yet exhausted and this is a topic that will be the object of further theoretical and experimental work in the years to come.

Phonons can couple to other elementary excitations in solids as well, in particular to magnetic excitations. The rich variety of physical phenomena that can arise from this interaction in paramagnetic and magnetically ordered crystals, the underlying theory, and the relevant experimental work is the subject of the chapter by Lüthi.

We wish to thank our friends from the North-Holland Publishing Company, Dr. W. H. Wimmers, Dr. P. S. H. Bolman, Dr. William

Montgomery, and Mrs. J. Kuurman for their cooperation and supervision of the publishing process.

It is also a pleasure to thank Mrs. Kathryn L. Roberts for her help in the preparation of the Subject Index for this volume.

January 1980

G. K. Horton
New Brunswick, N. J.
USA

A. A. Maradudin
Irvine, California
USA

Contents

Volume 3

List of Contributors

P. B. Allen, Dept. of Physics, State University of New York, Stony Brook, New York 11794, USA

P. A. Fleury, Bell Laboratories, Murray Hill, New Jersey 07974, USA

B. Lüthi, Physikalisches Institut der Universität Frankfurt, Frankfurt, F.R. Germany

S. K. Sinha, Argonne National Laboratory, Argonne, Illinois 60439, USA

CHAPTER 1

Phonons in Transition Metals*

S. K. SINHA

Solid State Science Division
Argonne National Laboratory
Argonne, Illinois 60439, USA

*Work performed under the auspices of the US Department of Energy.

Dynamical Properties of Solids, edited by
G. K. Horton and A. A. Maradudin

Contents

1. *Introduction*

Transition metals and their alloys and compounds provide one of the most fascinating areas in the study of phonons in crystal lattices. This is because, in addition to the richness and variety of structure of their phonon dispersion curves, these materials also often exhibit the phenomena of lattice instabilities and relatively high-temperature (8 K–23 K) superconductivity, in which phonons play a vital role. Most of the information on the phonon spectra of these materials has come in recent years from the results of inelastic neutron scattering experiments, and in some cases, also from ultrasonic measurements of elastic constants, from heat capacity measurements and from superconducting junction tunnelling experiments.

Fig. 1 shows the phonon dispersion curves along the symmetry directions of Nb. Historically, these measurements (Nakagawa and Woods 1965, Powell et al. 1968) provided the first direct evidence of the existence of structure or "anomalies" in the dispersion curves of a transition metal and indicated the presence, in these metals, of more complicated long-range forces than are found in the "simple" free-electron-like metals. These anomalies are primarily due to softening of the phonon frequencies of particular branches in particular regions of reciprocal space below their "normal" behavior, and have since come to be recognized as incipient lattice instabilities. (In a naive picture, such a lattice instability would occur in the frequency of a particular mode actually softened all the way to zero.)

Further experimental work has shown that the presence of anomalies in the phonon spectra of transition metals, alloys and compounds is fairly common. The identification of these anomalies as incipient lattice instabilities has been borne out by the fact that many of these materials *are* actually unstable, i.e. at certain temperatures or alloy concentrations they undergo actual crystallographic phase transformations. In addition, there are some intriguing correlations between the existence of these anomalies or instabilities, the density of d-like states around the Fermi level and the occurrence of relatively high superconducting transition temperatures.

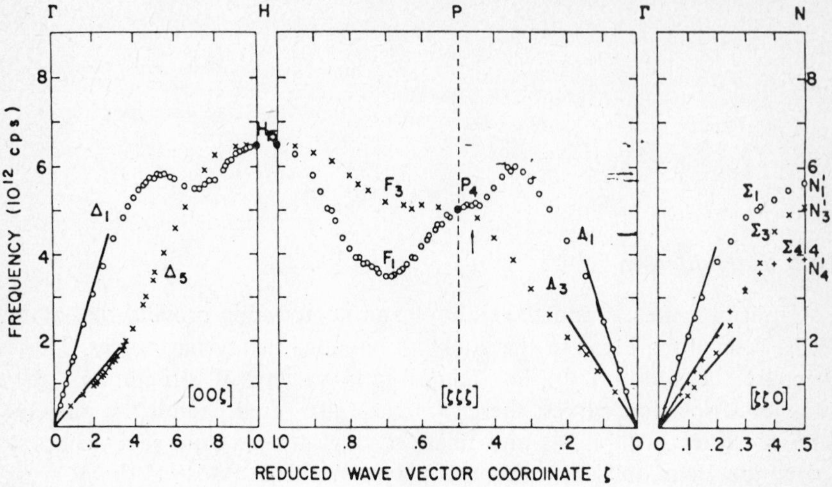

Fig. 1. Dispersion curves for Nb from Powell et al. (1968). The straight lines are the velocity of sound lines calculated from the measured elastic constants.

Thus, for instance, alloying the element Zr into Nb produces eventually (at roughly 80% Zr concentration) an instability from the body centered cubic structure to the so-called "ω-phase" (hexagonal) structure, which may be regarded as arising from an instability with regard to the $(2\pi/a)(\frac{2}{3},\frac{2}{3},\frac{2}{3})$ longitudinal acoustic mode of the bcc structure (de Fontaine et al. 1971, de Fontaine and Buck 1973). This transition appears to take place in a continuous fashion with critical overdamping of the "soft" phonon and an evolution of a "central" peak at the superlattice sites corresponding to the ω-phase structure (Moss et al. 1973, 1975), as shown in fig. 2. Note that alloying Zr into Nb decreases the electron/atom ratio and (at least in the rigid band picture) increases the density of states $n(E_F)$ at the Fermi level, as may be seen from fig. 3. The superconducting transition temperature T_c is also increased. Increasing the electron/atom ratio by alloying Mo into Nb tends to stabilize the lattice and both $n(E_F)$ and T_c go down. The anomalies present in the phonon spectra of pure Nb continuously disappear (Powell et al. 1968) and are instead replaced by an anomaly at the point $(2\pi/a)(1,0,0)$, as may be seen from fig. 4 where the dispersion curves of pure Mo (Woods and Chen 1964, Powell et al. 1968) are shown.

The transition metal compounds of the rocksalt structure also show the same phenomenon. Thus the dispersion curves of NbC and TaC (which have superconducting transition temperatures of 11.5 and 10.35 K respectively) show anomalies in certain longitudinal acoustic (LA) and transverse

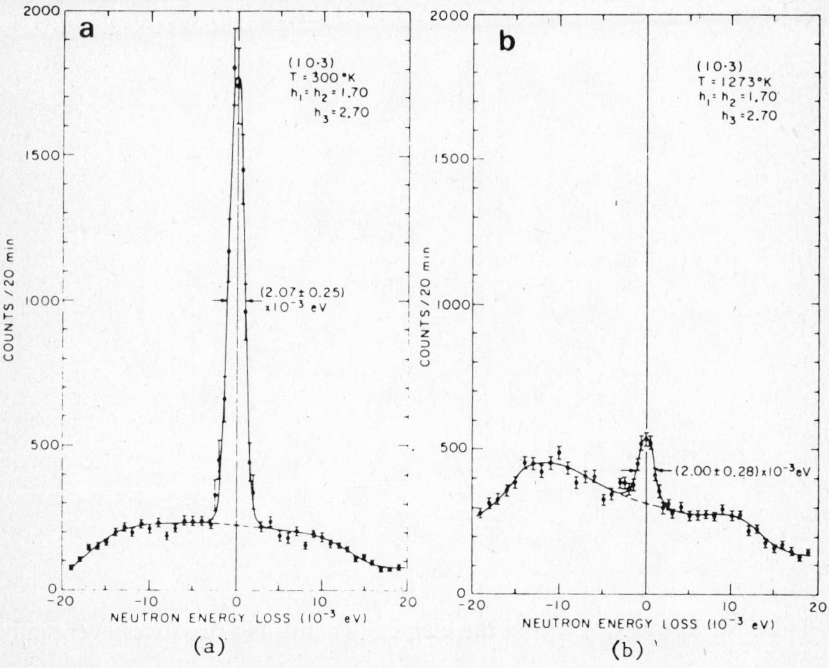

Fig. 2. Inelastic neutron scattering spectrum from $Zr_{0.8}Nb_{0.2}$ at $T = 1273$ K and $T = 300$ K at the bcc reciprocal lattice position (1.7, 1.7, 2.7) $(2\pi/a)$ showing the overdamped phonon "wings" and the appearance of a central component which grows as the temperature is decreased. From Moss et al. (1973).

acoustic (TA) modes (Smith and Gläser 1970, 1971, Smith 1972, Smith et al. 1976) whereas the dispersion curves of ZrC and HfC (with superconducting transition temperatures of < 0.05 K) show none (see fig. 5). Recently, similar anomalies have been found in the LA modes of YS, which has the same structure (Roedhammer et al. 1978). Fig. 6 shows the electronic density of states for NbC (Gupta and Freeman 1976c) with the position of the Fermi level indicated. Note that for ZrC, with one electron/atom less, the rigid band model would put E_F at the minimum in the density of states. On the other hand, the dispersion curves of the rocksalt structure uranium compounds UC and UN show no such anomalies in the acoustic branches (Smith and Gläser 1971, Dolling et al. 1978), although the behavior of some of the optic modes is anomalous.

The highest T_c's currently known are possessed by the A-15 structure family of compounds of the form A_3B, where A is usually Nb or V and B is Si, Ge, Sn, Ga or Al. Owing to the lack of availability of single crystals of sufficient size, detailed measurements of the phonon dispersion curves have hitherto not been carried out by inelastic neutron scattering, although

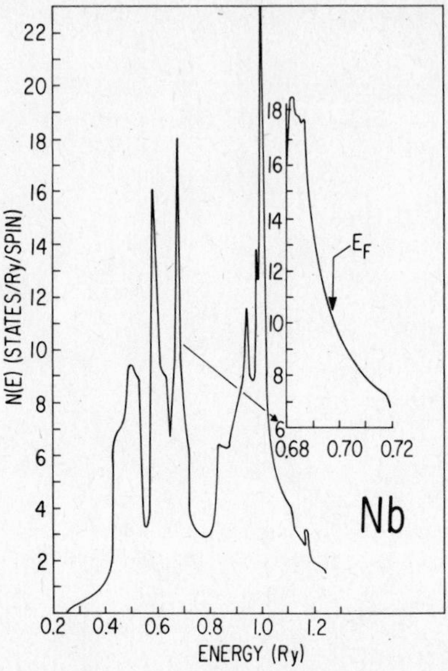

Fig. 3. Density of states for Nb as calculated by Elyashar and Koelling (1977). The position of the Fermi energy is indicated in the inset denoted by the arrow.

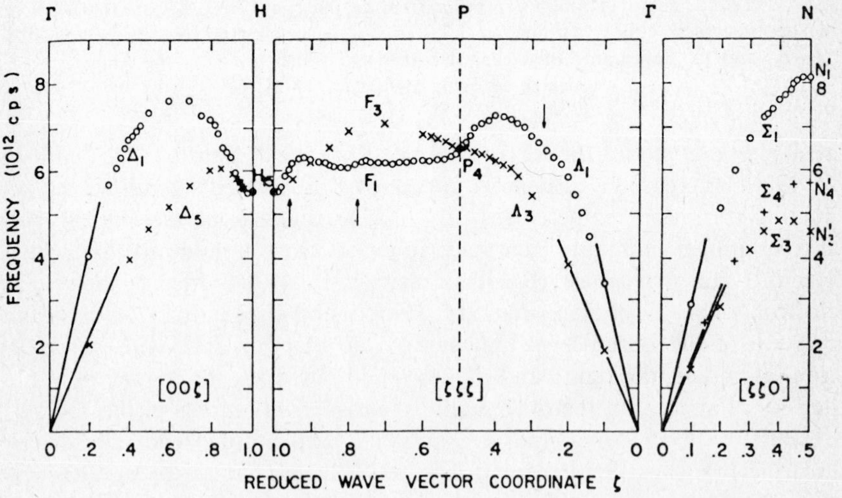

Fig. 4. Dispersion curves for molybdenum from Powell et al. (1968). The straight lines are the velocity of sound lines calculated from the measured elastic constants.

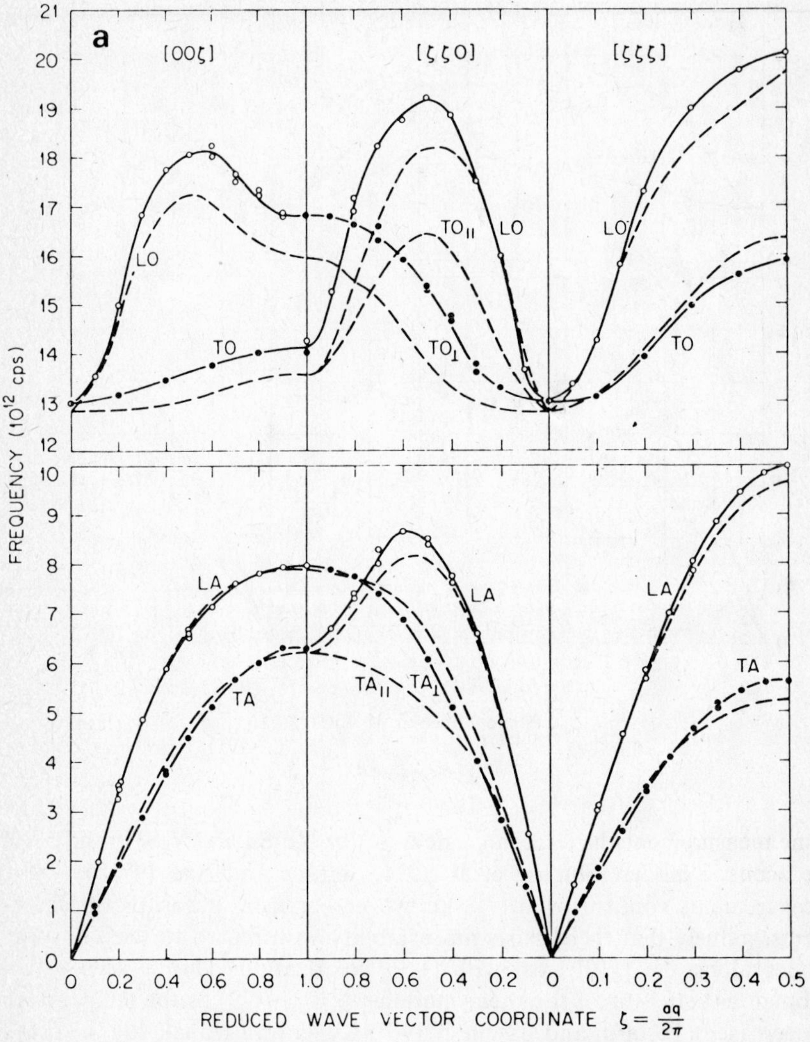

Fig. 5. (a) Dispersion curves for ZrC at room temperature from Smith et al. (1976). The dashed curves represent a "screened-shell model" fit by Weber (1973). (b) Dispersion curves for TaC. The full (open) symbols are the experimental points at room temperature (4.2 K). The full curves represent a "double-shell model" fit of Weber (1973) as discussed in §5. From Smith et al. (1976).

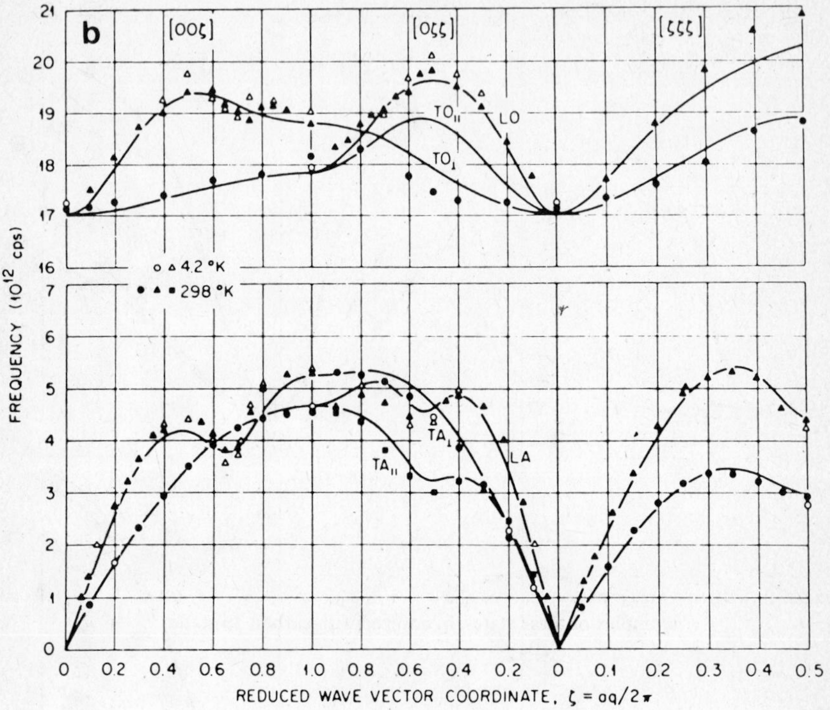

Phonon Spectra of TaC.

Fig. 5. (cont.)

some measurements have been carried out for Nb_3Sn and V_3Si for a few of the acoustic modes (Shirane et al. 1971, Shirane and Axe 1971b). These measurements confirm what was known earlier from ultrasonic measurements, namely that there exists an instability with regard to the TA shear modes along the [110] axis, with $[1\bar{1}0]$ polarization. This is seen as an anomalous softening of the shear modulus $\frac{1}{2}(C_{11} - C_{12})$ as the temperature is decreased (Testardi and Bateman 1967, Keller and Hanak 1967). At the phase transition, there is a crystallographic phase transformation to a tetragonally distorted structure, as was also known earlier from X-ray diffraction experiments (Mailfert et al. 1967). The neutron measurements show that the mode softening is not restricted to just the long-wavelength region but in fact persists for most of the measured q-values. Schweiss et al. (1976) have used the technique of inelastic neutron scattering from powdered samples to measure the "neutron-weighted" density of phonon states in Nb_3Sn, V_3Si, V_3Ge, V_3Ga and Nb_3Al. Their results show that there is an appreciable softening of the lower-frequency part of the phonon spectrum as the temperature is lowered in the high-T_c materials such as

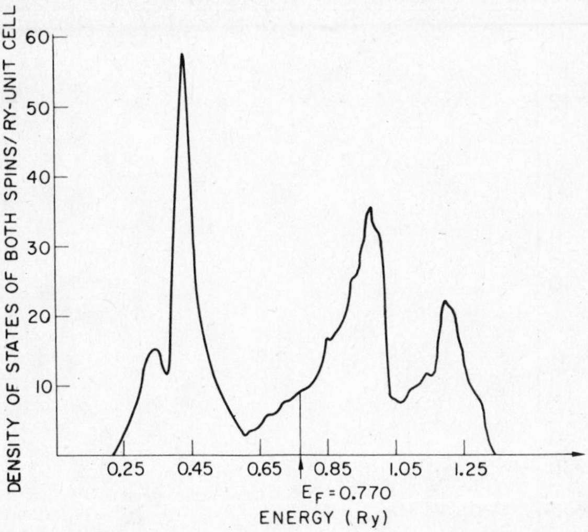

Fig. 6. Electronic density of states for NbC, from Gupta and Freeman (1976b). The double peak below E_F is due mainly to carbon 2p-states (with some admixture of Nb 4d-states), while the higher energy peaks are due mainly to Nb 4d-states. E_F lies within a "t_{2g}" band complex of these states, hybridized with carbon 2p-states.

Nb_3Sn and V_3Si, implying that a large number of modes must be softening anomalously. Knapp et al. (1975, 1976) also found from a careful analysis of heat capacity data on the A-15 compounds in terms of moments of the frequency spectrum, that V_3Si, V_3Ga and Nb_3Sn showed a strong decrease of the geometric mean frequency with temperature, while low-temperature superconducting compounds such as Nb_3Sb showed the more normal mode *stiffening* at lower temperatures. These authors also showed an interesting approximately linear correlation of the fractional frequency shift per degree with the density of states at the Fermi level for the A-15 compounds. Owing to the lack of detailed inelastic neutron measurements on single crystals, it is not yet known whether further specific anomalies or instabilities occur at shorter wavelengths in the acoustic modes or in optic modes in these materials, or whether the softening takes place more or less uniformly. Recently, Wipf et al. (1978) have observed the Γ_{25} optic spin-center made by Raman scattering. They observe only slight softening as the temperature is lowered, but pronounced line-broadening effects which they ascribe to the electron–phonon interaction. Shirane and Axe (1971a) have shown by neutron diffraction that the martensitic phase transformation in Nb_3Sn is accompanied by a sublattice distortion corresponding to a dimerization of Nb-atom chains, so undoubtedly some optic modes are involved in the instability at large wavelengths.

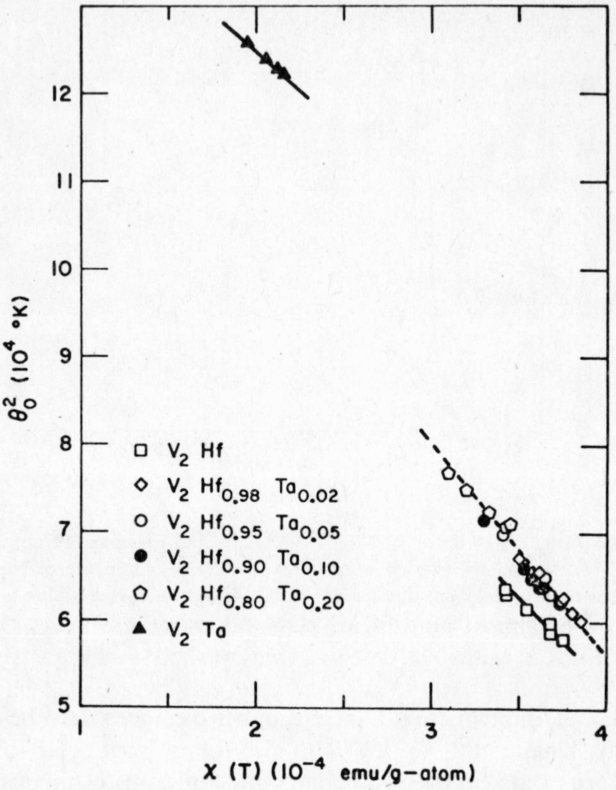

Fig. 7. Plot of Θ_0^2, i.e. $\chi_g^2(T)$ versus $\chi(T)$ with temperature as an implicit parameter, from Hafstrom et al. (1978). Data are plotted for the temperature range 100–300 K which includes the extremes in $\omega_g(T)$ and $\chi(T)$, and, in two cases, the lattice transformation temperature.

More recently, Hafstrom et al. (1978) have shown that the same effect exists in the C-15 superconductors $V_2Hf_{1-x}Ta_x$. For $x<0.1$, they find evidence from heat capacity data of a structural transformation, and for these compounds find a decrease of the geometric mean phonon frequency $\omega_g(T)$ as the temperature is decreased towards the phase transition temperature, mirroring an increase in the magnetic susceptibility $\chi(T)$ of these materials which of course is related to the thermally smeared-out electronic density of states at the Fermi level. In fact, they find a universal correlation between $\omega_g(T)$ and $\chi(T)$ for all these compounds (see fig. 7).

Another group of important transition metal compounds exhibiting phonon anomalies and lattice instabilities are the so-called "charge density wave" group of layered dichalcogenides (for a review, see for example,

Wilson et al. 1975). Typical such materials are $NbSe_2$ and $TaSe_2$. At temperatures below room temperature, electron diffraction patterns revealed the existence of satellites (Wilson et al. 1974) which were ascribed to "charge density waves" with periodicities incommensurate with the lattice periodicity, although close to three times the lattice periodicity along the *a*-axis. Neutron diffraction investigations (Moncton et al. 1975) revealed that the satellites were associated with periodic lattice distortions (PLD), i.e. the phase transitions were really structural phase transformations. Inelastic neutron scattering measurements (Wakabayashi et al. 1974, Moncton et al. 1975, 1977) have shown the existence of a dip in the Σ_1 (LA) branch along the $\boldsymbol{a}^*$ reciprocal space axis in the region of the superlattice positions for the distorted structure. The dip increases somewhat as the temperature is lowered towards the transition temperature, but instead of softening all the way to zero at the transition (as one might expect on the basis of a naive soft mode theory), saturates and instead a "central peak" (centered at zero frequency) is observed which grows in intensity and becomes the superlattice Bragg reflection in the low-temperature phase (Moncton et al. 1975, 1977). These materials are reasonably good superconductors ($T_c = 7\,K$ for $NbSe_2$ and $=6\,K$ for NbS_2) and have high densities of almost pure d-like states at this Fermi level (Mattheiss 1973, Myron and Freeman 1975). Similar layered compounds with one less electron, such as MoS_2 and TiS_2, are semiconductors and show no such charge-density-wave-like behavior or dips in their dispersion curves (Wakabayashi et al. 1975) although the semimetal $TiSe_2$ exhibits a zone boundary instability (Wilson and Mahajan 1972, Di Salvo et al. 1976; Moncton et al. 1978).

Two other transition metal compounds of interest here are the oxides NbO_2 and VO_2. At high temperatures these are metallic and possess the rutile structure. At 1083 K for NbO_2 and 341 K for VO_2, a metal–insulator transition occurs, accompanied by a structural transformation, which in the case of NbO_2 corresponds to instability against a mode at the reciprocal space point $P(\frac{1}{4}, \frac{1}{4}, \frac{1}{2})$ (Shapiro et al. 1974, Pynn et al. 1978) and in VO_2 at the point $R(\frac{1}{2}, \frac{1}{2}, 0)$ (Westman 1961, Brews 1970) in the Brillouin zone. For NbO_2, the phase transition is second order, although no mode has been observed to soften at the point P (Pynn et al. 1978), while for VO_2 the transition is first order. Band structure calculations for VO_2 (Gupta et al. 1977) show again that for these solids the Fermi level sits on a sharp large peak in the density of states. Another interesting lattice instability which appears to be driven by the electron–phonon interaction is the one taking place at the so-called Verwey transition at 123 K in magnetite (Fe_3O_4), which has recently been studied by neutron diffraction and inelastic scattering (Fujii et al. 1975, Iizumi and Shirane 1975, Yamada 1975). It has been established that in this material the Verwey transition corresponds to an instability against the Δ_5 phonon mode at the reciprocal space point $(0,0,\frac{1}{2})$. A model proposed for this transition (Yamada 1975) involves charge fluctuations between the Fe-atom sites coupling to the phonons via the electron–phonon interaction. We shall have more to say about the relationship between charge fluctuations and lattice instabilities in §5 of this chapter.

A group of transition metal alloys which show instabilities and phonon mode softening connected with the appearance of *magnetic* ordering are the so-called Invar alloys. Neutron scattering studies of the alloy $Fe_{0.65}Ni_{0.35}$ (Endoh et al. 1977), Fe_3Pt (Tajima et al. 1976), and γ-phase MnNi alloys (Harley et al. 1978), and elastic constant measurements in these alloys (for instance Hausch 1974), show considerable mode softening of the TA mode along the [100] axis, and the TA(T_1) mode along the [110] axis with decreasing temperature *below* the magnetic ordering temperature. This is also reflected in softening of the corresponding shear moduli, namely, C_{44} and $\frac{1}{2}(C_{11}-C_{12})$, although the neutron results show that, for the $Fe_{0.65}Ni_{0.35}$ and the Fe_3Pt alloys, the mode softening occurs for large wavevectors as well. Both these alloys are ferromagnetic, the former being a disordered alloy while Fe_3Pt froms an ordered alloy with the Cu_3Au structure. For these alloys the mode softening saturates at low temperatures. Results for Fe_3Pt are shown in fig. 8. For the Mn–Ni alloy, which is *antiferromagnetic*, the [110]T_1 TA branch shows softening predominantly in the small wavevector (elastic constant) region until eventually a tetragonal distortion occurs (see fig. 9). These mode softening effects have been ascribed to magnetoelastic coupling effects, since the effects appear to be proportional to the square of the magnetic order parameter (Hausch 1974), but from a microscopic point of view the explanation must lie in the change in the electronic response to a phonon when a magnetic splitting of the electronic energy bands exists.

We may thus summarize the experimental situation by saying that (a) it is the characteristic of most of these materials that they have a high density of states of predominantly d-like character at the Fermi level, (b) their dispersion curves often show interesting "dips" or anomalies, (c) many of these materials tend to be unstable and undergo structural transformations to a lower symmetry phase which may be metallic or semiconducting, sometimes with periodicities which are incommensurate with the original lattice periodicity and that (d) quite often these materials are also good superconductors. As we shall see, these rather fascinating correlations have inspired a sizeable number of theoretical models attempting to account, in particular, for the phonon anomalies.

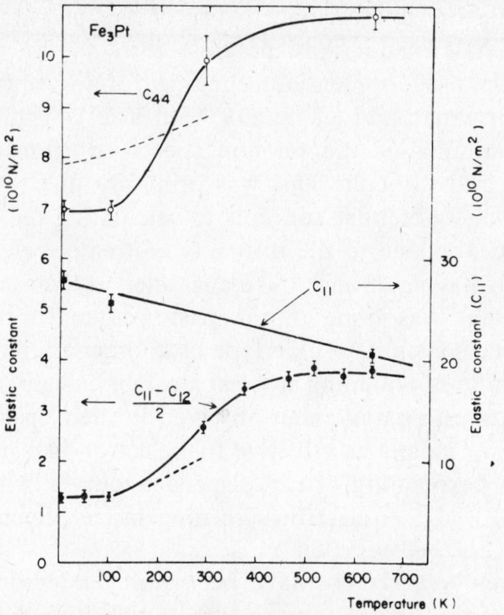

Fig. 8. Values of elastic constants deduced from the initial slopes of the measured acoustic phonon branches in Fe_3Pt plotted versus temperature. The dotted lines represent the results obtained by ultrasonic techniques. The magnetic ordering temperature is 500 K. From Tajima et al. (1976).

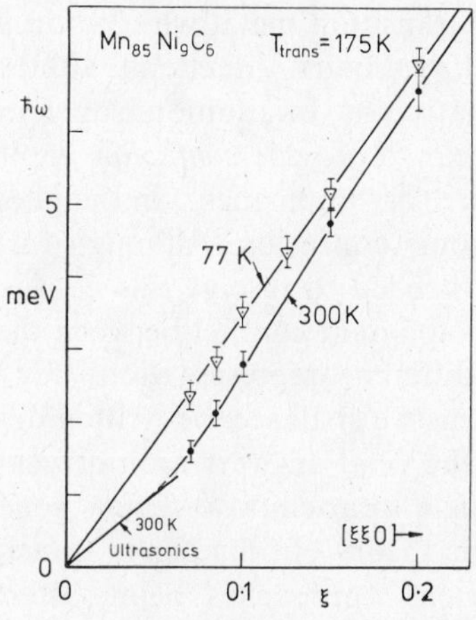

Fig. 9. Dispersion curves for the T_1 [110] branch of $Mn_{0.85}$ $Ni_{0.09}$ $C_{0.06}$ as measured by Harley et al. (1978). The martensitic cubic-to-tetragonal structural transformation is at 175 K.

Initially, while the microscopic theory of the phonon spectra of free-electron-like metals made rapid progress through the development and application of the pseudopotential technique (for a review, see for example, the article by Brovman and Kagan (1974) in Vol. 1 of this series), progress in our understanding of the phonon spectra of transition metals was relatively slow and difficult. This was primarily due to the difficulty of applying the pseudopotential concepts to calculating the electron–phonon interaction, and also due to the difficulty of treating electronic screening effects in metals having strong d-like character to their conduction bands. As a result, what was done in the past was to fit phenomenological interatomic force constants of one type or another to the observed dispersion curves (without obtaining a great deal of insight into the physical origin of the interesting anomalies observed in their dispersion curves), or to treat transition metals as effective free-electron-like metals and persist with fitting or "explaining" their dispersion curves with some kind of pseudopotential and free-electron-screening model. Such procedures are, of course, not very well justified.

In the last few years there have been many attempts to develop the microscopic theory of the phonon spectra of transition metals, both from a rigorous formal point of view, and also through elaborate numerical computations, and these have led to a much greater degree of understanding and insight into what the experimental phonon spectra have been trying to tell us all these years. However, at the time of writing, there has as yet been no completely first-principles parameterless calculation of the phonon spectrum of a transition metal which is in satisfactory agreement with experiment in all symmetry directions, although it appears we are rapidly approaching that point. In addition, superficially at least, many of the recent theories appear to provide *conflicting* explanations of the origins of the observed structure (or "anomalies") in the phonon dispersion curves, and this has created some confusion. Although further work in the next few years is obviously needed to resolve this confusion, it is this author's belief that much of the apparent conflict between the various viewpoints is due to a choice of different representations for the electron–phonon interaction, which can, as we shall see, be written down in a multiplicity of ways. Thus, although the final answers are not yet completely clear, it is the purpose of this article to attempt to review generally the basic framework of the microscopic theory of phonons in transition metals, to present and compare the various approaches being attempted, and finally to review some of the principal unanswered questions at this point.

A recent review of the experimental situation with regard to phonon spectra of the transition metals may be found in the article by Smith et al.

(1976) and further recent results may be found in the proceedings of the Paris International Conference on Lattice Dynamics (Balkanski 1978) and the Toronto Conference on the Physics of Transition Metals (Fawcett 1977).

2. *Phonon spectra and the electron density response function*

There have been a number of slightly different but closely related methods of deriving the general formal expressions for the phonon spectrum of a crystal. (See for instance the articles by Sham (1974), Brovman and Kagan (1974) and Bilz et al. (1974) in Vol. 1 of this series.) We state at the outset that one method commonly used for free-electron-like models, namely that of replacing the actual potential by a weak pseudopotential and calculating by second-order perturbation theory (in the pseudopotential) the energy of the metal for the ions in arbitrarily displaced configurations, (Harrison 1966) is not applicable to transition metals. Even in the case of free-electron-like metals, the effect of higher-order terms in the pseudopotential (and the corresponding problem of non-linear screening of the pseudopotential) has considerably complicated the theory for all but the simplest metals (Brovman and Kagan 1974). For all crystals, however, one may perform a systematic perturbation theory in powers of the ionic *displacements* (Born–Oppenheimer perturbation theory), since, in the spirit of the harmonic and adiabatic approximations, one then rigorously obtains the harmonic phonon frequencies by calculating the energy of the crystal to second order in the ionic displacements.

Let us suppose that the ionic displacements are given by $u(l,\alpha)$, where l denotes the unit cell in which the atom is situated and α the Cartesian component of the displacement. (For simplicity of notation, we restrict the mathematical treatment in this chapter to monatomic lattices, although the generalization to complex crystals is easy to carry out.) Although it is not strictly necessary, we shall follow the usual policy of dividing the electrons into "core" electrons, moving rigidly with the ions, and "valence" or "conduction" electrons which include all the electrons which deform in a more complicated way. For transition metals, the latter group must include the d-electrons, Now, by the Born–Oppenheimer theorem (Born and Huang 1954), the harmonic potential energy function for the vibrating ions consists of a part which is due to the direct Coulomb interaction between the ion cores, and a part which is the second-order change in the total energy of the valence electrons due to the ionic displacements, representing the ion–ion interaction transmitted through the valence electrons. We now calculate the latter contribution.

Let $\delta V_l^{(1)}$ denote the change in "bare" potential experienced by the electron system to first order in $u(l,\alpha)$ when the lth ion is displaced, and $\delta V_l^{(2)}$ the second-order change. Let $|n\rangle$ denote the *full many-electron state* of the valence electrons, with corresponding energy E_n, and let us assume for simplicity that in the unperturbed crystal, the electrons are in their ground state $|0\rangle$. (The treatment may be trivially extended to finite temperatures.) Born–Oppenheimer perturbation theory yields for the total second-order change in the electron energy

$$\Phi_{\mathrm{el}}^{(2)} = \sum_{ll'} \sum_{n} \langle 0|\delta V_l^{(1)}|n\rangle\langle n|\delta V_{l'}^{(1)}|0\rangle/(E_0 - E_n) + \sum_{l} \langle 0|\delta V_l^{(2)}|0\rangle. \quad (2.1)$$

However, the first-order change in the ground state electron wavefunction is given by

$$|0^{(1)}\rangle = \sum_{l} \sum_{n} |n\rangle\langle n|\delta V_l^{(1)}|0\rangle/(E_0 - E_n). \quad (2.2)$$

Thus

$$\Phi_{\mathrm{el}}^{(2)} = \frac{1}{2} \sum_{l} \left[\langle 0|\delta V_l^{(1)}|0^{(1)}\rangle + \text{comp. conj.}\right] + \sum_{l} \langle 0|\delta V_l^{(2)}|0\rangle. \quad (2.3)$$

This may be written in terms of the first-order change in electron density $\delta n(\boldsymbol{r})$, as

$$\Phi_{\mathrm{el}}^{(2)} = \frac{1}{2} \sum_{l} \int \mathrm{d}\boldsymbol{r}\, \delta n(\boldsymbol{r})\, \delta V_l^{(1)}(\boldsymbol{r}) + \sum_{l} \int \mathrm{d}\boldsymbol{r}\, n^0(\boldsymbol{r})\, \delta V_l^{(2)}(\boldsymbol{r}), \quad (2.4)$$

where $n^0(\boldsymbol{r})$ is the electron density of the unperturbed crystal.

Now the first-order change in $n(r)$ due to the first-order change $\delta V_l^{(1)}(\boldsymbol{r})$ may be calculated by linear response theory. (Note that the validity of using linear response does not now depend on the "weakness" of $\delta V_l^{(1)}$.) We have the relation (see for instance Sham 1974)

$$\delta n(\boldsymbol{r}) = \sum_{l'} \int \mathrm{d}\boldsymbol{r}'\, \chi(\boldsymbol{r},\boldsymbol{r}')\, \delta V_{l'}^{(1)}(\boldsymbol{r}'), \quad (2.5)$$

where $\chi(\boldsymbol{r},\boldsymbol{r}')$ is the electron density response function, discussed in more detail later in this chapter. If $V_{\mathrm{b}}(\boldsymbol{r}-\boldsymbol{r}_l)$ is the potential at the point $\boldsymbol{r}$ due to the "bare" ion core at $\boldsymbol{r}_l$, we have

$$\delta V_l^{(1)}(\boldsymbol{r}) = -\sum_{\alpha} \nabla_\alpha V_{\mathrm{b}}(\boldsymbol{r}-\boldsymbol{r}_l) u(l,\alpha) \tag{2.6a}$$

and

$$\delta V_l^{(2)}(\boldsymbol{r}) = \frac{1}{2} \sum_{\alpha\beta} \nabla_\alpha \nabla_\beta V_l(\boldsymbol{r}-\boldsymbol{r}_l) u(l,\alpha) u(l,\beta). \tag{2.6b}$$

Substituting the results (2.5) and (2.6) in eq. (2.4) we obtain

$$\Phi_{\mathrm{el}}^{(2)} = \frac{1}{2} \sum_{ll'} \left\{ \int\int \mathrm{d}\boldsymbol{r}\,\mathrm{d}\boldsymbol{r}' \nabla_\alpha V_{\mathrm{b}}(\boldsymbol{r}-\boldsymbol{r}_l) \chi(\boldsymbol{r},\boldsymbol{r}') \nabla_\beta V_{\mathrm{b}}(\boldsymbol{r}-\boldsymbol{r}_{l'}) \right.$$
$$\left. + \delta_{ll'} \int \mathrm{d}\boldsymbol{r}\, n^0(\boldsymbol{r}) \nabla_\alpha \nabla_\beta V_{\mathrm{b}}(\boldsymbol{r}-\boldsymbol{r}_l) \right\} u(l,\alpha) u(l',\beta). \tag{2.7}$$

Adding to this the direct ion–ion interaction, and noting that

$$\Phi_{\mathrm{ion}}^{(2)} + \Phi_{\mathrm{el}}^{(2)} = \frac{1}{2} \sum_{ll'} \phi_{\alpha\beta}(ll') u(l,\alpha) u(l',\beta), \tag{2.8}$$

where $\Phi_{\alpha\beta}(l,l')$ is the interatomic force constant tensor between the atoms at $\boldsymbol{r}_l, \boldsymbol{r}_{l'}$, we obtain

$$\phi_{\alpha\beta}(l,l') \underset{(l\neq l')}{=} \int\int \mathrm{d}\boldsymbol{r}\,\mathrm{d}\boldsymbol{r}' \nabla_\alpha V_{\mathrm{b}}(\boldsymbol{r}-\boldsymbol{r}_l) \chi(\boldsymbol{r},\boldsymbol{r}') \nabla_\beta V_{\mathrm{b}}(\boldsymbol{r}-\boldsymbol{r}_{l'})$$
$$- Z^2 \frac{\partial^2}{\partial x_\alpha \partial x_\beta} \left(\frac{e^2}{r} \right) \bigg|_{\boldsymbol{r}=\boldsymbol{r}_{l'}-\boldsymbol{r}_l} \tag{2.9a}$$

and

$$\phi_{\alpha\beta}(l,l) = \int\int \mathrm{d}\boldsymbol{r}\,\mathrm{d}\boldsymbol{r}' \nabla_\alpha V_{\mathrm{b}}(\boldsymbol{r}-\boldsymbol{r}_l) \chi(\boldsymbol{r},\boldsymbol{r}') \nabla_\beta V_{\mathrm{b}}(\boldsymbol{r}-\boldsymbol{r}_l)$$
$$+ \int \mathrm{d}\boldsymbol{r}\, n^{(0)}(\boldsymbol{r}) \nabla_\alpha \nabla_\beta V_{\mathrm{b}}(\boldsymbol{r}-\boldsymbol{r}_l) + Z^2 \sum_{l'\neq l} \frac{\partial^2}{\partial x_\alpha \partial x_\beta} \left(\frac{e^2}{r} \right) \bigg|_{\boldsymbol{r}=\boldsymbol{r}_{l'}-\boldsymbol{r}_l}. \tag{2.9b}$$

It may be shown (see below) that the condition of translational invariance requires that the relation

$$\sum_{l'} \phi_{\alpha\beta}(l,l')=0 \tag{2.10}$$

be satisfied, so in practice $\phi_{\alpha\beta}(l,l)$ does not need to be calculated explicitly.

The phonon frequencies and eigenvectors may be obtained from the eigenvalues and eigenvectors of the dynamical matrix given by (Maradudin 1974, Venkataraman et al. 1975)

$$D_{\alpha\beta}(\boldsymbol{q})=M^{-1}\sum_{l} \exp(\mathrm{i}\boldsymbol{q}\cdot\boldsymbol{r}_l)\phi_{\alpha\beta}(0,l), \tag{2.11}$$

where M is the atomic mass.

In practice, instead of calculating the interatomic force constants first, it is often convenient to calculate directly the dynamical matrix in $\boldsymbol{q}$-space in terms of the Fourier transforms of the quantities appearing in eqs. (2.9). One obtains (Sham 1974)

$$D_{\alpha\beta}(\boldsymbol{q})=M^{-1}\left[C_{\alpha\beta}(\boldsymbol{q})+E_{\alpha\beta}(\boldsymbol{q})\right], \tag{2.12}$$

where $C_{\alpha\beta}(\boldsymbol{q})$ is the Coulomb coupling coefficient arising from the second term in eq. (2.9a) and may be evaluated by standard methods (Kellerman 1940, Born and Huang 1954, Venkataraman et al. 1975), and

$$E_{\alpha\beta}(\boldsymbol{q})=\sum_{\boldsymbol{G}\boldsymbol{G}'} E_{\alpha\beta}(\boldsymbol{q}+\boldsymbol{G},\boldsymbol{q}+\boldsymbol{G}')-E_{\alpha\beta}(\boldsymbol{G},\boldsymbol{G}'), \tag{2.13}$$

where

$$\begin{aligned} E_{\alpha\beta}(\boldsymbol{q}+\boldsymbol{G},\boldsymbol{q}+\boldsymbol{G}')=(\boldsymbol{q}+\boldsymbol{G})_{\alpha}V_b^*(\boldsymbol{q}+\boldsymbol{G})\chi(\boldsymbol{q}+\boldsymbol{G},\boldsymbol{q}+\boldsymbol{G}') \\ \times V_{\mathrm{b}}(\boldsymbol{q}+\boldsymbol{G}')(\boldsymbol{q}+\boldsymbol{G}')_{\beta}, \end{aligned} \tag{2.14}$$

where $V_{\mathrm{b}}(\boldsymbol{K})$ is the Fourier transform of $V_{\mathrm{b}}(\boldsymbol{r})$ and $\chi(\boldsymbol{K},\boldsymbol{K}')$ is the double Fourier transform of $\chi(\boldsymbol{r},\boldsymbol{r}')$.

In this formulation, the central problem then is that of calculating the density response matrix $\chi(\boldsymbol{K},\boldsymbol{K}')$ for the conduction electrons of the system. In the random phase approximation (RPA) we may write (Sham 1974) the Fourier transform of eq. (2.5) in terms of a self-consistent field formulation

$$\begin{aligned} \delta n(\boldsymbol{K}) &= \sum_{\boldsymbol{K}'} \chi(\boldsymbol{K},\boldsymbol{K}')\delta V_{\mathrm{ext}}(\boldsymbol{K}') \\ &= \sum_{\boldsymbol{K}'} \chi^0(\boldsymbol{K},\boldsymbol{K}')\left[\delta V_{\mathrm{ext}}(\boldsymbol{K}')+\delta V_{\mathrm{s}}(\boldsymbol{K}')\right], \end{aligned} \tag{2.15}$$

where χ^0 is the response function of the *non-interacting* electron system and δV_s is the self-consistent screening potential set up by δn.

Thus,

$$\delta V_s(\boldsymbol{K}) = v(\boldsymbol{K})\,\delta n(\boldsymbol{K}), \tag{2.16}$$

where

$$v(\boldsymbol{K}) = 4\pi e^2/\Omega_0 K^2 \tag{2.17}$$

is the Fourier transform of the electron–electron interaction (Ω_0 is the unit cell volume). From these equations, we obtain

$$\delta V_s(\boldsymbol{K}) = \sum_{\boldsymbol{K}'} \varepsilon^{-1}(\boldsymbol{K},\boldsymbol{K}')v(\boldsymbol{K}')\sum_{\boldsymbol{K}''}\chi_0(\boldsymbol{K}',\boldsymbol{K}'')\,\delta V_{\text{ext}}(\boldsymbol{K}''),$$

where

$$\varepsilon(\boldsymbol{K},\boldsymbol{K}') = \delta_{\boldsymbol{K}\boldsymbol{K}'} - v(\boldsymbol{K})\chi^0(\boldsymbol{K},\boldsymbol{K}'). \tag{2.18}$$

Eqs. (2.16) and (2.17) yield

$$\chi(\boldsymbol{K},\boldsymbol{K}') = \frac{1}{v(K)}\left[\varepsilon^{-1}(\boldsymbol{K},\boldsymbol{K}') - \delta_{\boldsymbol{K}\boldsymbol{K}'}\right] \tag{2.19a}$$

which may also be written as

$$\chi(\boldsymbol{K},\boldsymbol{K}') = \sum_{\boldsymbol{K}''}\chi^0(\boldsymbol{K},\boldsymbol{K}'')\varepsilon^{-1}(\boldsymbol{K}'',\boldsymbol{K}'). \tag{2.19b}$$

For Bloch electrons, $\boldsymbol{K},\boldsymbol{K}'$ must always be related by a reciprocal lattice vector of the crystal, and χ^0 is easily obtained by perturbation theory as

$$\chi^0(\boldsymbol{K},\boldsymbol{K}') = \frac{1}{N}\sum_{k\lambda\lambda'}\frac{n(\boldsymbol{k}\lambda) - n(\boldsymbol{k}+\boldsymbol{q}\lambda')}{E(\boldsymbol{k}\lambda) - E(\boldsymbol{k}+\boldsymbol{q}\lambda')}\langle \boldsymbol{k}\lambda|\exp(-\mathrm{i}\boldsymbol{K}\cdot\boldsymbol{r})|\boldsymbol{k}+\boldsymbol{q}\lambda'\rangle \\ \times\langle \boldsymbol{k}+\boldsymbol{q}\lambda'|\exp(\mathrm{i}\boldsymbol{K}'\cdot\boldsymbol{r})|\boldsymbol{k}\lambda\rangle, \tag{2.20}$$

where $\boldsymbol{q}$ is the wavevector $\boldsymbol{K}$ or $\boldsymbol{K}'$ reduced to the first zone, and $n_{\boldsymbol{k}\lambda}, E_{\boldsymbol{k}\lambda}$ are the occupation number and energy respectively of the Bloch state with wavevector $\boldsymbol{k}$ and band index λ in the reduced zone scheme. N is the total number of unit cells in the crystal.

We note that while for free-electron-like metals, the dielectric matrix $\boldsymbol{\varepsilon}$ is approximately diagonal, for transition metals the inversion of $\varepsilon(\boldsymbol{q}+\boldsymbol{G}, \boldsymbol{q}+\boldsymbol{G}')$ becomes a formidable problem owing to the atomic-like nature of the d-wavefunctions near the ion cores. Additional problems in connection with exchange and correlation effects are discussed in §4.

Thus we see that there are some serious difficulties associated with actual calculations using the expressions (2.12)–(2.14). In the first place, both terms in eq. (2.12) are large but tend to cancel each other. As a consequence $E_{\alpha\beta}(\boldsymbol{q})$ must be calculated quite accurately to obtain reasonable values for the phonon frequencies. However, since $V_{\mathrm{b}}(\boldsymbol{q}+\boldsymbol{G})$ is the Fourier transform of the "bare" ion potential, which is strong and rapidly varying near the core, and since as we have seen $\chi(\boldsymbol{q}+\boldsymbol{G},\boldsymbol{q}+\boldsymbol{G}')$ has to be obtained in terms of the inverse of a very large dielectric matrix, an accurate calculation of $E_{\alpha\beta}(\boldsymbol{q})$ using eq. (2.14) as it stands is indeed a formidable task. However, the theory may be reformulated in a way that partly alleviates some of these problems, as we shall see in the next section.

Before closing this section, we point out that the Fourier representation is not the only one for expressing $E_{\alpha\beta}(\boldsymbol{q})$, as given by eqs. (2.13) and (2.14). In fact often it may be more convenient to work in terms of an orbital representation. Let ϕ_ξ denote a complete set of functions. We write

$$\chi(\boldsymbol{q}+\boldsymbol{G},\boldsymbol{q}+\boldsymbol{G}')=\sum_{\xi_1\xi_2\xi_3\xi_4}\langle\xi_1|\exp[-\mathrm{i}(\boldsymbol{q}+\boldsymbol{G})\cdot\boldsymbol{r}]|\xi_2\rangle\chi_{\xi_1\xi_3;\,\xi_2\xi_4}(\boldsymbol{q})\times\langle\xi_3|\exp[\mathrm{i}(\boldsymbol{q}+\boldsymbol{G}')\cdot\boldsymbol{r}]|\xi_4\rangle. \tag{2.21}$$

$\chi_{\xi_1\xi_2;\,\xi_3\xi_4}(\boldsymbol{q})$ is the density response function in the orbital representation. It denotes the amplitude of a charge fluctuation of type $(\phi^*_{\xi_1}\phi_{\xi_2})$ excited in response to a perturbation which couples to a charge fluctuation of type $(\phi^*_{\xi_3}\phi_{\xi_4})$.

Substituting eq. (2.21) in eqs. (2.13) and (2.14) we obtain

$$E_{\alpha\beta}(\boldsymbol{q})=E'_{\alpha\beta}(\boldsymbol{q})-E'_{\alpha\beta}(0), \tag{2.22}$$

where

$$E'_{\alpha\beta}(\boldsymbol{q})=\sum_{\xi_1\xi_2\xi_3\xi_4}\langle\xi_1|\delta V_{-\boldsymbol{q}}|\xi_2\rangle\chi_{\xi_1\xi_3;\,\xi_2\xi_4}(\boldsymbol{q})\langle\xi_3|\delta V_{\boldsymbol{q}}|\xi_4\rangle \tag{2.23}$$

and $\delta V_{\boldsymbol{q}}$ is given by

$$\delta V_{\boldsymbol{q}}=-\sum_{l,\alpha}\nabla_\alpha V_{\mathrm{b}}(\boldsymbol{r}-\boldsymbol{r}_l)\exp(\mathrm{i}\boldsymbol{q}\cdot\boldsymbol{r}_l). \tag{2.24}$$

3. *Reformulation of the electron–phonon interaction*

It was pointed out several years ago by the present author (Sinha 1968) that a large part of $\delta V_l^{(1)}(\boldsymbol{r})$ (given in eq. (2.6a)), in particular the strong rapidly-varying part near the core, simply had the effect of moving the atomic-like part of the wavefunction near the core rigidly with the ion, and thus did not really contribute to the phonon frequencies. This implies that some large part of the contributions of the $V_b(\boldsymbol{q}+\boldsymbol{G})$ to $E_{\alpha\beta}(\boldsymbol{q})$ via eqs. (2.13) and (2.14) may be separated to explicitly cancel out most of the $C_{\alpha\beta}(\boldsymbol{q})$ term in eq. (2.12), thus removing the problem referred to in §2, and also hopefully leaving eq. (2.13) in a more rapidly convergent form in reciprocal space. This may in fact be achieved by explicitly building into the problem from the beginning the rigid motion of some part of the conduction electron, which, together with the ion core, resembles a "pseudo-atom" which is neutral or at least has a much reduced charge compared to the bare ion core. This part may then be taken to contribute to the *first* term $C_{\alpha\beta}(\boldsymbol{q})$ in eq. (2.12). The second term involves now only the further distortion of the electron density. This may be described in terms of a reduced number of Fourier components, or charge fluctuation components (in the representation of eq. (2.21)), thus reducing the size of the matrix which must be inverted to obtain $\boldsymbol{\chi}$ accurately enough to yield reasonable phonon frequencies. (It should be pointed out that there is yet a different method of reformulating explicitly the cancellation between $C_{\alpha\beta}(\boldsymbol{q})$ and $E_{\alpha\beta}(\boldsymbol{q})$ in eq. (2.12) which is not equivalent to the one just described. We shall return to it in §4.)

The rest of this section is devoted to some formal results. First we show that use of the Schrödinger Equation ensures a relation between the electron density response function and the bare ionic potential such that the rigid "pseudo-atom" response to the movement of the ion-case appears explicitly, together with a formal expression for the "deformation" part of the response. This corresponds to a microscopic proof of the "acoustic sum rule" (Pick et al. 1970, Shaw 1974). Secondly, we show that use of this result in eq. (2.14) for $E_{\alpha\beta}$ directly renormalizes the ion–ion Coulomb contribution $C_{\alpha\beta}$. While these derivations constitute the formal proofs of the remarks made above, they will not actually be used to discuss detailed calculations for the phonon spectrum as the Bloch representation turns out to be rather clumsy for this purpose. Since a more elegant and compact derivation of the dynamical matrix for transition metals may be obtained in terms of atomic orbitals we provide in the next section an alternating Green's function formulation of both the density response function and dynamical matrix starting from the full electron–phonon Hamiltonian. Therefore, the reader who is more interested in the results of the microscopic theory for yielding phonon dispersion curves may go directly to §4.

Let us choose an arbitrary volume $V_0(l)$ in the lth unit cell, which for convenience we may assume to be spherical. (Some convenient choices for $V_0(l)$ are the muffin-tin sphere commonly used in energy band calculations or even the "atomic sphere" which is the spherical approximation to the symmetric Wigner–Seitz atomic unit cell.) We define a potential

$$U(\boldsymbol{r}-\boldsymbol{r}_l)=V(\boldsymbol{r}) \qquad (\boldsymbol{r}\text{ inside } V_0(l))$$
$$=0 \qquad (\boldsymbol{r}\text{ outside } V_0(l)), \tag{3.1}$$

where $V(r)$ is the *actual* crystal potential.

We define an operator A such that

$$\langle \boldsymbol{k}\lambda|A|\boldsymbol{k}'\lambda'\rangle=\frac{1}{2}\sum_{l,\alpha}u(l,\alpha)\int_{V_0(l)}\mathrm{d}\boldsymbol{r}'\{\psi^*_{\boldsymbol{k}\lambda}\nabla_\alpha\psi_{\boldsymbol{k}'\lambda'}-(\nabla_\alpha\psi^*_{\boldsymbol{k}\lambda})\psi_{\boldsymbol{k}'\lambda'}\}. \tag{3.2}$$

The Schrödinger equation for the $\psi_{\boldsymbol{k}\lambda}$ may be used to prove the following identities (Sinha 1968)

$$-\sum_{l,\alpha}\langle \boldsymbol{k}\lambda|\nabla_\alpha U(\boldsymbol{r}-\boldsymbol{r}_l)|\boldsymbol{k}'\lambda'\rangle u(l,\alpha)=M^{l,\alpha}_{\boldsymbol{k}\lambda,\boldsymbol{k}'\lambda'}u(l,\alpha)+\sum_{l,\alpha}u(l,\alpha)(E_{\boldsymbol{k}\lambda}-E_{\boldsymbol{k}'\lambda'})$$
$$\times\int_{V_0(l)}\mathrm{d}\boldsymbol{r}\,\psi^*_{\boldsymbol{k}\lambda}\nabla_\alpha\psi_{\boldsymbol{k}'\lambda'}, \tag{3.3}$$

where (in units such that $\hbar^2/2m=1$)

$$M^{l,\alpha}_{\boldsymbol{k}\lambda,\boldsymbol{k}'\lambda'}=\int_{V_0(l)}\mathrm{d}\boldsymbol{r}\big[(\nabla^2\psi^*_{\boldsymbol{k}\lambda})(\nabla_\alpha\psi_{\boldsymbol{k}'\lambda'})-\psi^*_{\boldsymbol{k}\lambda}\nabla^2(\nabla_\alpha\psi_{\boldsymbol{k}'\lambda'})\big], \tag{3.4}$$

which may also be expressed, using Green's theorem, as a surface integral over the surface of $V_0(l)$,

$$M^{l,\alpha}_{\boldsymbol{k}\lambda,\boldsymbol{k}'\lambda'}=\int_{S_0(l)}\mathrm{d}S\big[(\nabla_\mathrm{n}\psi^*_{\boldsymbol{k}\lambda})(\nabla_\alpha\psi_{\boldsymbol{k}'\lambda'})-\psi^*_{\boldsymbol{k}\lambda}\nabla_\mathrm{n}(\nabla_\alpha\psi_{\boldsymbol{k}'\lambda'})\big], \tag{3.5}$$

where ∇_n denotes differentiation normal to the surface. The following identity may also be established (Sinha 1968)

$$M^{*l,\alpha}_{\boldsymbol{k}'\lambda',\boldsymbol{k}}=M^{l,\alpha}_{\boldsymbol{k}\lambda,\boldsymbol{k}'\lambda'}+u(l,\alpha)(E_{\boldsymbol{k}\lambda}-E_{\boldsymbol{k}'\lambda'})\int_{V_0(l)}\mathrm{d}\boldsymbol{r}\nabla_\alpha(\psi^*_{\boldsymbol{k}\lambda}\psi_{\boldsymbol{k}'\lambda'}). \tag{3.6}$$

Using eqs. (3.3) and (3.6) we obtain

$$-\sum_{l,\alpha}\langle \boldsymbol{k}\lambda|\nabla_\alpha U(\boldsymbol{r}-\boldsymbol{r}_l)|\boldsymbol{k}'\lambda'\rangle u(l,\alpha)$$
$$=(E_{\boldsymbol{k}\lambda}-E_{\boldsymbol{k}'\lambda'})\langle \boldsymbol{k}\lambda|A|\boldsymbol{k}'\lambda'\rangle+\sum_{l,\alpha}u(l,\alpha)Q^{l,\alpha}_{\boldsymbol{k}\lambda,\boldsymbol{k}'\lambda'}, \tag{3.7}$$

where

$$Q^{l,\alpha}_{\boldsymbol{k}\lambda,\boldsymbol{k}'\lambda'}=\frac{1}{2}(M^{l,\alpha}_{\boldsymbol{k}\lambda,\boldsymbol{k}'\lambda'}+M^{*l,\alpha}_{\boldsymbol{k}'\lambda',\boldsymbol{k}\lambda}). \tag{3.8}$$

Writing $u(l,\alpha)$ in the form of a running wave for a particular normal mode $(\boldsymbol{q}j)$,

$$u(l,\alpha)=A_{\boldsymbol{q}j}e_\alpha(\boldsymbol{q}j)\exp(\mathrm{i}\boldsymbol{q}\cdot\boldsymbol{r}_l), \tag{3.9}$$

where $e_\alpha(\boldsymbol{q}j)$ is the phonon eigenvector, and Fourier transforming $\nabla_\alpha U(\boldsymbol{r}-\boldsymbol{r}_l)$ in the left-hand side of eq. (3.7) we obtain

$$\begin{aligned}-\mathrm{i}A_{\boldsymbol{q}j}\sum_\alpha e_\alpha(\boldsymbol{q}j)\sum_{\boldsymbol{G}'}(\boldsymbol{q}+\boldsymbol{G}')_\alpha U(\boldsymbol{q}+\boldsymbol{G}')\langle\boldsymbol{k}\lambda|\exp[\mathrm{i}(\boldsymbol{q}+\boldsymbol{G}')\cdot\boldsymbol{r}]|\boldsymbol{k}'\lambda'\rangle\\ =\frac{1}{N}(E_{\boldsymbol{k}\lambda}-E_{\boldsymbol{k}'\lambda'})\langle\boldsymbol{k}\lambda|A|\boldsymbol{k}'\lambda'\rangle+\frac{1}{N}\sum_{l,\alpha}u(l,\alpha)Q^{l,\alpha}_{\boldsymbol{k}\lambda,\boldsymbol{k}'\lambda'}.\end{aligned} \tag{3.10}$$

Multiplying both sides of this equation by

$$\frac{1}{N}\langle\boldsymbol{k}'\lambda'|\exp[-\mathrm{i}(\boldsymbol{q}+\boldsymbol{G})\cdot\boldsymbol{r}]|\boldsymbol{k}\lambda\rangle(n_{\boldsymbol{k}\lambda}-n_{\boldsymbol{k}'\lambda'})/(E_{\boldsymbol{k}\lambda}-E_{\boldsymbol{k}'\lambda'})$$

and summing over $\boldsymbol{k},\boldsymbol{k}',\lambda,\lambda'$, we obtain, using eq. (2.20)

$$\begin{aligned}&-\mathrm{i}A_{\boldsymbol{q}j}\sum_\alpha e_\alpha(\boldsymbol{q}j)\sum_{\boldsymbol{G}'}\{\chi^0(\boldsymbol{q}+\boldsymbol{G},\boldsymbol{q}+\boldsymbol{G}')(\boldsymbol{q}+\boldsymbol{G}')_\alpha U(\boldsymbol{q}+\boldsymbol{G}')\}\\ &=\frac{1}{N^2}\sum_{l,\alpha}u(l,\alpha)\sum_{\substack{\boldsymbol{k}\boldsymbol{k}'\\ \lambda\lambda'}}Q^{l,\alpha}_{\boldsymbol{k}\lambda,\boldsymbol{k}'\lambda'}\langle\boldsymbol{k}'\lambda'|\exp[-\mathrm{i}(\boldsymbol{q}+\boldsymbol{G})\cdot\boldsymbol{r}]|\boldsymbol{k}\lambda\rangle\\ &\quad\times(n_{\boldsymbol{k}\lambda}-n_{\boldsymbol{k}'\lambda'})/(E_{\boldsymbol{k}\lambda}-E_{\boldsymbol{k}'\lambda'})\\ &\quad+\frac{1}{N^2}\sum_{\substack{\boldsymbol{k}\boldsymbol{k}'\\ \lambda\lambda'}}(n_{\boldsymbol{k}\lambda}-n_{\boldsymbol{k}'\lambda'})\langle\boldsymbol{k}\lambda|A|\boldsymbol{k}'\lambda'\rangle\langle\boldsymbol{k}'\lambda'|\exp[-\mathrm{i}(\boldsymbol{q}+\boldsymbol{G})\cdot\boldsymbol{r}]|\boldsymbol{k}\lambda\rangle.\end{aligned} \tag{3.11}$$

Now using the completeness properties of the $\psi_{\boldsymbol{k}\lambda}$, the last term in eq. (3.11) can be shown to yield

$$-\mathrm{i}A_{\boldsymbol{q}j}\sum_\alpha e_\alpha(\boldsymbol{q}j)(\boldsymbol{q}+\boldsymbol{G})_\alpha\tilde{n}^0(\boldsymbol{q}+\boldsymbol{G}),$$

where $\tilde{n}^0(\boldsymbol{K})$ is the Fourier transform of the unperturbed crystal electron density inside $V_0(l)$. Thus, finally, eq. (3.11) yields

$$\begin{aligned}&\sum_{\boldsymbol{G}'}\chi^0(\boldsymbol{q}+\boldsymbol{G},\boldsymbol{q}+\boldsymbol{G}')(\boldsymbol{q}+\boldsymbol{G}')_\alpha U(\boldsymbol{q}+\boldsymbol{G})=(\boldsymbol{Q}+\boldsymbol{G})_\alpha\tilde{n}^0(\boldsymbol{q}+\boldsymbol{G})\\ &\qquad+(\mathrm{i}/N^2)\sum_l\sum_{\substack{\boldsymbol{k}\boldsymbol{k}'\\ \lambda\lambda'}}Q^{l,\alpha}_{\boldsymbol{k}\lambda,\boldsymbol{k}'\lambda'}\exp(\mathrm{i}\boldsymbol{q}\cdot\boldsymbol{r}_l)\langle\boldsymbol{k}'\lambda'|\exp[-\mathrm{i}(\boldsymbol{q}+\boldsymbol{G})\cdot\boldsymbol{r}]|\boldsymbol{k}\lambda\rangle\\ &\qquad\times(n_{\boldsymbol{k}\lambda}-n_{\boldsymbol{k}'\lambda'})/(E_{\boldsymbol{k}\lambda}-E_{\boldsymbol{k}'\lambda'}).\end{aligned} \tag{3.12}$$

Let us define a column vector $\Delta\mathbf{U}^\alpha$ whose components are $(\boldsymbol{q}+\boldsymbol{G})_\alpha U(\boldsymbol{q}+\boldsymbol{G})$, a column vector $\Delta\mathbf{n}^\alpha$ whose components are $(\boldsymbol{q}+\boldsymbol{G})_\alpha \tilde{n}^0(\boldsymbol{q}+\boldsymbol{G})$ and a column vector $\Delta\mathbf{Q}^\alpha$ whose components are represented by the last term on the right of eq. (3.12). Then eq. (3.12) may be written more compactly in matrix form

$$\chi^0\Delta\mathbf{U}^\alpha=\Delta\mathbf{n}^\alpha+\Delta\mathbf{Q}^\alpha. \tag{3.13}$$

Neglecting exchange and correlation potentials (which may be formally incorporated without changing the basic results), we may write for the bare ion–electron potential

$$V_\mathrm{b}(\boldsymbol{r}-\boldsymbol{r}_l)=U(\boldsymbol{r}-\boldsymbol{r}_l)-\int \mathrm{d}\boldsymbol{r}'\,v(\boldsymbol{r}-\boldsymbol{r}')\tilde{n}^0(\boldsymbol{r}'-\boldsymbol{r}_l)+W(\boldsymbol{r}-\boldsymbol{r}_l), \tag{3.14}$$

where we note that *inside* $V_0(l)$, the potential $W(\boldsymbol{r}-\boldsymbol{r}_l)$ is the negative of the potentials due to the charges in all the *other* unit cells, and outside $V_0(l)$, it is the Coulomb tail of the potential due to the pseudo-atom inside $V_0(l)$. It thus behaves very much like a "local" pseudopotential. From eq. (3.14) we obtain after Fourier transforming and multiplying by $(q+G)_\alpha$,

$$\Delta\mathbf{V}_\mathrm{b}^\alpha=\Delta\mathbf{U}^\alpha-v\Delta\mathbf{n}^\alpha+\Delta\mathbf{W}^\alpha. \tag{3.15}$$

The dielectric and density response matrices are given respectively by

$$\varepsilon=\mathbf{1}-v\chi^0, \tag{3.16}$$

and

$$\chi=\chi^0\varepsilon^{-1}. \tag{3.17}$$

Then eqs. (3.13) and (3.15) may be combined to yield

$$\chi\Delta\mathbf{V}_\mathrm{b}^\alpha=\chi^0\Delta\mathbf{U}^\alpha+\chi\Delta\mathbf{W}^\alpha+\chi v\,\Delta\mathbf{Q}^\alpha. \tag{3.18}$$

Using eq. (3.13) we obtain

$$\chi\Delta\mathbf{V}_\mathrm{b}^\alpha=\Delta\mathbf{n}^\alpha+\chi\Delta\mathbf{W}^\alpha+(\mathbf{1}+\chi v)\Delta\mathbf{Q}^\alpha. \tag{3.19}$$

We may now write this explicitly as

$$\sum_{G'} \chi(q+G,q+G')(q+G')_\alpha V_b(q+G') = (q+G)_\alpha \tilde{n}^0(q+G)$$
$$+\sum_{G'} \chi(q+G,q+G')(q+G')_\alpha W(q+G')$$
$$+\sum_{G'} \{\delta_{GG'} + \chi(q+G,q+G')v(q+G')\}$$
$$\times \frac{1}{N} \sum_{\substack{kk' \\ \lambda\lambda'}} \langle k'\lambda'|\exp[-\mathrm{i}(q+G')\cdot r]|k\lambda\rangle (n_{k\lambda}-n_{k'\lambda'})/(E_{k\lambda}-E_{k'\lambda'})$$
$$\times \frac{1}{N} \sum_{l} \mathrm{i} Q^{l,\alpha}_{k\lambda,k'\lambda'} \exp(\mathrm{i}q\cdot r_l). \tag{3.20}$$

This important result shows that the *total* conduction electron density response to the displaced potentials of the *bare* ions consists of three terms: a rigid displacement of the unperturbed valence density inside each $V_0(l)$ with the corresponding ion core; a response to a set of weaker pseudopotential-like displaced potentials $W(r-r_l)$; and a term involving the surface integral matrix elements $Q^{l,\alpha}_{k\lambda,k'\lambda'}$ defined in eq. (3.8). The last two terms describe the deformation inside $V_0(l)$ of the electron density due to the phonon, as well as its total change outside the $V_0(l)$. An important special case arises when we take $V_0(l)$ to be the actual Wigner–Seitz (neutral polyhedron) cell and the limit $q\to 0$. It may be shown that in this case $W(q+G)$ and the $Q^{l,\alpha}_{k\lambda,k'\lambda'}$ vanish, and hence only the first term on the right side of eq. (3.20) survives. Eq. (3.20) then expresses the result that a uniform displacement of the bare ion potentials produces the response which is a uniform displacement of the unperturbed valence density by the corresponding amount. This is closely related to the acoustic sum rule in insulators (Pick et al. 1970, Sham 1974), which is related to the $G=0$, $q\to 0$ limit of eq. (3.20). Although the result is physically obvious, the above derivation emphasizes how the Schrödinger equation restricts the electronic states such that their density response function is related to the ionic potential in just this manner.

From eqs. (3.13) and (3.15) we may also derive the result

$$\varepsilon^{-1}\Delta \mathbf{V}^\alpha_b = \Delta \mathbf{U}^\alpha + \varepsilon^{-1} v \Delta \mathbf{Q}^\alpha + \varepsilon^{-1}\Delta \mathbf{W}^\alpha. \tag{3.21}$$

The left-hand side is a column vector which consists of the elements $\Delta V^\alpha_s(q+G)$, which are the components of the *screened* potential resulting from the application of the bare potential components $(q+G)_\alpha V_b(q+G)$.

The matrix elements of the *screened* potential between the states $\boldsymbol{k}\lambda$ and $\boldsymbol{k}'\lambda'$ are (using eq. (3.9))

$$\langle \boldsymbol{k}\lambda|\delta V_{\mathrm{s}}|\boldsymbol{k}'\lambda'\rangle = -\mathrm{i}\sum_{\boldsymbol{G},\alpha}\langle \boldsymbol{k}\lambda|\exp[\mathrm{i}(\boldsymbol{q}+\boldsymbol{G})\cdot\boldsymbol{r}]|\boldsymbol{k}'\lambda'\rangle\,\Delta V_{\mathrm{s}}^{\alpha}(\boldsymbol{q}+\boldsymbol{G})\mathrm{e}_{\alpha}(\boldsymbol{q}j)A_{\boldsymbol{q}j}. \tag{3.22}$$

Thus by eqs. (3.19) and (3.7) we obtain for the screened electron–phonon matrix element

$$\begin{aligned}\langle \boldsymbol{k}\lambda|\delta V_{\mathrm{s}}|\boldsymbol{k}'\lambda'\rangle &= (E_{\boldsymbol{k}\lambda}-E_{\boldsymbol{k}'\lambda'})\langle \boldsymbol{k}\lambda|A|\boldsymbol{k}'\lambda'\rangle \\ &\quad+\sum_{l,\alpha}u(l,\alpha)Q_{\boldsymbol{k}\lambda,\boldsymbol{k}'\lambda'}^{l,\alpha}-\mathrm{i}\sum_{\boldsymbol{G},\alpha}e_{\alpha}(\boldsymbol{q}j)A_{\boldsymbol{q}j} \\ &\quad\times\langle \boldsymbol{k}\lambda|\exp[\mathrm{i}(\boldsymbol{q}+\boldsymbol{G})\cdot\boldsymbol{r}]|\boldsymbol{k}'\lambda'\rangle\sum_{\boldsymbol{G}'}\varepsilon^{-1}(\boldsymbol{q}+\boldsymbol{G},\boldsymbol{q}+\boldsymbol{G}') \\ &\quad\times[v(\boldsymbol{q}+\boldsymbol{G}')\Delta Q^{\alpha}(\boldsymbol{q}+\boldsymbol{G}')+(\boldsymbol{q}+\boldsymbol{G}')_{\alpha}W(\boldsymbol{q}+\boldsymbol{G}')].\end{aligned} \tag{3.23}$$

For energy-conserving transitions, such as those across the Fermi surface, the first term on the right-hand side vanishes. The second term corresponds to the surface integral in eqs. (3.5) and (3.8) and may be expressed in terms of logarithmic derivatives (Golibersuch 1967, Sinha 1968) or phase shifts (Gaspari and Gyorffy 1972) of the radial wavefunctions, as we shall see in more detail in the next section. Keeping only this term corresponds to the so-called "Rigid Muffin Tin Approximation (RMTA)" for the screened electron–phonon matrix element, so commonly used in superconductivity calculations. (For a more detailed discussion see §4, and also the chapter by Allen in this volume.) We thus see that it corresponds to neglecting (a) the tails of the potentials of the "pseudo-atoms" inside $V_0(l)$ as represented by $W(\boldsymbol{q}+\boldsymbol{G})$ and (b) screening of the $Q_{\boldsymbol{k}\lambda,\boldsymbol{k}'\lambda'}^{l,\alpha}$ as embodied in the term involving $\Delta Q^{\alpha}(\boldsymbol{q}+\boldsymbol{G})$. We note that the complicated form of $\Delta Q^{\alpha}(\boldsymbol{q}+\boldsymbol{G})$ (see eq. 3.12) and the somewhat messy expressions involving this term which have appeared in this section are due to the fact that $Q_{\boldsymbol{k}\lambda,\boldsymbol{k}'\lambda'}^{l,\alpha}$ cannot be expressed in terms of Fourier components of a local potential. In fact a much more compact and elegant formulation may be obtained if one changes from a Fourier representation to an orbital representation. This will be described in §4.

Returning to the problem of the phonon spectrum and the calculation of $E_{\alpha\beta}(\boldsymbol{q})$, we note that the first term on the right of eq. (2.13) may be written in our matrix notation as

$$E'_{\alpha\beta}(\boldsymbol{q}) = \Delta \mathbf{V}_{\mathrm{b}}^{\alpha\dagger} \cdot \boldsymbol{\chi} \cdot \Delta \mathbf{V}_{\mathrm{b}}^{\beta}. \tag{3.24}$$

Using eq. (3.19) it may be seen that this may be rewritten as

$$E'_{\alpha\beta}(\boldsymbol{q}) = \Delta \mathbf{V}_{\mathrm{b}}^{\alpha\dagger} \cdot \Delta \mathbf{n}^{\beta} + \Delta \mathbf{n}^{\alpha\dagger} \cdot \Delta \mathbf{W}^{\beta} + \Delta \mathbf{W}^{\alpha\dagger} \cdot \boldsymbol{\chi} \cdot \Delta \mathbf{W}^{\beta} + (\text{terms involving } \Delta \mathbf{Q}^{\alpha}). \tag{3.25}$$

The first term on the right of eq. (3.25) represents the interaction of the ion cores with the conduction electron densities inside the volumes $V_0(l)$, the second term that of the conduction electron density inside the volumes $V_0(l)$ with the cores *and* electrons inside other $V_0(l')$. When added to the $C_{\alpha\beta}(\boldsymbol{q})$ which represents the interaction between the ion cores alone, it may be seen that the total represents a modified $C'_{\alpha\beta}(\boldsymbol{q})$ corresponding to interactions between the *total* pseudo-atoms (almost neutral objects) inside the $V_0(l)$, thus alleviating the problem of the large cancellation between the "bare" $C_{\alpha\beta}(\boldsymbol{q})$ and the $E_{\alpha\beta}(\boldsymbol{q})$ alluded to previously. The problem then reduces to calculating the contributions of the terms involving $\Delta \mathbf{W}^{\alpha}$ and $\Delta \mathbf{Q}^{\alpha}$. We do not explicitly state these expressions here since they look complicated in the Fourier representation, but rather give the results in the orbital representation in the next section.

4. *The localized orbital representation*

We have hitherto skirted the problem of actually calculating the density response matrix $\chi(\boldsymbol{q}+\boldsymbol{G},\boldsymbol{q}+\boldsymbol{G}')$, or equivalently of inverting the dielectric function matrix $\chi(\boldsymbol{q}+\boldsymbol{G},\boldsymbol{q}+\boldsymbol{G}')$. It has been known for some time (Hayashi and Shimizu 1969, Sinha 1969) that if the non-interacting response matrix $\chi^0(\boldsymbol{q}+\boldsymbol{G},\boldsymbol{q}+\boldsymbol{G}')$ can be written in separable form, then an expression for $\varepsilon^{-1}(\boldsymbol{q}+\boldsymbol{G},\boldsymbol{q}+\boldsymbol{G}')$ and hence for χ can be written down explicitly. The original form chosen for this separable form of χ^0 was guided by the requirements of reproducing the phenomenological models used to "explain" dispersion curves, such as the shell models or bond charge models (Sinha et al. 1974, Price et al. 1974, Sinha 1973) and was in the nature of an "Ansatz". It was also shown (Hanke 1971, Pick 1971, Sham 1974) that a rigorously separable form of χ^0 could be written down if the Bloch wavefunctions were rewritten in a localized representation. Although the earlier formulations of this method were based on a representation in terms of Wannier functions, as we shall see below, *any* set of localized orbitals will do, provided that they form a good basis set for representation of the wavefunctions. In fact, the angular momentum orbital representations such as occur in the cellular methods of band theory, e.g. the linear combination of muffin-tin orbitals (LCMTO) (Andersen 1975) may also be conveniently used (Sinha and Harmon 1976). In this section, we shall indicate how the density response matrix may be formulated in such a localized representation and also consider the effect of exchange and correlation on χ.

It turns out to be convenient to use the same representation to discuss the expressions for the phonon frequencies in terms of the transformed electron–phonon matrix elements discussed in the last section and this will be done in §4.2.

4.1. Evaluation of the density response matrix

Hanke and Sham (1975) have given a detailed treatment of the problem of calculating χ using the localized orbital representation. They used a diagrammatic technique and included the leading exchange term but no correlation effects. In this section we briefly present an alternative derivation of their result using the Green's function method.

Let us write the spatial part of the Bloch wavefunctions in terms of orbitals centered on the lattice sites,

$$\psi_{\boldsymbol{k}\lambda} = N^{-1/2} \sum_{l} \exp(\mathrm{i}\boldsymbol{k}\cdot\boldsymbol{r}_l) \sum_{\xi} A^{\xi}_{\boldsymbol{k}\lambda} \phi_{\xi}(\boldsymbol{r}-\boldsymbol{r}_l), \tag{4.1.1}$$

where for the moment we leave the ϕ_ξ unspecified. We do not necessarily assume that these orbitals form a complete orthonormal set, only that they are sufficient to adequately represent, via eq. (4.1.1), the wavefunctions in the bands of importance (which are necessarily the ones around the Fermi level). By dropping the spin index, we have assumed that the wavefunctions are not spin-dependent, as in a paramagnetic crystal.

At this point, we have two choices: we may cut off the $\phi_\xi(\boldsymbol{r}-\boldsymbol{r}_l)$ *inside* the unit cell of $\boldsymbol{r}_l$ so there is *no* overlap of orbital between different cells (by definition) *or* we may adopt overlapping orbitals. The first method is well suited to using the results of cellular band structure calculations of the wavefunction but may require a larger number of orbitals to achieve a good representation of the wavefunction. (However, it avoids the necessity of calculating multicenter integrals.) The second method may be more economical with orbitals but requires calculation of overlap terms. In order to keep the treatment general, we shall include such overlap terms in the present discussion.

The electron Hamiltonian may be written as (Yamada and Shimizu 1967)

$$\mathcal{H}=\sum_{k\lambda\sigma}E_{k\lambda}C^{\dagger}_{k\lambda\sigma}C_{k\lambda\sigma}+\sum_{\substack{k_1k_2k_3k_4\\ \lambda_1\lambda_2\lambda_3\lambda_4}}\sum_{\sigma\sigma'}W_{k_1\lambda_1,k_2\lambda_2;k_4\lambda_4,k_3\lambda_3}$$
$$\times C^{\dagger}_{k_1\lambda_1\sigma}C^{\dagger}_{k_2\lambda_2\sigma'}C_{k_3\lambda_3\sigma'}C_{k_4\lambda_4\sigma}-\sum_{k_1k_2}\sum_{\sigma\sigma'}W_{k_1\lambda_1,k_2\lambda_2;k_1\lambda_3,k_2\lambda_2}$$
$$\times\langle n_{k_2\lambda_2\sigma}\rangle C^{\dagger}_{k_1\lambda_1\sigma'}C_{k_1\lambda_3\sigma'}+\sum_{\substack{k_1k_2\\ \lambda_1\lambda_2\lambda_3}}\sum_{\sigma}W_{k_1\lambda_1,k_2\lambda_2;k_2\lambda_2,k_1\lambda_3}$$
$$\times\langle n_{k_2\lambda_2\sigma}\rangle C^{\dagger}_{k_1\lambda_1\sigma}C_{k_1\lambda_3\sigma},\tag{4.1.2}$$

where $C^{\dagger}_{k\sigma}, C_{k\lambda\sigma}$ are the creation and annihilation operators respectively for the Bloch state $(k\lambda)$ with spin index σ, and

$$W_{k_1\lambda_1,k_2\lambda_2;k_4\lambda_4,k_3\lambda_3}=\int\int \mathrm{d}r\,\mathrm{d}r'\psi^{*}_{k_1\lambda_1}(r)\psi^{*}_{k_2\lambda_2}(r')v(r-r')$$
$$\times\psi_{k_3\lambda_3}(r')\psi_{k_4\lambda_4}(r),\tag{4.1.3}$$

where $v(r-r')$ is the Coulomb repulsion between electrons. The $E_{k\lambda}$ are the one-electron Hartree–Fock energies, and the third and fourth terms correct for the electron–electron contributions to the first term. $\langle n_{k\lambda\sigma}\rangle$ is the expectation value of the occupation of state $(k\lambda\sigma)$.

Substituting eq. (4.1.1) into eq. (4.1.3), we obtain

$$W_{k_1\lambda_1,k_2\lambda_2;k_4\lambda_4,k_3\lambda_3}=\frac{1}{N^2}\sum_{\xi_1\xi_2\xi_3\xi_4}A^{*\xi_1}_{k_1\lambda_1}A^{*\xi_2}_{k_2\lambda_2}A^{\xi_3}_{k_3\lambda_3}A^{\xi_4}_{k_4\lambda_4}$$
$$\times\sum_{ijkl}\exp\left[\mathrm{i}(k_3\cdot r_k+k_4\cdot r_l-k_1\cdot r_i-k_2\cdot r_j)\right]V_{\xi_1 i,\xi_2 j;\xi_4 l,\xi_3 k},\tag{4.1.4}$$

where

$$V_{\xi_1 i,\xi_2 j;\xi_4 l,\xi_3 k}=\int\int \mathrm{d}r\,\mathrm{d}r'\phi^{*}_{\xi_1}(r-r_i)\phi^{*}_{\xi_2}(r-r_j)v(r-r')$$
$$\times\phi_{\xi_4}(r-r_l)\phi_{\xi_3}(r'-r_k).\tag{4.1.5}$$

Eq. (4.1.4) may be rewritten as

$$W_{\boldsymbol{k}_1\lambda_1,\boldsymbol{k}_2\lambda_2;\boldsymbol{k}_4\lambda_4,\boldsymbol{k}_3\lambda_3}=\frac{1}{N}\delta_{\boldsymbol{k}_1+\boldsymbol{k}_2,\boldsymbol{k}_3+\boldsymbol{k}_4}\sum_{\xi_1\xi_2\xi_3\xi_4}\sum_{m,n}A^{*\xi_1}_{\boldsymbol{k}_1\lambda_1}A^{*\xi_2}_{\boldsymbol{k}_2\lambda_2}$$
$$\times\left(A^{\xi_3}_{\boldsymbol{k}_3\lambda_3}\exp(\mathrm{i}\boldsymbol{k}_3\cdot\boldsymbol{r}_m)\right)\left(A^{\xi_4}_{\boldsymbol{k}_4\lambda_4}\exp(\mathrm{i}\boldsymbol{k}_4\cdot\boldsymbol{r}_n)\right)$$
$$+\left[V^{\mathrm{c}}_{\xi_1,\xi_2;\xi_4 n,\xi_3 m}(\boldsymbol{k}_1-\boldsymbol{k}_4)+V^{x}_{\xi_1,\xi_2;\xi_4 n,\xi_3 m}(\boldsymbol{k}_1-\boldsymbol{k}_3)\right], \tag{4.1.6}$$

where $\boldsymbol{r}_m$, $\boldsymbol{r}_n$ are lattice vectors,

$$V^{\mathrm{c}}_{\xi_1,\xi_2;\xi_4 n,\xi_3 m}(\boldsymbol{q})=\sum_j\exp(\mathrm{i}\boldsymbol{q}\cdot\boldsymbol{r}_j)\int\int\mathrm{d}\boldsymbol{r}\,\mathrm{d}\boldsymbol{r}'\phi^*_{\xi_1}(\boldsymbol{r})\phi^*_{\xi_2}(\boldsymbol{r}'-\boldsymbol{r}_j)v(\boldsymbol{r}-\boldsymbol{r}')$$
$$\times\phi_{\xi_3}(\boldsymbol{r}'-\boldsymbol{r}_j-\boldsymbol{r}_m)\phi_{\xi_4}(\boldsymbol{r}-\boldsymbol{r}_n) \tag{4.1.7}$$

and

$$V^{\mathrm{x}}_{\xi_1,\xi_2;\xi_4 n,\xi_3 m}(\boldsymbol{q})=\sum_j{}'\exp(\mathrm{i}\boldsymbol{q}\cdot\boldsymbol{r}_j)\int\int\mathrm{d}\boldsymbol{r}\,\mathrm{d}\boldsymbol{r}'\phi^*_{\xi_1}(\boldsymbol{r})\phi^*_{\xi_2}(\boldsymbol{r}'-\boldsymbol{r}_j)v(\boldsymbol{r}-\boldsymbol{r}')$$
$$\times\phi_{\xi_3}(\boldsymbol{r}'-\boldsymbol{r}_m)\phi_{\xi_4}(\boldsymbol{r}-\boldsymbol{r}_j-\boldsymbol{r}_n), \tag{4.1.8}$$

where the prime on the summation sign means the term with $\boldsymbol{r}_j=0$ is excluded, since it is already included in V^{c}.

The frequency-dependent density response function matrix may be expressed in terms of the retarded thermal Green's function (Zubarev 1960) as

$$\chi(\boldsymbol{q}+\boldsymbol{G},\boldsymbol{q}+\boldsymbol{G}',\omega)=-2\pi\sum_{k_1k_2}\sum_{\sigma}\langle\boldsymbol{k}_1\lambda_1|\exp[-\mathrm{i}(\boldsymbol{q}+\boldsymbol{G})\cdot\boldsymbol{r}]|\boldsymbol{k}_1+\boldsymbol{q}\lambda_2\rangle$$
$$\times\langle\boldsymbol{k}_2+\boldsymbol{q}\lambda_3|\exp[\mathrm{i}(\boldsymbol{q}+\boldsymbol{G})\cdot\boldsymbol{r}]|\boldsymbol{k}_2\lambda_4\rangle$$
$$\times G_{\boldsymbol{k}_1\lambda_1\sigma,\boldsymbol{k}_1+\boldsymbol{q}\lambda_2\sigma;\boldsymbol{k}_2\boldsymbol{q}\lambda_3\sigma,\boldsymbol{k}_2\lambda_4\sigma}(\omega), \tag{4.1.9}$$

where $G_{\boldsymbol{k}_1\lambda_1\sigma,\boldsymbol{k}_1+\boldsymbol{q}\lambda_2\sigma;\boldsymbol{k}_2\boldsymbol{q}\lambda_3\sigma,\boldsymbol{k}_2\lambda_4\sigma}(\omega)$ is the time Fourier transform of the two-particle electron Green's function

$$G_{\boldsymbol{k}_1\lambda_1\sigma,\boldsymbol{k}_1+\boldsymbol{q}\lambda_2\sigma;\boldsymbol{k}_2+\boldsymbol{q}\lambda_3\sigma,\boldsymbol{k}_2\lambda_4\sigma}(t-t')$$
$$=\mathrm{i}\theta(t-t')\left\langle\left[C^{\dagger}_{\boldsymbol{k}_1\lambda_1\sigma}(t)C_{\boldsymbol{k}_1+\boldsymbol{q}\lambda_2\sigma}(t),C^{\dagger}_{\boldsymbol{k}_2+\boldsymbol{q}\lambda_3\sigma}(t')C_{\boldsymbol{k}_2\lambda_4\sigma}(t')\right]\right\rangle. \tag{4.1.10}$$

In the usual manner, we may write down the equation of motion for this Green's function, take the time Fourier transform and approximate the commutator $[C^{\dagger}_{k\lambda,\sigma}C_{k+q\lambda_2\sigma}, \mathcal{H}]$ by making the extended random phase approximation (Yamada and Shimizu 1967). That is, we reduce the resulting four-particle operators according to

$$C_1^{\dagger}C_2^{\dagger}C_3C_4 \simeq \langle n_1\rangle[\delta_{14}C_2^{\dagger}C_3-\delta_{13}C_2^{\dagger}C_4]+\langle n_2\rangle[\delta_{23}C_1^{\dagger}C_4-\delta_{24}C_1^{\dagger}C_3]. \tag{4.1.11}$$

The final result for the equation of motion is

$$\begin{aligned}
(\omega+E_{k_1\lambda_1}-E_{k_1+q\lambda_2})&G_{k_1\lambda_1\sigma,k_1+q\lambda_2\sigma;k_2+q\lambda_3\sigma,k_2\lambda_4\sigma}(\omega)\\
&=-\frac{1}{2\pi}[\langle n_{k_1\lambda_1\sigma}\rangle-\langle n_{k_1+q\lambda_2\sigma}\rangle]\delta_{k_1k_2}\delta_{\lambda_1\lambda_4}\delta_{\lambda_2\lambda_3}\\
&\quad+[\langle n_{k_1\lambda_1\sigma}\rangle-\langle n_{k_1+q\lambda_2\sigma}\rangle]\sum_{\xi_1\xi_2\xi_3\xi_4}\sum_{\substack{k_3\\ \mu_1\mu_2}}\sum_{\sigma'}\sum_{mn}A^{*\xi_1}_{k_1+q\lambda_2}A^{*\xi_2}_{k_3\mu_1}\\
&\quad\times(A^{\xi_3}_{k_3+q\mu_2}\exp(\mathrm{i}(k_3+q)\cdot r_m))(A^{\xi_4}_{k_1\lambda_1}\exp(\mathrm{i}k_1\cdot r_n))\\
&\quad\times\{V^{\mathrm{c}}_{\xi_1,\xi_2;\xi_4n,\xi_3m}(q)+V^{\mathrm{x}}_{\xi_1\xi_2;\xi_4n,\xi_3m(k_1-k_3)}-\delta_{\sigma'\sigma}V^{\mathrm{x}}_{\xi_1\xi_2;\xi_3m,\xi_4n}(q)\\
&\qquad-\delta_{\sigma'\sigma}V^{\mathrm{c}}_{\xi_1\xi_2;\xi_3m,\xi_4n}(k_1-k_3)\}\\
&\quad\times G_{k_3\mu_1\sigma',k_3+q\mu_2\sigma';k_2+q\lambda_3\sigma,k_2\lambda_4\sigma}(\omega).
\end{aligned} \tag{4.1.12}$$

The presence of the terms involving arguments (k_1-k_3) on the right-hand side in eq. (4.1.12) makes this integral equation difficult to solve. If we average these terms over k_1 and k_3, it may be seen from eqs. (4.1.7) and (4.1.8) that the only non-vanishing contribution from these terms comes from $r_j=0$, i.e. $-\frac{1}{2}V_{\xi_10,\xi_20;\xi_3m,\xi_4n}$ which we may denote by $-\frac{1}{2}V^{0}_{\xi_1,\xi_2;\xi_3m,\xi_4n}$. We may now combine these terms with the q-dependent terms to obtain

$$\begin{aligned}
(\omega+E_{k_1\lambda_1}-E_{k_1+q\lambda_2})&G_{k_1\lambda_1\sigma,k_1+q\lambda_2\sigma;k_2+q\lambda_3\sigma,k_2\lambda_4\sigma}(\omega)\\
&=-\frac{1}{2\pi}[\langle n_{k_1\lambda_1\sigma}\rangle-\langle n_{k_1+q\lambda_2\sigma}\rangle]\delta_{k_1k_2}\delta_{\lambda_1\lambda_4}\delta_{\lambda_2\lambda_3}\\
&\quad+[\langle n_{k_1\lambda_1\sigma}\rangle-\langle n_{k_1+q\lambda_2\sigma}\rangle]\sum_{\xi_1\xi_2\xi_3\xi_4}\sum_{k_3}A^{\xi_1}_{k_1+q\lambda_2}A^{*\xi_2}_{k_3\mu_1}\\
&\quad\times(A^{\xi_3}_{k_3+q\mu_2}\exp(\mathrm{i}(k_3+q)\cdot r_n))(A^{\xi_1}_{k_1\lambda_1}\exp(\mathrm{i}k_1\cdot r_n))\\
&\quad\times\tilde{V}_{\xi_1\xi_2;\xi_4n,\xi_3m}(q)G_{k_3\mu_1\sigma,k_3+q\mu_2\sigma;k_2+q\lambda_3\sigma,k_2\lambda_4\sigma}(\omega),
\end{aligned} \tag{4.1.13}$$

where

$$\tilde{V}_{\xi_1\xi_2;\xi_4 n,\xi_3 m}(\boldsymbol{q}) = V^{c}_{\xi_1\xi_2;\xi_4 n,\xi_3 m}(\boldsymbol{q}) - \frac{1}{2} V^{x}_{\xi_1\xi_2;\xi_3 m,\xi_4 n}(\boldsymbol{q}) - \frac{1}{2} V^{0}_{\xi_1;\xi_2;\xi_3 m,\xi_4 n}. \quad (4.1.14)$$

We may solve eq. (4.1.13) by defining a quantity

$$\chi_{\xi_1\xi_2;\xi_4 n,\xi_3 m}(q,\omega) = -2\pi \sum_{\substack{\boldsymbol{k}_1\boldsymbol{k}_2\\ \lambda_1\lambda_2\lambda_3\lambda_4}} \sum_{\sigma} A^{*\xi_1}_{\boldsymbol{k}_1\lambda_1} A^{*\xi_2}_{\boldsymbol{k}_2+\boldsymbol{q}\lambda_3} \times \left(A^{\xi_4}_{\boldsymbol{k}_1+\boldsymbol{q}\lambda_2}\exp(\mathrm{i}(\boldsymbol{k}_1+\boldsymbol{q})\cdot\boldsymbol{r}_n)\right)\left(A^{\xi_3}_{\boldsymbol{k}_2\lambda_4}\exp(\mathrm{i}\boldsymbol{k}_2\cdot\boldsymbol{r}_m)\right) \times G_{\boldsymbol{k}_1\lambda_1\sigma,\boldsymbol{k}_1+\boldsymbol{q}\lambda_2\sigma;\boldsymbol{k}_2+\boldsymbol{q}\lambda_3\sigma,\boldsymbol{k}_2\lambda_4\sigma}(\omega). \quad (4.1.15)$$

Then, multiplying both sides of eq. (4.1.13) by

$$-2\pi A^{*\xi_1}_{\boldsymbol{k}_1\lambda_1} A^{*\xi 2}_{\boldsymbol{k}_2+\boldsymbol{q}\lambda_3}\left(A_{\boldsymbol{k}_1+\boldsymbol{q}\lambda_2}\exp(\mathrm{i}(\boldsymbol{k}_1+\boldsymbol{q})\cdot\boldsymbol{r}_n)\right) \times \left(A_{\boldsymbol{k}_2\lambda_4}\exp(\mathrm{i}\boldsymbol{k}_2\cdot\boldsymbol{r}_m)\right)\left[E_{\boldsymbol{k}_1\lambda_1} - E_{\boldsymbol{k}_1+\boldsymbol{q}\lambda_2} + \omega\right]^{-1}$$

and summing over $\boldsymbol{k}_1$, $\boldsymbol{k}_2$, λ_1, λ_2, λ_3, λ_4 and σ we obtain

$$\chi_{\xi_1\xi_2;\xi_4 n,\xi_3 m}(\boldsymbol{q},\omega) = \chi^{0}_{\xi_1\xi_2;\xi_4 n,\xi_3 m}(\boldsymbol{q},\omega) + \sum_{\xi_2\xi_6\xi_7\xi_8} \sum_{m'n'} \chi^{0}_{\xi_1,\xi_5;\xi_4 n,\xi_6 m'}(\boldsymbol{q},\omega)\,\tilde{V}_{\xi_5,\xi_7;\xi_6 m',\xi_8 n'}(\boldsymbol{q}) \times \chi_{\xi_7\xi_2;\xi_8 n',\xi_3 m}(\boldsymbol{q},\omega), \quad (4.1.16)$$

where

$$\chi^{0}_{\xi_1,\xi_2;\xi_4 n,\xi_3 m}(\boldsymbol{q},\omega) = \sum_{\boldsymbol{k}\lambda_1\lambda_2} \sum_{\sigma} \frac{\langle n_{\boldsymbol{k}\lambda_1\sigma}\rangle - \langle n_{\boldsymbol{k}+\boldsymbol{q}\lambda_2\sigma}\rangle}{E_{\boldsymbol{k}\lambda} - E_{\boldsymbol{k}+\boldsymbol{q}\lambda_2} + \omega} \times A^{*\xi_1}_{\boldsymbol{k}\lambda_1} A^{*\xi_2}_{\boldsymbol{k}+\boldsymbol{q}\lambda_2}\left(A_{\boldsymbol{k}+\boldsymbol{q}\lambda_2}\exp(\mathrm{i}(\boldsymbol{k}+\boldsymbol{q})\cdot\boldsymbol{r}_n)\right) \times \left(A^{\xi_3}_{\boldsymbol{k}\lambda_1}\exp(\mathrm{i}\boldsymbol{k}\cdot\boldsymbol{r}_m)\right). \quad (4.1.17)$$

If we combine pairs of indices $(\xi_1,\xi_4 n)$, $(\xi_2,\xi_3 m)$, etc. as new *compound* indices μ_1, μ_2 etc., eq. (4.1.16) may be written compactly as

$$\chi_{\mu_1\mu_2}(\boldsymbol{q},\omega) = \chi^0_{\mu_1\mu_2}(\boldsymbol{q},\omega) + \sum_{\mu_3\mu_4} \chi^0_{\mu_1\mu_3}(\boldsymbol{q},\omega)\tilde{V}_{\mu_3\mu_4}(\boldsymbol{q})\chi_{\mu_4\mu_2}(\boldsymbol{q},\omega) \tag{4.1.18}$$

which has the solution

$$\chi_{\mu_1\mu_2}(\boldsymbol{q},\omega) = \left[(1-\chi^0\tilde{V})^{-1}\chi^0\right]_{\mu_1\mu_2}. \tag{4.1.19a}$$

An equivalent form is

$$\chi_{\mu_1\mu_2}(\boldsymbol{q},\omega) = \left[\chi^0(1-\tilde{V}\chi^0)^{-1}\right]_{\mu_1\mu_2}. \tag{4.1.19b}$$

Substituting in eq. (4.1.9) the representation (4.1.1) for the Bloch wavefunctions and using eq. (4.1.15) we obtain

$$\chi(\boldsymbol{q}+\boldsymbol{G},\boldsymbol{q}+\boldsymbol{G}',\omega) = \sum_{\xi_1\xi_2\xi_3\xi_4} \sum_{mn} \langle\xi_1|\exp[-\mathrm{i}(\boldsymbol{q}+\boldsymbol{G})\cdot\boldsymbol{r}]|\xi_4 n\rangle \times\langle\xi_2|\exp[\mathrm{i}(\boldsymbol{q}+\boldsymbol{G}')\cdot\boldsymbol{r}]|\xi_3 m\rangle\chi_{\mu_1\mu_2}(\boldsymbol{q},\omega), \tag{4.1.20}$$

where $\mu_1 \to (\xi_1, \xi_4 n)$, $\mu_2 \to (\xi_2, \xi_3 m)$ and

$$\chi_{\mu_1\mu_2}(\boldsymbol{q},\omega) = \chi_{\xi_1,\xi_2;\xi_4 n,\xi_3 m}(\boldsymbol{q},\omega).$$

Similarly in eq. (4.1.19), the matrix element $\tilde{V}_{\mu_1\mu_2}(\boldsymbol{q})$ stands for $\tilde{V}_{\xi_1,\xi_2;\xi_4 n,\xi_3 m}(\boldsymbol{q})$. Note that by deriving the frequency-dependent density response function $\chi(\boldsymbol{q}+\boldsymbol{G},\boldsymbol{q}+\boldsymbol{G}',\omega)$ we have to some extent gone beyond the adiabatic approximation. However, for most phonon spectrum calculations, the ω-dependence (which enters via the ω term in the denominator of the expression for χ^0 in eq. (4.1.17)) of χ is negligible. This is because phonon frequencies are small compared to the typical electronic energy differences $(E_{\boldsymbol{k}\lambda_1} - E_{\boldsymbol{k}+\boldsymbol{q}\lambda_2})$ involved.

Comparing eq. (4.1.19b) with eq. (2.19b) we see that what has been achieved is a change of representation from the Fourier representation to a localized orbital representation so that the matrix to be inverted is now in (μ_1, μ_2) space rather than $(\boldsymbol{G}, \boldsymbol{G}')$ space. The idea behind it is, of course, that if a small number of orbitals provide a good representation of the conduction wavefunctions of interest, then μ_1, μ_2 (which we must remember stand for pairs of such orbital indexes) can be kept to a manageable size and the inversion procedure carried out. Physically, as discussed at the end of §2, the index μ_1 labels a *charge fluctuation* corresponding to a virtual transition from an orbital ξ_1 on one lattice site to an orbital ξ_4 on some neighboring site r_n. Fig. 10 illustrates schematically such charge fluctuations and one notices that they may have monopole symmetry, dipole symmetry and so on. $\tilde{V}_{\mu_1\mu_2}(\boldsymbol{q})$ which is the Coulomb (and exchange) interaction in this charge fluctuation representation stands fo the $\boldsymbol{q}$-dependent interaction coefficient between unit amplitude charge fluctuations of types μ_1 and μ_2 propagating through the crystal with periodicity $\boldsymbol{q}$. $\chi_{\mu_1\mu_2}(\boldsymbol{q},\omega)$ is a measure of the amount of charge fluctuation of type μ_1 introduced by an external perturbation of frequency ω and wavevector $\boldsymbol{q}$ which couples to charge fluctuations of type μ_2.

Although the leading exchange contribution is rigorously taken into account in the above formalism, there is a problem arising from the exchange term $V^{\mathrm{x}}_{\mu_1,\mu_2;\mu_3 m,\mu_4 n}(\boldsymbol{q})$ in $\tilde{V}(\boldsymbol{q})$ (eq. (4.1.14)). If one examines the structure of eq. (4.1.8) one sees that because of the long-range nature of the Coulomb interaction, orbitals at quite distant sites $\boldsymbol{r}_m$ ($\simeq \boldsymbol{r}_j$ in eq. (4.1.8)) and $\boldsymbol{r}_n$ ($\simeq -\boldsymbol{r}_j$) can contribute to this term, even if they do not overlap the origin cell. Thus, by necessity the size of the $\tilde{V}$ matrix and hence of the matrix to be inverted becomes unmanageably large. In fact one knows that higher-order effects such as correlations will tend to screen out the long-ranged nature of the Coulomb interaction in eq. (4.1.8) so that the problem may not be a real one. Hanke and Sham (1975) arbitrarily

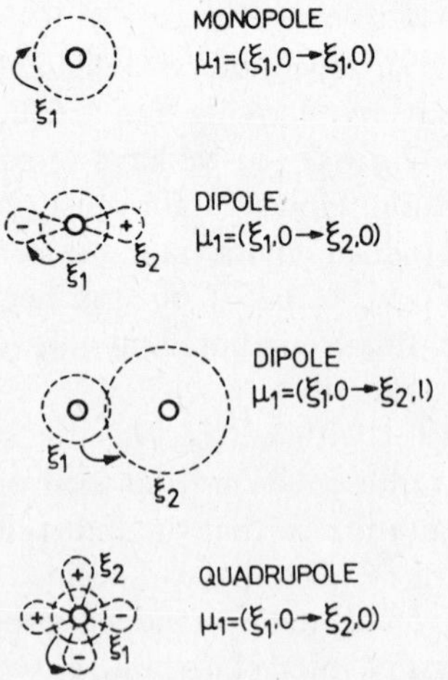

Fig. 10. Schematic representation of various kinds of charge fluctuations corresponding to virtual transitions from an orbital of type ξ_1 on site 0 to an orbital of type ξ_2 on site l.

restricted the m, n to nearest neighbors in calculating this exchange term, but the situation remains somewhat unsatisfactory from this point of view.

An alternative approach which might ultimately prove more satisfactory is to incorporate exchange and correlation effects in the response function using the local density functional theory (Sinha et al., to be published). The local density functional formalism (Hohenberg and Kohn 1964, Kohn and Sham 1965, Hedin and Lundqvist 1971, von Barth and Hedin 1972) appears to describe ground state properties of the electron system in most solids very well (Moruzzi et al. 1977). Since phonons cause essentially adiabatic perturbations of the ground state of the electron system, one might expect the method to work quite well for describing the static density response relevant to phonon calculations. (One could not of course expect the method to yield good results at frequencies comparable to *electronic* excitations.)

We proceed as follows. The effective one-electron potential is given by the sum of an ionic potential, the Hartree potential of the electron system $V_H(\boldsymbol{r})$ and an "exchange-correlation" potential $V_{xc}(\boldsymbol{r})$ which is a functional of the local density $n(\boldsymbol{r})$. If we consider a *weak* external potential acting on the electron system, then using linear response theory we have

$$\delta n(\boldsymbol{r}) = \int \mathrm{d}\boldsymbol{r}' \chi^0(\boldsymbol{r}, \boldsymbol{r}') \{ \delta V_{\text{ext}}(\boldsymbol{r}') + \delta V_H(\boldsymbol{r}') + \delta V_{xc}[n(\boldsymbol{r}')] \}, \tag{4.1.21}$$

where $\chi^0(\boldsymbol{r}, \boldsymbol{r}')$ is the non-interacting response function. Now

$$\delta V_H(\boldsymbol{r}') = \int \mathrm{d}\boldsymbol{r}'' v(\boldsymbol{r}' - \boldsymbol{r}'') \delta n(\boldsymbol{r}'') \tag{4.1.22}$$

and

$$\delta V_{xc}(\boldsymbol{r}') = \frac{\delta V_{xc}[n(\boldsymbol{r}')]}{\delta[n(\boldsymbol{r}')]} \delta n(\boldsymbol{r}') \equiv \mathfrak{F}(\boldsymbol{r}') \delta n(\boldsymbol{r}'), \tag{4.1.23}$$

where the *functional* derivative has been taken and denoted by $\mathfrak{F}(\boldsymbol{r}')$.

Fourier transformation of eq. (4.1.21) yields

$$\begin{aligned} \delta n(\boldsymbol{q}+\boldsymbol{G}) = {} & \sum_{\boldsymbol{G}'} \chi^0(\boldsymbol{q}+\boldsymbol{G}, \boldsymbol{q}+\boldsymbol{G}') [\delta V_{\text{ext}}(\boldsymbol{q}+\boldsymbol{G}') + v(\boldsymbol{q}+\boldsymbol{G}') \delta n(\boldsymbol{q}+\boldsymbol{G}')] \\ & + \sum_{\boldsymbol{G}'\boldsymbol{G}''} \chi^0(\boldsymbol{q}+\boldsymbol{G}, \boldsymbol{q}+\boldsymbol{G}') \mathfrak{F}(\boldsymbol{G}' - \boldsymbol{G}'') \delta n(\boldsymbol{q}+\boldsymbol{G}''), \end{aligned} \tag{4.1.24}$$

which yields for the density response matrix relating $\delta n(\boldsymbol{q}+\boldsymbol{G})$ to $\delta V_{\text{ext}}(\boldsymbol{q}+\boldsymbol{G}')$,

$$\chi(\boldsymbol{q}+\boldsymbol{G},\boldsymbol{q}+\boldsymbol{G}')=\sum_{\boldsymbol{G}''}T^{-1}(\boldsymbol{q}+\boldsymbol{G},\boldsymbol{q}+\boldsymbol{G}'')\chi^0(\boldsymbol{q}+\boldsymbol{G}'',\boldsymbol{q}+\boldsymbol{G}'), \tag{4.1.25}$$

where

$$T(\boldsymbol{q}+\boldsymbol{G},\boldsymbol{q}+\boldsymbol{G}')=\delta_{\boldsymbol{G}\boldsymbol{G}'}-\chi^0(\boldsymbol{q}+\boldsymbol{G},\boldsymbol{q}+\boldsymbol{G}')v(\boldsymbol{q}+\boldsymbol{G}')$$
$$-\sum_{\boldsymbol{G}''}\chi^0(\boldsymbol{q}+\boldsymbol{G},\boldsymbol{q}+\boldsymbol{G}'')\mathcal{F}(\boldsymbol{G}''-\boldsymbol{G}'). \tag{4.1.26}$$

Now in the orbital representation, we may write (see eq. (4.1.20)),

$$\chi^0(\boldsymbol{q}+\boldsymbol{G},\boldsymbol{q}+\boldsymbol{G}')=\sum_{\mu_1\mu_2}A_{\mu_1}(-\boldsymbol{q}-\boldsymbol{G})\chi^0_{\mu_1\mu_2}(\boldsymbol{q})A_{\mu_2}(\boldsymbol{q}+\boldsymbol{G}'), \tag{4.1.27}$$

where $A_{\mu_2}(\boldsymbol{q}+\boldsymbol{G}')$ stands for $\langle\xi_2|\exp[\mathrm{i}(\boldsymbol{q}+\boldsymbol{G}')\cdot\boldsymbol{r}]\|\xi_3 m\rangle$, etc. and $\chi^0_{\mu_1\mu_2}(\boldsymbol{q})$ is given by eq. (1.1.17) with $\omega=0$. Then eq. (4.1.26) may be written as

$$T(\boldsymbol{q}+\boldsymbol{G},\boldsymbol{q}+\boldsymbol{G}')=\delta_{\boldsymbol{G}\boldsymbol{G}'}-\sum_{\mu_1\mu_2}A_{\mu_1}(-\boldsymbol{q}-\boldsymbol{G})\chi^0_{\mu_1\mu_2}(\boldsymbol{q})B_{\mu_2}(\boldsymbol{q}+\boldsymbol{G}'), \tag{4.1.28}$$

where

$$B_{\mu_2}(\boldsymbol{q}+\boldsymbol{G})=A_{\mu_2}(\boldsymbol{q}+\boldsymbol{G})v(\boldsymbol{q}+\boldsymbol{G})+\sum_{\boldsymbol{G}''}A_{\mu_2}(\boldsymbol{q}+\boldsymbol{G}'')\mathcal{F}(\boldsymbol{G}''-\boldsymbol{G}'). \tag{4.1.29}$$

T is now of separable form and its inverse may be written down explicitly (see for instance Sinha 1973). The net result is

$$\chi(\boldsymbol{q}+\boldsymbol{G},\boldsymbol{q}+\boldsymbol{G}')=\sum_{\mu_1\mu_2}A_{\mu_1}(-\boldsymbol{q}-\boldsymbol{G})\chi_{\mu_1\mu_2}(\boldsymbol{q})A_{\mu_2}(\boldsymbol{q}+\boldsymbol{G}'), \tag{4.1.30}$$

where

$$\chi_{\mu_1\mu_2}(\boldsymbol{q})=\left[\left(\mathbf{1}-\boldsymbol{\chi}^0\mathbf{V}_{\text{eff}}\right)^{-1}\boldsymbol{\chi}^0\right]_{\mu_1\mu_2}(\boldsymbol{q}), \tag{4.1.31}$$

where

$$(V_{\text{eff}})_{\mu_1\mu_2}(\boldsymbol{q})=\sum_{\boldsymbol{G}}A_{\mu_1}(-\boldsymbol{q}-\boldsymbol{G})A_{\mu_2}(\boldsymbol{q}+\boldsymbol{G})v(\boldsymbol{q}+\boldsymbol{G})$$
$$+\sum_{\boldsymbol{G}\boldsymbol{G}'}A_{\mu_1}(-\boldsymbol{q}-\boldsymbol{G}')\mathcal{F}(\boldsymbol{G}'-\boldsymbol{G})A_{\mu_2}(\boldsymbol{q}+\boldsymbol{G}). \tag{4.1.32}$$

Comparing eq. (4.1.31) with (4.1.19a) we see that in this method we have yet another effective interaction matrix between charge fluctuations μ_1,μ_2.

The first term on the right of eq. (4.1.32) can be shown to be identical to the first term on the right of eq. (4.1.14), so that the difference here is that the "bare exchange" correction term obtained before has been replaced by another exchange and correlation correction term represented by the second term in eq. (4.1.32).

4.2. Derivation of the phonon spectrum in the localized representation

While a formal expression for the dynamical matrix was given in §2, it was pointed out in §3 that it was desirable to transform the electron–phonon interaction in a way that explicitly brought about the large cancellation between $C_{\alpha\beta}(\boldsymbol{q})+E_{\alpha\beta}(\boldsymbol{q})$ by introducing pseudo-neutral objects which moved rigidly with the ion cores. The expressions are most compactly obtained in terms of a canonical transformation of the Hamiltonian and the use of the orbital representation.

We write

$$\mathcal{H}=\mathcal{H}_{\mathrm{el}}+\sum_{\boldsymbol{q}j}\omega^0_{\boldsymbol{q}j}b^\dagger_{\boldsymbol{q}j}b_{\boldsymbol{q}j}+\mathcal{H}_{\mathrm{ep}},\tag{4.2.1}$$

where $\mathcal{H}_{\mathrm{el}}$ is the electronic part of the Hamiltonian as given in eq. (4.1.2), $\omega^0_{\boldsymbol{q}j}$ are the ion plasma frequencies, and

$$\begin{aligned}\mathcal{H}_{\mathrm{ep}}=&-\sum_{l,\alpha}\sum_{\substack{\boldsymbol{k}\boldsymbol{k}'\\ \lambda\lambda'}}\sum_{\sigma}\langle\boldsymbol{k}\lambda|\nabla_\alpha V_{\mathrm{b}}(\boldsymbol{r}-\boldsymbol{r}_l)|\boldsymbol{k}'\lambda'\rangle C^\dagger_{\boldsymbol{k}\lambda}C_{\boldsymbol{k}'\lambda'\sigma}u(l,\alpha)\\ &+\frac{1}{2}\sum_{l\alpha\beta}\sum_{\substack{\boldsymbol{k}\boldsymbol{k}'\\ \lambda\lambda'}}\sum_{\sigma}\langle\boldsymbol{k}\lambda|\nabla_\alpha\nabla_\beta V_{\mathrm{b}}(\boldsymbol{r}-\boldsymbol{r}_l)|\boldsymbol{k}'\lambda'\rangle C^\dagger_{\boldsymbol{k}\lambda\sigma}C_{\boldsymbol{k}'\lambda'\sigma}u(l,\alpha)u(l,\beta).\end{aligned}\tag{4.2.2}$$

We may use eqs. (3.14) and (3.7) to rewrite $\mathcal{H}_{\mathrm{ep}}$ as

$$\begin{aligned}\mathcal{H}_{\mathrm{ep}}=&\sum_{l,\alpha}\sum_{\substack{\boldsymbol{k}\boldsymbol{k}'\\ \lambda\lambda'}}\sum_{\sigma}\Big\{Q^{l,\alpha}_{\boldsymbol{k}\lambda,\boldsymbol{k}'\lambda'}-\langle\boldsymbol{k}\lambda|\nabla_\alpha W(\boldsymbol{r}-\boldsymbol{r}_l)\\ &\qquad-\int\mathrm{d}\boldsymbol{r}'v(\boldsymbol{r}-\boldsymbol{r}')\nabla_\alpha\tilde{n}^0(\boldsymbol{r}-\boldsymbol{r}_l)|\boldsymbol{k}'\lambda'\rangle\Big\}C^\dagger_{\boldsymbol{k}\lambda\sigma}C_{\boldsymbol{k}'\lambda'\sigma}u(l,\alpha)\\ &+\sum_{\substack{\boldsymbol{k}\boldsymbol{k}'\\ \lambda\lambda'}}(E_{\boldsymbol{k}\lambda}-E_{\boldsymbol{k}'\lambda'})\langle\boldsymbol{k}\lambda|A|\boldsymbol{k}'\lambda'\rangle C^\dagger_{\boldsymbol{k}\lambda\sigma}C_{\boldsymbol{k}'\lambda'\sigma}\\ &+\frac{1}{2}\sum_{l\alpha\beta}\sum_{\substack{\boldsymbol{k}\boldsymbol{k}'\\ \lambda\lambda'}}\sum_{\sigma}\langle\boldsymbol{k}\lambda|\nabla_\alpha\nabla_\beta V_{\mathrm{b}}(\boldsymbol{r}-\boldsymbol{r}_l)|\boldsymbol{k}'\lambda'\rangle C^\dagger_{\boldsymbol{k}\lambda\sigma}C^\dagger_{\boldsymbol{k}'\lambda'\sigma}u(l,\alpha)u(l',\beta).\end{aligned}\tag{4.2.3}$$

We now carry out a canonical transformation

$$\mathcal{H}' = \mathrm{e}^{\mathrm{i}S}\mathcal{H}\mathrm{e}^{-\mathrm{i}S} \tag{4.2.4}$$

where

$$S = -\mathrm{i}\sum_{\substack{\boldsymbol{k}_1\boldsymbol{k}_2\\ \lambda_1\lambda_2}}\sum_{\sigma'}\langle \boldsymbol{k}_1\lambda_1|A|\boldsymbol{k}_2\lambda_2\rangle C^{\dagger}_{\boldsymbol{k}_1\lambda_1\sigma'}C_{\boldsymbol{k}_2\lambda_2\sigma'}, \tag{4.2.5}$$

where the matrix element was defined in eq. (3.2). Note that S is first order in the $u(l,\alpha)$.

Keeping terms up to second order in the $u(l,\alpha)$, and making an RPA-like decoupling of the electron operators in terms such as $C^{\dagger}_{\boldsymbol{k}_1\lambda_1}C^{\dagger}_{\boldsymbol{k}_2\lambda_2}C_{\boldsymbol{k}_3\lambda_3}C_{\boldsymbol{k}_4\lambda_4}u(l,\alpha)$ in terms which arise from $\mathcal{H}_{\mathrm{el}}$, we obtain after some tedious algebra the result

$$\mathcal{H}' = \mathcal{H}_{\mathrm{el}} + \sum_{\boldsymbol{q}j}\tilde{\omega}_{\boldsymbol{q}j}b^{\dagger}_{\boldsymbol{q}j}b_{\boldsymbol{q}j} + \mathcal{H}'_{\mathrm{ep}}, \tag{4.2.6}$$

$$\mathcal{H}'_{\mathrm{ep}} = \sum_{l,\alpha}\sum_{\substack{\boldsymbol{k}\boldsymbol{k}'\\ \lambda\lambda'}}\left[Q^{l,\alpha}_{\boldsymbol{k}\lambda,\boldsymbol{k}'\lambda'} - \langle \boldsymbol{k}\lambda|\nabla_{\alpha}W(\boldsymbol{r}-\boldsymbol{r}_l)|\boldsymbol{k}'\lambda'\rangle\right]C^{\dagger}_{\boldsymbol{k}\lambda\sigma}C_{\boldsymbol{k}'\lambda'\sigma}u(l,\alpha), \tag{4.2.7}$$

where $\mathcal{H}_{\mathrm{el}}$ is the same operator as given in eq. (4.1.2), $\tilde{\omega}_{\boldsymbol{q}j}$ are not the "bare" ion plasma frequencies, but frequencies associated with mutual interaction of the pseudo-neutral objects or pseudo-atoms consisting of the (cores and conduction charge inside the volume V_0 around the cores). $W(\boldsymbol{r}-\boldsymbol{r}_l)$ formally should have corrections for the exchange and correlation potentials due to the valence charge inside $V_0(l)$, but this vanishes in the *local* density functional formalism as may be seen from its definition following eq. (3.14).

In deriving eqs. (4.2.6) and (4.2.7), we have neglected terms which arose from $\omega^0_{\boldsymbol{q}j}b^{\dagger}_{\boldsymbol{q}j}b_{\boldsymbol{q}j}$ in eq. (4.2.1) since these are of order (phonon energy)/(electron energy) relative to the terms in eq. (4.2.6). We have also averaged directly over the electron operators in terms involving terms of second order in the displacements such as $u(l,\alpha)u(l,\beta)$.

The ionic displacements are given by

$$u(l,\alpha) = \sum_{\boldsymbol{q}j}\left(\frac{1}{2MN\tilde{\omega}_{\boldsymbol{q}j}}\right)^{1/2}e_{\alpha}(\boldsymbol{q}j)\exp(\mathrm{i}\boldsymbol{q}\cdot\boldsymbol{r}_l)\left(b_{\boldsymbol{q}j}+b^{\dagger}_{-\boldsymbol{q}j}\right) \tag{4.2.8}$$

so that we may write

$$\mathcal{H}'_{\mathrm{ep}} = \sum_{\boldsymbol{q}j}\sum_{\substack{\boldsymbol{k}\boldsymbol{k}'\\ \lambda\lambda'}}\sum_{\sigma}\left(\frac{1}{2NM\tilde{\omega}_{\boldsymbol{q}j}}\right)^{1/2}\sum_{\alpha}P_{\boldsymbol{k}\lambda,\boldsymbol{k}'\lambda'}(\boldsymbol{q}j)e_{\alpha}(\boldsymbol{q}j)C^{\dagger}_{\boldsymbol{k}\lambda\sigma}C_{\boldsymbol{k}'\lambda'\sigma}\left(b_{\boldsymbol{q}j}+b^{\dagger}_{-\boldsymbol{q}j}\right), \tag{4.2.9}$$

where

$$P^{\alpha}_{k\lambda,k'\lambda'}(qj)=\sum_{l}\left[Q^{l,\alpha}_{k\lambda,k'\lambda'}-\langle k\lambda|\nabla_{\alpha}W(r-r_l)|k'\lambda'\rangle\right]\exp(iq\cdot r_l)\qquad(4.2.10)$$

is the electron–phonon matrix element. Expressing the Bloch wavefunctions in the localized representation we may write

$$P^{\alpha}_{k\lambda,k'\lambda'}(q)=\sum_{\xi_1\xi_2 m}A^{*\xi_1}_{k\lambda}(A^{\xi_2}_{k'\lambda'}\exp(ik\cdot r_m))I^{\alpha}_{\xi_1,\xi_2 m}(q)\,\delta_{k,k'+q}.\qquad(4.2.11)$$

where, using eqs. (3.5), (3.8) and (4.1.1)

$$\begin{aligned}I^{\alpha}_{\xi_1,\xi_2^m}(q)=\tfrac{1}{2}\int_{S_0}\mathrm{d}S\Big[&(\nabla_n\phi^{*}_{\xi_1}(r))(\nabla_{\alpha}\phi_{\xi_1}(r-r_m))-\phi^{*}_{\xi_1}(r)\nabla_n(\nabla_{\alpha}\phi_{\xi_2}(r-r_m))\\&+(\nabla_n\phi_{\xi}(r-r_m))(\nabla_{\alpha}\phi^{*}_{\xi_1}(r))-\phi_{\xi_2}(r-r_n)\nabla_n(\nabla_{\alpha}\phi^{*}_{\xi_1}(r))\Big]\\&-i\sum_{G}(q+G)_{\alpha}W(q+G)\int dr\phi^{*}_{\xi_1}(r)\exp(i(q+G)\cdot r)\phi_{\xi_2}(r-r_m)\end{aligned}\qquad(4.2.11a)$$

$I^{\alpha}_{\xi_1,\xi_2 m}(q)$ plays the role of the electron–phonon matrix element in the localized orbital representation.

We may now discuss the expression for the screened electron–phonon matrix element. By eq. (2.19a) we have

$$\varepsilon^{-1}(q+G,q+G')=\delta_{GG'}+v(q+G)\chi(q+G,q+G').\qquad(4.2.12)$$

We shall for the moment neglect exchange and correlation effects. Using eq. (4.1.20), we may write this as

$$\begin{aligned}\varepsilon^{-1}(q+G,q+G')=\delta_{GG'}+v(q+G)&\sum_{\xi_1\xi_2\xi_3\xi_4}\sum_{mn}\langle\xi_1|\exp[-i(q+G)]|\xi_4 n\rangle\\&\times\langle\xi_2|\exp[i(q+G')\cdot r]|\xi_3 m\rangle\chi_{\xi_1\xi_2,\xi_4 n,\xi_3 m}(q).\end{aligned}\qquad(4.2.13)$$

If $\Delta\phi(q+G)$ represent the Fourier components of *any* "bare" potential, the Fourier components of the corresponding screened potential are

$$\sum_{G'}\varepsilon^{-1}(q+G,q+G')\Delta\phi(q+G'),$$

so that, by eq. (4.2.13)

$$\begin{aligned}\langle k\lambda|\Delta\phi_s|k'\lambda'\rangle=\langle k\lambda|\Delta\phi|k'\lambda'\rangle+&\sum_{\xi_1\xi_2\xi_3\xi_4}\sum_{mn}\sum_{GG'}\langle k\lambda|\exp[i(q+G)\cdot r]|k'\lambda'\rangle\\&\times v(q+G)\langle\xi_1|\exp[-i(q+G)\cdot r]|\xi_4 n\rangle\\&\times\Delta\phi(q+G')\langle\xi_2|\exp[i(q+G')\cdot r]|\xi_3 m\rangle\chi_{\xi_1\xi_2;\xi_4 n,\xi_3 m}(q).\end{aligned}\qquad(4.2.14)$$

Writing $\psi_{k\lambda}, \psi_{k'\lambda'}$ in terms of the orbitals, and rewriting eq. (4.1.7) for $V(\boldsymbol{q})$ in terms of a reciprocal space sum instead of a real space sum, we obtain

$$\langle k\lambda|\Delta\phi_s|k'\lambda'\rangle = \langle k\lambda|\Delta\phi k'\lambda'\rangle + \sum_{\xi_1\xi_2\xi_3\xi_4}\sum_{mn} A^{*\xi_1}_{k\lambda}\left(A^{\xi_4}_{k'\lambda'}\exp(\mathrm{i}k'\cdot r_n)\right) \times (\mathbf{V}\chi)_{\xi_1\xi_2;\,\xi_4 n,\,\xi_3 m}\langle\xi_2|\Delta\phi|\xi_3 m\rangle. \tag{4.2.15}$$

Now using the relation $\chi = \chi_0(\mathbf{1}-\mathbf{V}\chi_0)^{-1}$ (eq. (4.1.19b)), this may be rewritten

$$\langle k\lambda|\Delta\phi_s|k'\lambda'\rangle = \sum_{\xi_1\xi_2\xi_3\xi_4}\sum_{mn} A^{*\xi_1}_{k\lambda}\left(A^{\xi_4}_{k'\lambda'}\exp(\mathrm{i}k'\cdot r_n)\right) \times (\mathbf{1}-\mathbf{V}\chi^0)^{-1}_{\xi_1\xi_2;\,\xi_4 n,\,\xi_3 m}\langle\xi_2|\Delta\phi|\xi_3 m\rangle. \tag{4.2.16}$$

Note that $(\mathbf{1}-\mathbf{V}\chi^0)$ plays the role in the localized representation which ε plays in the Fourier representation. We could at this stage put $\Delta\phi(q+G) = \Delta V_b(q+G)$, but we do not wish direct matrix elements of ΔV_b to appear in the expression for the screened electron–phonon matrix element, but rather the transformed expressions obtained earlier. For this purpose, we use the expression (3.19) to eliminate $\Delta V_b(q+G)$, and then use the result (3.7). After some manipulation the final result is for a normal mode (qj)

$$\begin{aligned}\langle k\lambda|\delta V_s|k'\lambda'\rangle &= (E_{k\lambda}-E_{k'\lambda'})\langle k\lambda|A|k'\lambda'\rangle \\ &+ \left(\frac{1}{2NM\tilde{\omega}_{qj}}\right)^{1/2}\sum_{\xi_1\xi_2\xi_3\xi_4}\sum_{mn} A^{*\xi_1}_{k\lambda}\left(A^{\xi_4}_{k'\lambda'}\exp(\mathrm{i}k'\cdot r_n)\right) \\ &\times \sum_{\alpha}(\mathbf{1}-\mathbf{V}\chi^0)^{-1}_{\xi_1\xi_2;\,\xi_4 n,\,\xi_3 m} I^{\alpha}_{\xi_2,\xi_3 m}(q) \\ &\times e_\alpha(qj)\,\delta_{k,k'+q}.\end{aligned} \tag{4.2.17}$$

(Compare eq. (3.21), which was the result in the Fourier representation.) Note that the first term on the right of eq. (4.2.17) vanishes for energy-conserving transitions. Finally, note the exchange and correlation effects may be included by replacing $\mathbf{V}$ in the above by $\mathbf{V}_{\mathrm{eff}}$ (as in eq. (4.1.32)).

The operators $u(l,\alpha)$ are not affected by the canonical transformation since they commute with S. Thus the phonon Green's function (which is essentially the displacement–displacement Green's function) may be obtained from the transformed Hamiltonian $\mathcal{H}'$ by the usual equation of motion method. The new Hamiltonian has the desired property of partially renormalized "ionic" frequencies $\tilde{\omega}_{qj}$ and a transformed electron–phonon matrix element which has the effect of the deep potential well in

$\nabla_\alpha V_b(\boldsymbol{r}-\boldsymbol{r}_l)$ near the ion cores removed. The solution of the phonon Green's function follows methods similar to those discussed by Takahashi (1968), Rajagopal and Cohen (1972) and Gliss and Bilz (1968). It has been carried out in detail by Sinha et al. (unpublished). We shall sketch the method here.

The phonon Green's function $G_{\boldsymbol{q}j}(t-t')$ is defined by

$$G_{\boldsymbol{q}j}(t-t')=\mathrm{i}\theta(t-t')\langle[A_{\boldsymbol{q}j}(t),A^{\dagger}_{\boldsymbol{q}j}(t')]\rangle, \tag{4.2.18}$$

where

$$A_{\boldsymbol{q}j}=b_{\boldsymbol{q}j}+b^{\dagger}_{-\boldsymbol{q}j}. \tag{4.2.19}$$

Using eqs. (4.2.5)–(4.2.9), we have

$$[b_{\boldsymbol{q}j},\mathcal{H}']=\tilde{\omega}_{\boldsymbol{q}j}b_{\boldsymbol{q}j}+\sum_{\boldsymbol{k}\lambda_1\lambda_2}\sum_{\sigma}\sum_{\alpha}\left(\frac{1}{2NM\tilde{\omega}_{\boldsymbol{q}j}}\right)^{1/2} \times P^{\alpha}_{\boldsymbol{k}\lambda_1,\boldsymbol{k}+\boldsymbol{q}\lambda_2}(\boldsymbol{q})e^{*}_{\alpha}(\boldsymbol{q}j)C^{\dagger}_{\boldsymbol{k}\lambda_1}C_{\boldsymbol{k}+\boldsymbol{q}\lambda_2}, \tag{4.2.20}$$

$$[b^{\dagger}_{-\boldsymbol{q}j},\mathcal{H}']=-\tilde{\omega}_{\boldsymbol{q}j}b^{\dagger}_{-\boldsymbol{q}j}-\sum_{\boldsymbol{k}\lambda_1\lambda_2}\sum_{\sigma}\sum_{\alpha}\left(\frac{1}{2NM\tilde{\omega}_{\boldsymbol{q}j}}\right)^{1/2}P^{\alpha}_{\boldsymbol{k}\lambda_1,\boldsymbol{k}+\boldsymbol{q}\lambda_2}(-\boldsymbol{q}) \times e^{*}_{\alpha}(\boldsymbol{q}j)C^{\dagger}_{\boldsymbol{k}\lambda_1\sigma}C_{\boldsymbol{k}+\boldsymbol{q}\lambda_2\sigma}. \tag{4.2.21}$$

The Fourier transform of $G_{\boldsymbol{q}j}(t)$ can be shown, using the above two equations to obey the equation of motion

$$(\omega^2-\omega^2_{\boldsymbol{q}j})G_{\boldsymbol{q}j}(\omega)=-\frac{1}{\pi}\tilde{\omega}_{\boldsymbol{q}j}+2\tilde{\omega}_{\boldsymbol{q}j}\sum_{\boldsymbol{k}\lambda_1\lambda_2}\sum_{\sigma}\sum_{\alpha}P^{\alpha}_{\boldsymbol{k}\lambda_1,\boldsymbol{k}+\boldsymbol{q}\lambda_2}(-\boldsymbol{q}) \times e^{*}_{\alpha}(\boldsymbol{q}j)G_{\boldsymbol{k}\lambda_1\sigma,\boldsymbol{k}+\boldsymbol{q}\lambda_2\sigma;\boldsymbol{q}j}(\omega), \tag{4.2.22}$$

where $G_{\boldsymbol{k}\lambda_1\sigma,\boldsymbol{k}+\boldsymbol{q}\lambda_2\sigma}(\omega)$ is the Fourier transform of the Green's function

$$G_{\boldsymbol{k}\lambda_1\sigma,\boldsymbol{k}+\boldsymbol{q}\lambda_2\sigma;\boldsymbol{q}j}(t-t')=\mathrm{i}\theta(t-t')\langle[C^{\dagger}_{\boldsymbol{k}\lambda_1\sigma}(t)C_{\boldsymbol{k}+\boldsymbol{q}\lambda_2\sigma}(t),A^{\dagger}_{\boldsymbol{q}j}(t')]\rangle. \tag{4.2.23}$$

The equation of motion of this Green's function and its solution may be obtained by methods almost identical to those used in deriving the result for the two-electron Green's function in §4.1, utilizing the orbital representation, and the result is

$$G_{k\lambda_1\sigma,k+q\lambda_2\sigma;qj}(\omega)=\sum_{\xi_1\xi_2\xi_3\xi_4}\sum_{mn}A^{*\xi_1}_{k+q\lambda_2}\left(A^{\xi_2}_{k\lambda_1}\exp(\mathrm{i}k\cdot r_m)\right)$$
$$\times\frac{\langle n_{k\lambda_1\sigma}\rangle-\langle n_{k+q\lambda_2\sigma}\rangle}{E_{k\lambda_1\sigma}-E_{k+q\lambda_2\sigma+\omega}}$$
$$\times\left[\mathbf{1}-\tilde{\mathbf{V}}\boldsymbol{\chi}^0\right]^{-1}_{\xi_1\xi_3;\,\xi_2 m,\xi_4 n}\left(\frac{1}{2NM\tilde{\omega}_{qj}}\right)^{1/2}$$
$$\times\sum_{\alpha}I^{\alpha}_{\xi_3,\xi_4 n}(q)e_{\alpha}(qj)G_{qj}(\omega).\tag{4.2.24}$$

In deriving this result, we have decoupled operators of the form $C^{\dagger}_{k\lambda\sigma}C_{k'\lambda'\sigma}A_{qj}$ in the equations of motion as $\langle C^{\dagger}_{k\lambda\sigma}C_{k'\lambda'\sigma}\rangle A_{qj}$ i.e. we have at this stage decoupled the electrons from the phonons. Substituting the results in eq. (4.2.24) in eq. (4.2.22), and using eq. (4.2.10) we obtain finally after summing over spins

$$G_{qj}(\omega)=-\frac{1}{\pi}\tilde{\omega}_{qj}\left[\omega^2-\tilde{\omega}^2_{qj}+2\tilde{\omega}_{qj}\Sigma_{qj}(\omega)\right]^{-1},\tag{4.2.25}$$

where

$$\Sigma_{qj}(\omega)=\frac{1}{2M\tilde{\omega}_{qj}}\sum_{\alpha\beta}\sum_{\mu_1\mu_2\mu_3}e^{*}_{\alpha}(qj)I^{\alpha}_{\mu_1}(-q)\chi^{0}_{\mu_1\mu_2}(q,\omega)$$
$$\times\left[\mathbf{1}-\mathbf{V}\boldsymbol{\chi}^0\right]^{-1}_{\mu_2\mu_3}I^{\beta}_{\mu_3}(q)e_{\beta}(qj),\tag{4.2.26}$$

where we have again adopted the compound index μ_1 to stand for $(\xi_1,\xi_2 m)$ etc. Eqs. (4.2.25) and (4.2.26) show that the phonon frequencies are given by

$$\omega^2_{qj}=\tilde{\omega}^2_{qj}-\frac{1}{M}\sum_{\alpha\beta}e^{*}_{\alpha}(qj)\sum_{\mu_1\mu_2}I^{\alpha}_{\mu_1}(-q)\chi_{\mu_1\mu_2}(q)I^{\beta}_{\mu_2}(q)e_{\beta}(qj),\tag{4.2.27}$$

where we have neglected the ω-dependence of $\chi_{\mu_1\mu_2}(q,\omega)$ in accordance with the adiabatic approximation. Eq. (4.2.27) shows that the ω_{qj} are the eigenvalues of the dynamical matrix

$$D_{\alpha\beta}(q)=C'_{\alpha\beta}(q)-\frac{1}{M}\sum_{\mu_1\mu_2}I^{\alpha}_{\mu_1}(-q)\chi_{\mu_1\mu_2}(q)I^{\beta}_{\mu_2}(q),\tag{4.2.28}$$

where $C'_{\alpha\beta}(\boldsymbol{q})$ is the " pseudo-atom" part which yields $\tilde{\omega}^2_{qj}$. (This term also contains the constant term which cancels the second term in the limit $\boldsymbol{q}\to 0$ in order to satisfy translational invariance.) The result (4.2.28) may be compared with eq. (2.23). The only difference is a transformation of the matrix elements and a change of $C_{\alpha\beta}(\boldsymbol{q})$ to $C'_{\alpha\beta}(\boldsymbol{q})$.

Let us examine the small-$\boldsymbol{q}$ behavior of the second term on the right of eq. (4.2.28). Let (Z^*e) be the total charge inside the volume $V_0(l)$. Then at large r, $W(r)$ will behave as $-Z^*e/r$. For simplicity consider only one orbital of importance in the bands at the Fermi level. The dominant (singular) terms in eq. (4.2.28) as $\boldsymbol{q}\to 0$ will arise from the monopole charge fluctuations, i.e. where μ_1, μ_2 correspond to transitions from one orbital to the same orbital on the same site. $Q^{l,\alpha}_{k\lambda,k'\lambda'}$ stays finite as $\boldsymbol{q}\to 0$. Thus by eqs. (4.2.11) and (4.2.11a), the dominant term in $I^{\alpha}_{\mu_1}(\boldsymbol{q})$ as $\boldsymbol{q}\to 0$ arises from the term involving $\nabla_\alpha W(\boldsymbol{r}-\boldsymbol{r}_l)$ and is $(4\pi Z^* e^2/\Omega_0 q^2)q_\alpha$($\Omega_0$=volume of unit cell). The monopole term in the response function $\chi_{\mu_1\mu_1}(\boldsymbol{q})$ tends to $(4\pi e^2/\Omega_0 q^2)^{-1}$ as $\boldsymbol{q}\to 0$. Thus the second term on the right in eq. (4.2.28) goes as $-(1/M)(4\pi e^2 Z^{*2}/\Omega_0)(q_\alpha q_\beta/q^2)$ which is exactly cancelled by the singular behavior of $C'_{\alpha\beta}(\boldsymbol{q})$ as $\boldsymbol{q}\to 0$, since the pseudo-atoms have a charge Z^*.

Eq. (4.2.28) is the central result of the development of our theory of phonon spectra in terms of localized orbitals and we shall now discuss its physical significance. $I^{\alpha}_{\mu_2}$ represents the coupling of the lattice to the deformation part (i.e. *apart* from the rigid motion of the "pseudo-atom") of the electron density fluctuations or "dynamic charge density waves" of type μ_1, $\chi_{\mu_1\mu_2}$ represents the self-consistent response exciting μ_1 type of charge fluctuations in response to the coupling of the lattice to the μ_2-type charge fluctuations, and finally $I^{\alpha}_{\mu_1}$ represents the coupling of these charge fluctuations back to the lattice, thus giving the usual type of second-order result for the energy.

Let us write the lattice equations of motion (4.2.27) in terms of extra internal degrees of freedom w_{μ_1}, representing the amplitudes of the charge fluctuations of type μ_1, etc. Using the result in eq. (4.1.31), we have

$$\omega^2 e_\alpha = \sum_\beta C'_{\alpha\beta}(\boldsymbol{q})e_\beta + \sum_{\mu_2} I^{\alpha}_{\mu_2}(-\boldsymbol{q})w_{\mu_2},$$

$$0 = \sum_\beta I^{\beta}_{\mu_2}(\boldsymbol{q})e_\beta + \sum_{\mu_2}\left[(V_{\text{eff}})_{\mu_1\mu_2}(\boldsymbol{q}) + \left(\chi^0_{\mu_1\mu_2}\right)^{-1}(\boldsymbol{q})\right]w_{\mu_2}. \qquad (4.2.29)$$

The similarity of structure here to the equations of motion of the various model theories, such as the shell model, breathing shell model, etc., is obvious. These model theories essentially pick out some particular subset of charge fluctuations μ_2 of monopolar type, dipolar type, etc. on some physical grounds.

4.3. Representation in terms of (l,m) orbitals

The original formulations of the localized orbital method conceived of the $\phi_\xi(\boldsymbol{r}-\boldsymbol{r}_l)$ in terms of the Wannier function formed from the Bloch orbitals. In general, however, this leads to certain practical difficulties, namely the problem of obtaining well-localized Wannier functions from the band wavefunctions, and the necessity for calculating multicenter overlap integrals. Therefore, it is worth examining an expansion of the band wavefunctions in terms of angular momentum orbitals localized in each cell. We call these (l,m) orbitals.

All cellular methods such as the augmented plane wave (APW) method or the Kohn–Korringa–Rostoker (KKR) method yield a form for the wavefunctions inside the muffin-tin sphere which is of the form

$$\psi_{\boldsymbol{k}\lambda}=N^{-1/2}\exp(\mathrm{i}\boldsymbol{k}\cdot\boldsymbol{r}_j)\sum_{lm}A_{\boldsymbol{k}\lambda}^{lm}Y_{lm}(\theta,\phi)R_l(|\boldsymbol{r}-\boldsymbol{r}_j|,E_{\boldsymbol{k}\lambda}), \tag{4.3.1}$$

where θ,ϕ represent the polar angles of the vector $(\boldsymbol{r}-\boldsymbol{r}_j)$ and eq. (4.3.1) stands for $\psi_{\boldsymbol{k}\lambda}$ inside the jth unit cell. The radial function R_l depends on the energy of the state $(\boldsymbol{k}\lambda)$. In the spirit of Andersen's atomic sphere approximation (ASA) (Andersen 1975) we assume that eq. (4.3.1) provides an adequate representation of $\psi_{\boldsymbol{k}\lambda}$ throughout the unit cell (approximated by a Wigner–Seitz sphere) rather than just inside the muffin-tin sphere. Next we examine the energy dependence of the functions R_l. For transition metals, the energy dependence of R_l for $l=0$ and $l=1$ is very small (involving a maximum change of 5% in the function) across all the bands of importance around the Fermi level. This is not true for $l\geqslant 2$. As in Andersen's linear combination of muffin-tin orbitals (LCMTO) method (Andersen 1975, Pettifor 1972) we may write

$$R_l(E,r)=R_{l,1}(E_0,r)+(E-E_0)R_{l,2}(E_0,r), \tag{4.3.2}$$

where E_0 is the resonance energy.

Combining this with eq. (4.3.1) we see that we may write $\psi_{\boldsymbol{k}\lambda}$ in the general form

$$\psi_{\boldsymbol{k}\lambda}=N^{-1/2}\exp(\mathrm{i}\boldsymbol{k}\cdot\boldsymbol{r}_j)\sum_{lms}A_{\boldsymbol{k}\lambda}^{lm;s}\phi_{lm,s}(\boldsymbol{r}-\boldsymbol{r}_j), \tag{4.3.3}$$

where the index s is only different from 1 for $l\geqslant 2$, in which case it runs over 1 and 2 denoting the orbitals given in eq. (4.3.2).

In eq. (4.3.3),

$$\begin{aligned}\phi_{lm,s}(\boldsymbol{r}) &= Y_{lm}(\theta,\phi)R_{l,s}(\boldsymbol{r}), && \boldsymbol{r} \text{ inside unit cell}\\ &=0, && \boldsymbol{r} \text{ outside unit cell.}\end{aligned} \tag{4.3.4}$$

We may combine (l,m,s) into a single index ξ.

With this choice of the ϕ_ξ, we may neglect all the multicenter overlap integrals in eq. (4.1.7), and the indices m,n may be dropped. We now consider the evaluation of the $I^{\alpha}_{\xi_1\xi_2}(\boldsymbol{q})$ given by eq. (4.2.11). It may be shown using the gradient operator theorem that the surface integral given in eq. (3.5) can be reduced to give (Sinha 1968)

$$M^{l,\alpha}_{\boldsymbol{k}\lambda,\boldsymbol{k}'\lambda'} = N^{-1}\sum_{\xi_1\xi_2} A^{*\xi_1}_{\boldsymbol{k}\lambda} A^{\xi_2}_{\boldsymbol{k}'\lambda'} \exp\left[\mathrm{i}(\boldsymbol{k}'-\boldsymbol{k})\cdot\boldsymbol{r}_l\right] N^{\alpha}_{\xi_1\xi_2}, \tag{4.3.5}$$

where (in units such that $\hbar^2/2m=1$)

$$\begin{aligned} N^{\alpha}_{\xi_1\xi_2} = 4\pi r_s^2 \Bigg[& \delta_{l_1,l_2-1}\left(\frac{l_2}{2l_2+1}\right)^{1/2} C^{l_2-1,1,l_2}_{m_1,m_2-m_1}\Bigg\{\frac{l_2+1}{r_s^2}R_{l_1s_1}(r_s)R_{l_2s_2}(r_s)\\ &+\frac{l_2+1}{r_s}\left(R'_{l_1s_1}(r_s)R_{l_2s_2}(r_s)-R_{l_1s_1}(r_s)R'_{l_2s_2}(r_s)\right)\\ &+\left(R'_{l_1s_1}(r_s)R'_{l_2s_2}(r_s)-R_{l_1s_1}(r_s)R''_{l_2s_2}(r_s)\right)\Bigg\}\\ &+\delta_{l_1,l_2+1}\left(\frac{l_2+1}{2l_2+1}\right)^{1/2}\\ &\times C^{l_2+1,1,l_2}_{m_1,m_2-m_1}\Bigg\{\frac{l_2}{r_s^2}R_{l_1s_1}(r_s)R_{l_2s_2}(r_s)+\frac{l_2}{r_s}\big(R'_{l_1s_1}(r_s)\\ &\times R_{l_2s_2}(r_s)-R_{l_1s_1}(r_s)R'_{l_2s_2}(r_s)\big)\\ &-\left(R'_{l_1s_1}(r_s)R'_{l_2s_2}(r_s)-R_{l_1s_1}(r_s)R''_{l_2s_2}(r_s)\right)\Bigg\}\Bigg]A_{\alpha,m_2-m_1}, \end{aligned} \tag{4.3.6}$$

where r_s is the muffin-tin radius, ξ_1 stands for (l_1,m_1,s_1), etc. the C's are the Clebsch–Gordan coefficients and m_1,m_2 can differ by only 0, 1 or -1. A_{α,m_2-m_1} is the matrix which transforms vectors from a spherical to a Cartesian basis (Sinha 1968).

Similarly,

$$M^{*l,\alpha}_{\boldsymbol{k}'\lambda',\boldsymbol{k}\lambda} = N^{-1}\sum_{\xi_1\xi_2} A^{*\xi_1}_{\boldsymbol{k}\lambda} A^{\xi_2}_{\boldsymbol{k}'\lambda'} \exp\left[\mathrm{i}(\boldsymbol{k}-\boldsymbol{k})\cdot\boldsymbol{r}_l\right] N^{*\alpha}_{\xi_2\xi_1}. \tag{4.3.7}$$

Thus we may write, using eqs. (3.8) and (4.2.10)

$$P^{\alpha}_{kv,k'\lambda'}(q)=\delta_{k,k'+q}\sum_{\xi_1\xi_2} A^{*\xi_1}_{k\lambda} A^{\xi_2}_{k'\lambda'}\left\{\frac{1}{2}\left[N^{\alpha}_{\xi_1\xi_2}+N^{*\alpha}_{\xi_2\xi_1}\right]\right.$$

$$\left.-\mathrm{i}\sum_{G}(q+G)_{\alpha}W(q+G)\langle\xi_1|\exp[-\mathrm{i}(q+G)\cdot r]|\xi_2\rangle\right\}. \tag{4.3.8}$$

Comparing with eq. (4.2.11) we see that

$$I^{\alpha}_{\xi_1\xi_2}(q)=\frac{1}{2}\left[N^{\alpha}_{\xi_1\xi_2}+N^{*\alpha}_{\xi_1\xi_2}\right]-\mathrm{i}\sum_{G}(q+G)_{\alpha}W(q+G)$$

$$\times(\xi_1|\exp[-\mathrm{i}(q+G)\cdot r]|\xi_2), \tag{4.3.9}$$

with the $N^{\alpha}_{\xi_1\xi_2}$ defined in eq. (4.3.6). Note that these may be expressed in terms of the logarithmic derivative of the radial wavefunction obtained in the course of a regular band calculation. (The second derivates $R''_{l_2s_2}$ appearing in eq. (4.3.6) may be eliminated by use of the radial Schrödinger equation.)

Eq. (4.3.9) thus provides an explicit expression in terms of which the effective electron–phonon matrix element may be calculated and thus used to calculate phonon spectra using eq. (4.2.28). From the result in eq. (4.2.17), we see that for energy-conserving transitions, the screened electron–phonon matrix element is obtained by replacing $I^{\alpha}_{\xi_1\xi_2}(q)$ by

$$\sum_{\xi_3\xi_4}\left[\mathbf{1}-\mathbf{V}\chi^0\right]^{-1}_{\xi_1\xi_3;\,\xi_2\xi_4} I^{\alpha}_{\xi_3\xi_4}(q).$$

We shall now investigate this further with a view to establishing a rigorous result for the electron–phonon coupling strength λ in the (l,m) representation. In the strong-coupling theory of superconductivity (McMillan 1968), the electron–phonon coupling strength λ is given by

$$\lambda=\frac{2}{n(E_{\mathrm{F}})(2\pi)^3\Omega_0}\sum_{\lambda\lambda'}\sum_{j}\int\int\frac{\mathrm{d}S_k}{|\nabla E_{k\lambda}|}\frac{\mathrm{d}S_{k'}}{|\nabla E_{k'\lambda'}|}\frac{1}{\omega_{qj}}$$

$$\times I^{\mathrm{sc}}_{k\lambda,k'\lambda'}(qj)I^{\mathrm{sc}}_{k'\lambda',k\lambda}(-qj), \tag{4.3.10}$$

where $n(E_{\mathrm{F}})$ is the density of states at the Fermi level and the double integral is over the Fermi surface, q stands for the vector $(k-k')$, and

$I^{sc}_{k\lambda,k'\lambda'}(qj)$ is the coefficient of $C^{\dagger}_{k\lambda\sigma}C_{k'\lambda'\sigma}(b_{qj}+b^{\dagger}_{-qj})$ in the expression for $\mathcal{H}'_{ep}$ (eq. (4.2.9)) but with $P^{\alpha}_{k\lambda,k'\lambda'}$ replaced by the screened matrix element. Using the result derived above, we finally obtain

$$\lambda=\frac{1}{(2\pi)^3\Omega_0 n(E_F)M}\sum_{\lambda\lambda'}\sum_{j}\sum_{\xi_1\ldots\xi_4}\sum_{\xi'_1\ldots\xi'_4}\int\frac{dS_k}{|\nabla E_{k\lambda}|}A^{*\xi_1}_{k\lambda}A^{\xi'_1}_{k\lambda}$$
$$\times\int\frac{dS_{k'}}{|\nabla E_{k'\lambda'}|}A^{*\xi'_2}_{k'\lambda'}A^{\xi_2}_{k'\lambda'}\left[1-\mathbf{V}_{eff}(q)\boldsymbol{\chi}^0(q)\right]^{-1}_{\xi_1\xi_2;\,\xi_2\xi_4}\left[\sum_{\alpha}e_{\alpha}(qj)I^{\alpha}_{\xi_3\xi_4}(q)\right]$$
$$\times\left[1-\mathbf{V}_{eff}(q)\boldsymbol{\chi}^0(q)\right]^{-1}_{\xi'_2\xi'_4;\,\xi'_1\xi'_3}\left[\sum_{\beta}e^{*}_{\beta}(qj)I^{*\beta}_{\xi'_3\xi'_4}(q)\right]\frac{1}{\omega^2_{qj}}. \tag{4.3.11}$$

Note that the so-called rigid-muffin-tin approximation (RMTA) corresponds to (a) neglecting the second term in eq. (4.3.9), i.e. the outer tails of the potentials due to the pseudo-atoms inside $V_0(l)$. (This error may be minimized by extending the muffin-tin radius to the "atomic sphere radius") and (b) neglecting the "screening factors", i.e. putting $[\mathbf{1}-\mathbf{V}_{eff}(q)\boldsymbol{\chi}^0(q)]\simeq\mathbf{1}$.) In this representation, it is not a complete neglect of screening effects (as far as the bare ion potential is concerned) since eqs. (3.19) and (3.21) show that a large portion of the screening is included in replacing ΔV_b by ΔU, i.e. replacing the displaced bare ion potential by the displaced *crystal* (muffin-tin) potential. We also note that $I^{\alpha}_{\xi_3\xi_4}(q)$ has *no* diagonal elements, i.e. it connects only ξ_3 and ξ_4 having l values which differ by ±1 (see eq. (4.3.6)). In the (l,m) representation, we may write $\boldsymbol{\chi}^0$ (eq. (4.1.17)) as

$$\chi^0_{l_1m_1s_1,l_2m_2s_2;\,l_4m_4s_4,l_3m_3s_3}(q)=\sum_{k\lambda_1\lambda_2}\frac{\langle n_{k\lambda_1}-n_{k+q\lambda_2}\rangle}{E_{k\lambda}-E_{k+q\lambda_2}}\cdot$$
$$\times A^{*l_1m_1s_1}_{k\lambda_1}A^{*lm_2s_2}_{k+q\lambda_2}A^{l_4m_4s_4}_{k+q\lambda_2}A^{l_3m_3s_3}_{k\lambda_1}. \tag{4.3.12}$$

Thus the elements of $(\mathbf{1}-\mathbf{V}_{eff}\boldsymbol{\chi}^0)^{-1}$ coupling to $l\to l\pm1$ transitions will be different from unity only for appreciable joint s–p, p–d or d–f densities of states in the bands around the Fermi energy. The RMTA has by now been widely applied in attempts to calculate λ and T_c of transition metal superconductors from first principles (Gaspari and Gyorffy 1972, Butler et al. 1976, Butler 1977, Papaconstantopoulos et al. 1977) with varying degrees of success. In order to make contact with these calculations, some further approximations are necessary.

Let us assume that $1/\omega_{qj}$ can be removed from inside the double integral in eq. (4.3.11) and replaced (McMillan 1968) by

$$1/\bar{\omega}^2=\langle\omega^{-1}\rangle/\langle\omega\rangle, \tag{4.3.13}$$

where $\langle\ \rangle$ denotes an average over the phonon spectrum. Then the closure property of the eigenvectors yields from eq. (4.3.11)

$$\lambda = n(E_F)\langle I^2\rangle / M\bar{\omega}^2, \tag{4.3.14}$$

$$\langle I^2\rangle = \frac{1}{(2\pi)^3\Omega_0 n^2(E_F)} \sum_{\lambda\lambda'} \sum_{\xi_1\xi_2\xi_3\xi_4} \int\int \frac{dS_k}{|\nabla E_{k\lambda}|} \frac{dS_{k'}}{|\nabla E_{k'\lambda'}|} \times A^{*\xi_1}_{k\lambda} A^{\xi_2}_{k\lambda} A^{*\xi_3}_{k'\lambda'} A^{\xi_4}_{k'\lambda'} \sum_\alpha I^\alpha_{\xi_1\xi_4} I^{*\alpha}_{\xi_3\xi_2}. \tag{4.3.15}$$

Since we are working here only at the Fermi energy, and thus do not have to deal with radial functions over *all* the energy bands, we may drop the s subscript and take the R_l to be radial wavefunctions at the Fermi energy. Thus for the above calculation of λ, ξ_1 simply stands for (l_1, m_1), etc.

Now in eq. (4.3.6), we may, since we are dealing with wavefunctions at the muffin-tin radius, replace the radial functions by their form outside the muffin-tin radius r_s; i.e.

$$R_l(r_s, E_F) = N_l^{-1/2}\{ j_l(\kappa r_s)\cos\delta_l - n_l(\kappa r_s)\sin\delta_l\}, \tag{4.3.16}$$

where $\kappa = (E_F)^{1/2}$, and δ_l is the phase shift for angular momentum l, and N_l is a normalization constant.

However, since $R_l''(r)$ appears in the expression (4.3.6), and since the usual muffin-tin potential has a discontinuity at r_s, one must be careful in using the phase-shift expression to put in a term which corrects for the corresponding discontinuity in the second derivative, or equivalently which evaluates $R_l''(r)$ just *inside* the muffin-tin radius rather than outside. If one does this, and uses the properties of the j_l and n_l, one may rewrite eq. (4.3.6) in the form

$$\begin{aligned} N^\alpha_{l_1m_1,l_2m_2} = \frac{4\pi}{2m} \Bigg\{ & -\delta_{l_1+1,l_2}\left(\frac{l_1+1}{2l_1+3}\right)^{1/2} C^{l_1,1,l_1+1}_{m_1,m_2-m_1m_2} \\ & \times\left[\sin(\delta_{l_1}-\delta_{l_1+1}) N_{l_1}^{-1/2} N_{l_1+1}^{-1/2} + 2m(\Delta V) R_{l_1} R_{l_1+1} r_s^2\right] \\ & + \delta_{l_1,l_2+1}\left(\frac{l_2+1}{2l_2+1}\right)^{1/2} C^{l_2+1,1,l_2}_{m_1,m_2-m_1m_2} \\ & \times\left[\sin(\delta_{l_2}-\delta_{l_2+1}) N_{l_2+1}^{-1/2} N_{l_2}^{-1/2} + 2m(\Delta V) R_{l_2} R_{l_2+1} r_s^2\right]\Bigg\} \\ & \times A_{\alpha,m_2-m_1}, \end{aligned} \tag{4.3.17}$$

where ΔV is the difference between the *inside* value of the muffin potential at r_s and the *outside* value (usually taken as zero). The terms involving ΔV were omitted by Gaspari and Gyorffy (1972) in their derivation. For cubic crystals,

$$\langle (I^x)^2 \rangle = \langle (I^y)^2 \rangle = \langle (I^z)^2 \rangle = \frac{1}{3}\langle I^2 \rangle, \tag{4.3.18}$$

so we need simply evaluate $I^z_{l_1m_1,l_2m_2}$. Using eq. (4.3.17) and symmetry properties of the Clebsch–Gordan coefficients, we obtain

$$\begin{aligned} I^z_{l_1m_1,l_2m_2} = 4\pi\, \delta_{m_1,m_2} \Bigg\{ & \delta_{l_1,l_2+1} \left(\frac{l_2+1}{2l_2+1} \right)^{1/2} C^{l_2+1\,1\,l_2}_{m_1\,0\,m_1} \\ & \times N_{l_2}^{-1/2} N_{l_2+1}^{-1/2} (\sin[\delta_{l_2+1} - \delta_{l_2}] + B(l_2)) \\ & - \delta_{l_2,l_1+1} \left(\frac{l_1+1}{2l_1+1} \right)^{1/2} C^{l_1+1\,1\,l_1}_{m_1\,0\,m_1} N_{l_1}^{-1/2} N_{l_1+1}^{-1/2} (\sin[\delta_{l_1} - \delta_{l_1+1}] \\ & + B(l_1)) \Bigg\}, \end{aligned} \tag{4.3.19}$$

where

$$\begin{aligned} B(l) = r_s^2(\Delta V)(j_l(\kappa r_s)\cos\delta_l - n_l(\kappa r_s)\sin\delta_l)(j_{l+1}(\kappa r_s)\cos\delta_{l+1} \\ - n_{l+1}(\kappa r_s)\sin\delta_{l+1}). \end{aligned} \tag{4.3.20}$$

Let us assume that

$$(2l_1+1)\sum_\lambda \int \frac{\mathrm{d}S_k}{|\nabla E_{k\lambda}|} A^{*l_1m_1}_{k\lambda} A^{*l_2m_2}_{k\lambda} = \delta_{l_1l_2}\delta_{m_1m_2} n_{l_1}, \tag{4.3.21}$$

where n_l is the partial density of states at E_F for angular momentum l. Then, substituting in eq. (4.3.15), and using the orthonormality relations for the Clebsch–Gordan coefficients, we obtain finally

$$\langle I^2 \rangle = \frac{2}{(2m^2)\pi\Omega_a n^2(E_F)} \sum_l 2(l+1)[\sin(\delta_l - \delta_{l+1}) + B(l)]^2 \frac{n_l n_{l+1}}{N_l N_{l+1}}, \tag{4.3.22}$$

which is exactly the Gaspari–Gyorffy formula except for the term $B(l)$ in the square brackets.

The assumption (4.3.21) may seem unnecessarily restrictive. However, following John (1973) one may show for crystals with cubic symmetry and provided only the $A_{k\lambda}^{lm}$ up to $l=2$ are important that the result (5.16) is still valid. In practice, the off-diagonal terms contributing to λ, (i.e. terms containing $(I_{p-d}I_{d-f}^*)$ in eq. (4.3.15), and the so-called "non-spherical" terms tend to give overall a numerically small correction to the result in eq. (4.3.22) (Butler et al. 1976).

This is not the proper place for a critical evaluation of the various calculations of λ (or the quantity $\alpha^2F(\omega)$) defined by

$$\lambda=2\int \alpha^2(\omega)\,F(\omega)/\omega\,d\omega \tag{4.3.23}$$

using the RMTA. A detailed discussion will be found in the chapter by Allen. Some general comments may be order, however. There has been a great deal of effort expanded on the calculation of λ using the phase-shift method (Evans et al. 1973, Klein and Papaconstantopoulos 1974, Papaconstantopoulos and Klein 1975, Papaconstantopoulos et al. 1977, Butler et al. 1976, Butler 1977) for a variety of pure transition metals, as well as transition metal compounds such as NbC, PdH and the A-15 compounds. An interesting feature of these calculations (based on the Gaspari–Gyorffy formula) is that they indicate that $l=2$ to $l=3$ (i.e. d−f) scattering processes to be dominant in their contributions to λ for the 4d transition metal series. (The $l=3$ component arises from the tails of overlapping d-orbitals from neighboring sites.) There also appears to be some sensitivity of the results to the choice of potential used for the band calculation. Calculations have also been performed for the spectral function α^2F for Nb (Harmon and Sinha 1977, Butler et al. 1979) and Pd (Allen, to be published), and for λ_k, i.e. the Fermi-surface *orbit*-averaged λ for particular orbits on the Fermi surface (rather than an overall Fermi-surface average), which may be compared directly with electron–phonon enhancement of orbital masses obtained from de Haas–van Alphen experiments (Karim et al. 1978, Dye et al. 1978). On the whole, the calculations predict the correct *trends* for λ across the transition metal series, and often give reasonable quantitative agreement with experiments. Nevertheless, several important questions remain unanswered. The first question relates to the tail of the potentials due to the charges inside the muffin-tin spheres (which are not neutral). We also note that the term B_l which must be added to the $\sin(\delta_l-\delta_{l+1})$ term in eq. (4.3.22) cancels the effect of the discontinuity in the muffin-tin potential at the sphere radius. Numerical calculations (Harmon and Sinha 1977) have shown that this term can make a significant difference in the value for λ for Nb, increasing the s−p and p−d contributions by 30% relative to the Gaspari–Gyorffy formula but decreasing the d−f contribution by 25%, resulting in an overall 10% increase in λ. Butler et al. (1976) have argued that keeping the effect of the muffin-tin discontinuity in the expression for $\langle I^2 \rangle$ (i.e. using the original Gaspari–Gyorffy formula) tends to allow for the tail of the muffin-tin potential in the interstitial regions. This is approximately true, since if one increases the muffin-tin radius to the atomic sphere radius, the muffin-tins *do* approximate neutral objects, the interstitial tail of the potential therefore decreases, and so does the muffin-tin discontinuity at the sphere radius, but it is hardly a rigorous argument. Note that if one does go beyond the RMTA and put in the interstitial potential tails as represented by the second term in eq. (4.3.9), one must put it in *screened* in the total electron–phonon matrix element, or else it will diverge at small $\boldsymbol{q}$-values. RMTA calculations of the electron mass-enhancement in Nb (Harmon and Sinha 1977) yielded opposite trends for the mass-enhancements experimentally observed by the de Haas–van Alphen effects (Dye et al. 1978) on the Γ-centered octahedron, jungle gym and N-centered ellipsoids, respectively. Preliminary calculations introducing the interstitial tails of the muffin-tin potentials, screened in a crude fashion, indicated that these could account for the right trends in the mass-enhancement on different portions of the Fermi surface, but the value of λ then came out too large. Thus we are led back to the question of screening, i.e. what effect the factors $[\mathbf{1}-\mathbf{V}_{\text{eff}}(\boldsymbol{q})\boldsymbol{\chi}^0(\boldsymbol{q})]^{-1}$ have in eq. (4.3.11).

In an interesting recent calculation, Pettifor (1977) has performed an RMTA calculation of the quantity of $(=n(E_F)\langle I^2 \rangle)$ for a number of elements in the 4d-series by expressing the logarithmic derivatives of the radial wavefunction (eq. 4.3.6) in terms of energy band parameters. He also finds very large d–f contributions, and overall of values appreciably larger than the experimental values. He then shows that introduction of screening of the matrix element via a simple Thomas–Fermi screening model decreases η substantially. In particular the d–f contribution is more heavily screened (by a factor of ~2) since it depends on the wavefunctions in the outer parts of the Wigner–Seitz cell where the screening is more effective.

Until the full screened matrix element (including the interstitial tail part) is calculated for a few cases at least and compared with the RMTA result, we shall not know for sure how much the reliability of the latter depends on a somewhat fortuitous cancellation of errors. At the time of writing, further calculations of $\alpha^2 F$ and λ_k are being done for a number of metals by several groups and so further insight should be available soon.

4.4 Tight-binding representation of the electron–phonon matrix elements

In §§4.2 and 4.3, we have developed expressions for the electron–phonon matrix element in terms of localized representations of the Bloch wavefunctions. However, the transformation of the matrix element to a representation in terms of pseudo-atoms inside muffin-tin spheres (and the consequent elimination of the deep potential wells near the ion cores) is mainly suited to a representation in terms of (l, m) or muffin-tin orbitals. An alternative way to proceed is to use the tight-binding form of the wavefunctions (including overlap) and transform the electron–phonon matrix element without introducing muffin tins at all. We shall outline this procedure here.

Let us consider again the representation (4.1.1) for the Bloch wavefunctions, where the ϕ_ξ are now atomic orbitals which are *not* necessarily orthogonal to those on neighboring sites. Let $V_{\rm el}(\boldsymbol{r}-\boldsymbol{r}_l)$ be the potential due to the electrons in the occupied orbitals on the site $\boldsymbol{r}_l$. Then

$$V_{\rm a}(\boldsymbol{r}-\boldsymbol{r}_l)=V_{\rm b}(\boldsymbol{r}-\boldsymbol{r}_l)+V_{\rm el}(\boldsymbol{r}-\boldsymbol{r}_l) \tag{4.4.1}$$

is the "atomic" potential due to the site $\boldsymbol{r}_l$. When the ions are displaced, the "atoms" (i.e. the ion cores plus the electrons in the occupied atomic orbitals) will move as a unit, and *in addition* there will be some distortion of the atomic orbitals. Thus the total self-consistent first-order change in potential at the point $\boldsymbol{r}$ will be

$$\delta V_{\rm s}(\boldsymbol{r})=-\sum_{l,\alpha}\nabla_\alpha V_{\rm a}(\boldsymbol{r}-\boldsymbol{r}_l)u(l,\alpha)+\delta V_{\rm d}(\boldsymbol{r}), \tag{4.4.2}$$

where $\delta V_{\rm d}(\boldsymbol{r})$ is the screening potential set up by the distortions of the atomic orbitals. We may at this stage, as has been done by all authors so far, ignore the term $\delta V_{\rm d}(\boldsymbol{r})$ and make the "rigid atom approximation" (RAA) which is the analogue here of the RMTA for the muffin-tin representations. Once one has made the RAA, one may derive an expression for the electron–phonon matrix element in two ways. One may use the so-called Frölich or modified tight-binding approach, where one obtains the matrix element by expansion of the displaced, wavefunction in powers of the ionic displacements. This method has been used by several authors (Barisic et al. 1970, Pelir et al. 1977, Birnboim and Gutfreund 1974 1975, Varma et al. 1979). The other alternative is the Bloch approach, which we have taken earlier in this article, where one expands the *potential* in powers of the ionic displacements, and uses wavefunctions constructed from atomic orbitals centered or equilibruim sites, as in eq. (4.1.1). In fact the two methods give matrix elements which differ by terms proportional to $(E_{k\lambda}-E_{k'\lambda'})$, and are thus the same for energy-conserving transitions. Ashkenazi et al. (1979) have also shown that the total second-order change in energy (given by eq. (2.1)), which involves also the matrix element of the *second-order* change in the lattice potential, is in fact the same by either method. Thus for any physical property it is immaterial which method one uses to calculate the electron–phonon matrix element. We give below the less usual derivation using the Bloch approach.

Writing the ionic displacements in the form of running waves (eq. (3.9)), we then obtain

$$\langle \boldsymbol{k}'\lambda'|\delta V_{\mathrm{s}}|\boldsymbol{k}\lambda\rangle_{\mathrm{RAA}} = -A_{\boldsymbol{q}j}\sum_{l,\alpha} e_\alpha(\boldsymbol{q}j)\langle \boldsymbol{k}'\lambda'|\nabla_\alpha V_{\mathrm{a}}(\boldsymbol{r}-\boldsymbol{r}_l)|\boldsymbol{k}\lambda\rangle \exp(\mathrm{i}\boldsymbol{q}\cdot\boldsymbol{r}_l). \tag{4.4.3}$$

We now write $\psi_{k\lambda}$ and $\psi_{k'\lambda'}$ in tight-binding form (eq. (4.1.1)) and make the further approximation of keeping *only* one- and two-center integrals. We then have

$$\begin{aligned}\langle \boldsymbol{k}'\lambda'|\delta V_{\mathrm{s}}|\boldsymbol{k}\lambda\rangle_{\mathrm{RAA}} = {} & -N^{-1}A_{\boldsymbol{q}j}\sum_{\alpha} e_\alpha(\boldsymbol{q}j)\sum_{\xi_1\xi_2}\sum_{l_1,l_2}{}' A^{*\xi_1}_{\boldsymbol{k}'\lambda'}A^{\xi_2}_{\boldsymbol{k}\lambda} \\ & \times\big(\langle \xi_1 l_1|\nabla_\alpha V_{\mathrm{a}}(\boldsymbol{r}-\boldsymbol{r}_{l_1})|\xi_2 l_2\rangle \exp(-\mathrm{i}(\boldsymbol{k}'-\boldsymbol{q})\cdot\boldsymbol{r}_{l_1})\exp(\mathrm{i}\boldsymbol{k}\cdot\boldsymbol{r}_{l_2}) \\ & +\langle \xi_1 l_1|\nabla_\alpha V_{\mathrm{a}}(\boldsymbol{r}-\boldsymbol{r}_{l_2})|\xi_2 l_2\rangle \exp(-i\boldsymbol{k}'\cdot\boldsymbol{r}_{l_1})\exp(\mathrm{i}(\boldsymbol{k}+\boldsymbol{q})\cdot\boldsymbol{r}_{l_2})\big) \\ & -N^{-1}A_{\boldsymbol{q}j}\sum_{\alpha} e_\alpha(\boldsymbol{q}j)\sum_{\xi_1\xi_2}\sum_{l} A^{*\xi_1}_{\boldsymbol{k}'\lambda'}A^{\xi_2}_{\boldsymbol{k}\lambda}\langle \xi_1 l|\nabla_\alpha V_{\mathrm{a}}(\boldsymbol{r}-\boldsymbol{r}_l)|\xi_2 l\rangle \\ & \times\exp[\mathrm{i}(\boldsymbol{k}+\boldsymbol{q}-\boldsymbol{k}')\cdot\boldsymbol{r}_l].\end{aligned} \tag{4.4.4}$$

In eq. (4.4.4), we have made the further approximation of neglecting the so-called "crystal field" terms, arising from the gradients of the tails of the potentials due to other atoms on a particular atomic site. Now, if we neglect three-center integrals, the matrix elements of the Hamiltonian in the tight-binding representation are

$$\mathcal{H}_{\xi_1 l_1,\xi_2 l_2} = \langle \xi_1 l_1| -\nabla^2 + V_{\mathrm{a}}(\boldsymbol{r}-\boldsymbol{r}_{l_1}) + V_{\mathrm{a}}(\boldsymbol{r}-\boldsymbol{r}_{l_2})|\xi_2 l_2\rangle. \tag{4.4.5}$$

Then,

$$\begin{aligned}(\partial/\partial r_{l_2,\alpha})\mathcal{H}_{\xi_1 l_1,\xi_2 l_2} & \equiv \nabla_\alpha \mathcal{H}_{\xi_1 l_1,\xi_2 l_2} \\ & = -\langle \xi_1 l_1|\mathcal{H}\nabla_\alpha|\xi_2 l_2\rangle - \langle \xi_1 l_1|\nabla_\alpha V_{\mathrm{a}}(\boldsymbol{r}-\boldsymbol{r}_2)|\xi_2 l_2\rangle,\end{aligned} \tag{4.4.6}$$

where inside the integrals we have substituted $-\nabla_\alpha$ for $(\partial/\partial r_{l_2,\alpha})$, when operating on functions of $(\boldsymbol{r}-\boldsymbol{r}_l)$. Thus

$$\langle \xi_1 l_1 | \nabla_\alpha V_a(\boldsymbol{r}-\boldsymbol{r}_{l_2}) | \xi_2 l_2 \rangle = -\langle \xi_1 l_1 | \mathcal{H} \nabla_\alpha | \xi_2 l_2 \rangle - \nabla_\alpha \mathcal{H}_{\xi_1 l_1, \xi_2 l_2}. \tag{4.4.7}$$

Now using the result (in the two-center approximation)

$$\langle \xi_1 l_1 | [\nabla_\alpha, \mathcal{H}] | \xi_2 l_2 \rangle = \langle \xi_1 l_1 | \nabla_\alpha V_a(\boldsymbol{r}-\boldsymbol{r}_{l_1}) + \nabla_\alpha V_a(\boldsymbol{r}-\boldsymbol{r}_{l_2}) | \xi_2 l_2 \rangle \tag{4.4.8}$$

we obtain from eq. (4.4.6)

$$\langle \xi_2 l_1 | \nabla_\alpha V_a(\boldsymbol{r}-\boldsymbol{r}_{l_1}) | \xi_2 l_2 \rangle = \langle \xi_1 l_1 | \nabla_\alpha \mathcal{H} | \xi_2 l_2 \rangle + \nabla_\alpha \mathcal{H}_{\xi_1 l_1, \xi_2 l_2}. \tag{4.4.9}$$

Substituting eqs. (4.4.7) and (4.4.9) in eq. (4.4.3) and using the Schrödinger equation for the $\psi_{\boldsymbol{k}\lambda}$, $\psi_{\boldsymbol{k}'\lambda'}$, we obtain

$$\langle \boldsymbol{k}'\lambda' | \delta V_s | \boldsymbol{k}\lambda \rangle_{\mathrm{RAA}} = -N^{-1} A_{\boldsymbol{q}j} \sum_\alpha e_\alpha(\boldsymbol{q}j) \sum_{\xi_1 \xi_2} A^{*\xi_1}_{\boldsymbol{k}'\lambda'} A^{\xi_2}_{\boldsymbol{k}\lambda} \times \left[I_{\xi_1,\xi_2}(\boldsymbol{k}\lambda) - I_{\xi_1,\xi_2}(\boldsymbol{k}'\lambda') \right] \delta_{\boldsymbol{k}',\boldsymbol{k}+\boldsymbol{q}} \tag{4.4.10}$$

where

$$I_{\xi_1 l_1, \xi_2 l_2}(\boldsymbol{k}\lambda) = \sum_{l_1, l_2}{}' \left[\nabla_\alpha S_{\xi_1 l_1, \xi_2 l_2} - E_{\boldsymbol{k}\lambda} \nabla_\alpha S_{\xi_1 l_1, \xi_2 l_2} \right] \exp\left[\mathrm{i}\boldsymbol{k}\cdot(\boldsymbol{r}_{l_2} - \boldsymbol{r}_{l_1}) \right] \tag{4.4.11}$$

where

$$S_{\xi_1 l_1, \xi_2 l_2} = \langle \xi_1 l_1 | \xi_2 l_2 \rangle \tag{4.4.12}$$

is the overlap matrix element, so that

$$\nabla_\alpha S_{\xi_1 l_1, \xi_2 l_2} \equiv (\partial / \partial r_{l_2,\alpha}) S_{\xi_1 l_1, \xi_2 l_2} = -\langle \xi_1 l_1 | \nabla_\alpha | \xi_2 l_2 \rangle. \tag{4.4.13}$$

This result differs from the expression given by Varma et al. (1979) in that their expression the same $E_{\boldsymbol{k}\lambda}$ (referring to the state on the right of the matrix element) appears in both terms on the right of eq. (4.4.10). Our expression is symmetric between $(\boldsymbol{k},\lambda)$ and $(\boldsymbol{k}',\lambda')$. The difference is proportional to the energy difference $(E_{\boldsymbol{k}\lambda} - E_{\boldsymbol{k}'\lambda'})$ and its effect may be incorporated in the "bound" part of the electron response, as shown below. In order to discuss the screening effects, we use eq. (4.4.2) to write

$$\langle \boldsymbol{k}'\lambda' | \delta V_s | \boldsymbol{k}\lambda \rangle = \langle \boldsymbol{k}'\lambda' | \delta V_s | \boldsymbol{k}\lambda \rangle_{\mathrm{RAA}} + \langle \boldsymbol{k}'\lambda' | \delta V_d | \boldsymbol{k}\lambda \rangle. \tag{4.4.14}$$

We may rewrite eq. (4.4.10) as

$$\langle \boldsymbol{k}'\lambda'|\delta V_{\mathrm{s}}|\boldsymbol{k}\lambda\rangle_{\mathrm{RAA}} = g_{\boldsymbol{k}'\lambda',\boldsymbol{k}\lambda} - N^{-1}A_{\boldsymbol{q}j}(E_{\boldsymbol{k}\lambda}-E_{\boldsymbol{k}'\lambda'})\sum_{\alpha} e_{\mathrm{a}}(\boldsymbol{q}j)$$
$$\times \sum_{\xi_1\xi_2}\sum_{l_1l_2} A^{*\xi_1}_{\boldsymbol{k}'\lambda'}A^{\xi_2}_{\boldsymbol{k}\lambda}\exp\left[\mathrm{i}\boldsymbol{k}'\cdot(\boldsymbol{r}_{l_2}-\boldsymbol{r}_{l_1})\right]$$
$$\times \langle \xi_1 l_1|\nabla_\alpha|\xi_2 l_2\rangle \delta_{\boldsymbol{k}',\boldsymbol{k}+\boldsymbol{q}}, \tag{4.4.15}$$

where

$$g_{\boldsymbol{k}'\lambda',\boldsymbol{k}\lambda} = -N^{-1}A_{\boldsymbol{q}j}\sum_{\alpha} e_\alpha(\boldsymbol{q}j)\sum_{\xi_1\xi_2} A^{*\xi_1}_{\boldsymbol{k}'\lambda'}A^{\xi_2}_{\boldsymbol{k}\lambda}\left\{\sum_{l_1l_2}{}' (\nabla_\alpha \mathcal{H}_{\xi_1l_1,\xi_2l_2}\right.$$
$$\left. - E_{\boldsymbol{k}\lambda}\nabla_\alpha S_{\xi_1l_1,\xi_2l_2})\left\{\exp\left[\mathrm{i}\boldsymbol{k}\cdot(\boldsymbol{r}_{l_2}-\boldsymbol{r}_{l_1})\right]-\exp\left[\mathrm{i}\boldsymbol{k}'\cdot(\boldsymbol{r}_{l_2}-\boldsymbol{r}_{l_1})\right]\right\}\right\}. \tag{4.4.16}$$

Thus the first-order change in the wavefunction $\psi_{\boldsymbol{k}\lambda}$ is (using (eq. 4.4.14))

$$\delta\psi_{\boldsymbol{k}\lambda} = \sum_{\boldsymbol{k}'\lambda'} \langle \boldsymbol{k}'\lambda'|\delta V_{\mathrm{s}}|\boldsymbol{k}\lambda\rangle (E_{\boldsymbol{k}\lambda}-E_{\boldsymbol{k}'\lambda'})^{-1}\psi_{\boldsymbol{k}'\lambda'}$$
$$= \sum_{\boldsymbol{k}'\lambda'} (g_{\boldsymbol{k}'\lambda',\boldsymbol{k}\lambda} + \langle \boldsymbol{k}'\lambda'|\delta V_{\mathrm{d}}|\boldsymbol{k}\lambda\rangle)(E_{\boldsymbol{k}\lambda}-E_{\boldsymbol{k}'\lambda'})^{-1}\psi_{\boldsymbol{k}'\lambda'}$$
$$- N^{-3/2}A_{\boldsymbol{q}j}\sum_{\alpha} e_\alpha(\boldsymbol{q}j)\sum_{\xi_2l_2} A^{\xi_2}_{\boldsymbol{k}\lambda}\exp\left[\mathrm{i}(\boldsymbol{k}+\boldsymbol{q})\cdot\boldsymbol{r}_{l_2}\right]$$
$$\times \sum_{\xi_1l_1}\sum_{\boldsymbol{k}'\lambda'} A^{*\xi_1}_{\boldsymbol{k}'\lambda'}A^{\xi_3}_{\boldsymbol{k}'\lambda'}\exp\left[\mathrm{i}\boldsymbol{k}'\cdot(\boldsymbol{r}_{l_3}-\boldsymbol{r}_{l_1})\right]\langle \xi_1l_1|\nabla_\alpha|\xi_2l_2\rangle\phi_{\xi_3}(\boldsymbol{r}-\boldsymbol{r}_{l_3}). \tag{4.4.17}$$

Now, from the completeness property of the $\psi_{\boldsymbol{k}\lambda}$, it may be shown that

$$N^{-1}\sum_{\boldsymbol{k}'\lambda'} A^{*\xi_1}_{\boldsymbol{k}'\lambda'}A^{\xi_3}_{\boldsymbol{k}'\lambda'}\exp\left[\mathrm{i}\boldsymbol{k}'\cdot(\boldsymbol{r}_{l_3}-\boldsymbol{r}_{l_1})\right] = S^{-1}_{\xi_3l_3,\xi_1l_1} \tag{4.4.18}$$

and that

$$\sum_{\substack{\xi_1\xi_3\\ l_1l_3}} |\xi_3l_3\rangle S^{-1}_{\xi_3l_3,\xi_1l_1}\langle \xi_1l_1| = 1. \tag{4.4.19}$$

Using these results in the last term in eq. (4.4.17) we obtain

$$\delta\psi_{\boldsymbol{k}\lambda} = \sum_{\boldsymbol{k}'\lambda'} (g_{\boldsymbol{k}'\lambda',\boldsymbol{k}\lambda} + \langle \boldsymbol{k}'\lambda'|\delta V_{\mathrm{d}}|\boldsymbol{k}\lambda\rangle)(E_{\boldsymbol{k}\lambda}-E_{\boldsymbol{k}'\lambda'})^{-1}\psi_{\boldsymbol{k}'\lambda'}$$
$$- A_{\boldsymbol{q}j}\sum_{\alpha} e_\alpha(\boldsymbol{q}j)A^{\xi_2}_{\boldsymbol{k}\lambda_2}\nabla_\alpha\phi_{\xi_2}(\boldsymbol{r}-\boldsymbol{r}_{l_2})\exp(\mathrm{i}\boldsymbol{q}\cdot\boldsymbol{r}_{l_2})\exp(\mathrm{i}\boldsymbol{k}\cdot\boldsymbol{r}_{l_2}). \tag{4.4.20}$$

The last term corresponds to the rigid displacement with the ions of the orbitals associated with $\psi_{k\lambda}$. This portion of the electron density response, (neglecting certain interference terms), together with the displacement of the bare ion potentials, gives the "rigid atom" change in potential

$$-\sum_{l,\alpha} \nabla_\alpha(\boldsymbol{r}-\boldsymbol{r}_l)u(l,\alpha).$$

The first term may be used to obtain the distortion part of the electron density response, and a self-consistency equation set up in the usual manner for δV_{d}. We obtain for the Fourier components of δV_{d},

$$\begin{aligned}\delta V_{\mathrm{d}}(\boldsymbol{q}+\boldsymbol{G}) = A_{\boldsymbol{q}j}\sum_{\boldsymbol{G}'} \varepsilon^{-1}(\boldsymbol{q}+\boldsymbol{G},\boldsymbol{q}+\boldsymbol{G}')v(\boldsymbol{q}+\boldsymbol{G}')\\ \times \sum_{\substack{\boldsymbol{k}\boldsymbol{k}'\\ \lambda\lambda'}} \langle \boldsymbol{k}\lambda|\exp[-\mathrm{i}(\boldsymbol{q}+\boldsymbol{G}')\cdot\boldsymbol{r}]|\boldsymbol{k}'\lambda'\rangle\\ \times (E_{\boldsymbol{k}\lambda}-E_{\boldsymbol{k}'\lambda'})^{-1}\{\langle n_{\boldsymbol{k}\lambda}\rangle g_{\boldsymbol{k}'\lambda',\boldsymbol{k}\lambda}-\langle n_{\boldsymbol{k}'\lambda'}\rangle g^*_{\boldsymbol{k}\lambda,\boldsymbol{k}'\lambda'}\}.\end{aligned} \tag{4.4.21}$$

It is the quantity

$$\sum_{\boldsymbol{G}} \langle \boldsymbol{k}'\lambda'|\exp[\mathrm{i}(\boldsymbol{q}+\boldsymbol{G})\cdot\boldsymbol{r}]|\boldsymbol{k}\lambda\rangle \delta V_{\mathrm{d}}(\boldsymbol{q}+\boldsymbol{G})$$

which is being neglected in the RAA.

Note that it is the terms involving $E_{\boldsymbol{k}}, E_{\boldsymbol{k}'\lambda'}$ in eq. (4.4.16) which prevent the total screened matrix element being written in compact form, as in the muffin-tin orbital representation. Thus, if we neglected the dependence on $\boldsymbol{k}\lambda, \boldsymbol{k}'\lambda'$ in these terms, we could write approximately,

$$g_{\boldsymbol{k}'\lambda',\boldsymbol{k}\lambda} = N^{-1}A_{\boldsymbol{q}j}\sum_\alpha e_\alpha(\boldsymbol{q}j)\sum_{\xi_1\xi_2} A^{*\xi_1}_{\boldsymbol{k}'\lambda'}A^{\xi_2}_{\boldsymbol{k}\lambda}\exp[\mathrm{i}\boldsymbol{k}\cdot(\boldsymbol{r}_{l_2}-\boldsymbol{r}_{l_1})]I^\alpha_{\xi_1 l_1,\xi_2 l_2}(\boldsymbol{q}). \tag{4.4.22}$$

Using the same arguments used in eqs. (4.4.11)–(4.2.17), we obtain for the total screened electron–phonon matrix element (compare eq. (4.2.11)),

$$\begin{aligned}\langle \boldsymbol{k}'\lambda'|\delta V_{\mathrm{s}}|\boldsymbol{k}\lambda\rangle = -N^{-1}A_{\boldsymbol{q}j}(E_{\boldsymbol{k}\lambda}-E_{\boldsymbol{k}'\lambda'})\sum_\alpha e_\alpha(\boldsymbol{q}j)\sum_{\xi_1\xi_2}\sum_{l_1l_2} A^{*\xi_1}_{\boldsymbol{k}'\lambda'}\\ \times A^{\xi_2}_{\boldsymbol{k}\lambda}\exp[\mathrm{i}\boldsymbol{k}\cdot(\boldsymbol{r}_{l_2}-\boldsymbol{r}_{l_1})]\langle \xi_1 l_1|\nabla_\alpha|\xi_2 l_2\rangle \delta_{\boldsymbol{k}',\boldsymbol{k}+\boldsymbol{q}}\\ +A_{\boldsymbol{q}j}\sum_\alpha e_\alpha(\boldsymbol{q}j)\sum_{\substack{\xi_1\xi_2\\ \xi_3\xi_4}}\sum_{l_1l_2} A^{*\xi_1}_{\boldsymbol{k}'\lambda'}(A^{\xi_2}_{\boldsymbol{k}\lambda}\exp(\mathrm{i}\boldsymbol{k}\cdot\boldsymbol{r}_{l_1}))\\ \times(1-\mathbf{V}\boldsymbol{\chi}^0)^{-1}_{\xi_1\xi_3;\,\xi_2 l_1,\,\xi_4 l_2}I^\alpha_{\xi_3,\,\xi_4 l_2}(\boldsymbol{q})\delta_{\boldsymbol{k}',\boldsymbol{k}+\boldsymbol{q}},\end{aligned} \tag{4.4.23}$$

where we have written $I^{\alpha}_{\xi_3 l_1, \xi_4 l_2}(\boldsymbol{q})$ as $I^{\alpha}_{\xi_3, \xi_4(l_2-l_1)}(\boldsymbol{q})$. For energy-conserving transitions ($E_{\boldsymbol{k}\lambda} = E_{\boldsymbol{k}'\lambda'}$) such as are involved in superconductivity calculations, the first term vanishes. Thus $I^{\alpha}_{\xi_3, \xi_3 l_2}(\boldsymbol{q})$ vanishes if $l_2 = 0$, and $\xi_3 = \xi_4$. Thus only the off-diagonal elements of $(\mathbf{1} - \mathbf{V}\boldsymbol{\chi}^0)^{-1}$ serve to screen the electron–phonon matrix element in this representation, and the neglect of screening may not be as serious as it would be if one simply evaluated $\langle \boldsymbol{k}'\lambda' | \delta V_{\mathrm{b}} | \boldsymbol{k}\lambda \rangle$. This because some part of the screening is already put into the RAA in terms of rigid atomic orbital displacements. This is the analogue of what happens in the RMTA where the matrix element one has to "screen", i.e. $I^{\alpha}_{\xi_1 \xi_2}(\boldsymbol{q})$ vanishes for $\xi_1 = \xi_2$ and is thus screened by only the off-diagonal elements of $(\mathbf{1} - \mathbf{V}\boldsymbol{\chi}^0)^{-1}$.

Let us now return to the RAA expression (as given by eqs. (4.4.10) and (4.4.11) or eqs. (4.4.15) and (4.4.16)) for the matrix elements, still assuming $E_{\boldsymbol{k}\lambda} = E_{\boldsymbol{k}'\lambda'}$. If we choose atomic orbitals *orthogonalized* to each other, so that the overlap matrix elements $S_{\xi_1 l_1, \xi_2 l_2}$ vanish then the expression reduces to the form derived by Barisic et al. (1970). Peter et al. (1977) and also Birnboim and Gutfreund (1974, 1975) have parameterized the distance-dependence of the tight-binding integrals $\mathcal{H}_{\xi_1 l_1, \xi_2 l_2}$ (where only d and s orbitals were used) to evaluate λ and hence T_{c} for Nb by solution of the Eliashberg equations. These authors find that the matrix elements must be reduced considerably (by about 35%) in order to obtain agreement with experiment. It is not clear whether the errors are due to neglect of screening, as discussed above, or because of the two-center integral approximation, which may become appreciable if orthogonalized orbitals are used since these tend to be more widely extended spatially. The latter point of view is taken by Varma et al., who choose non-orthogonal orbitals and thus keep the terms involving $\nabla_{\alpha} S_{\xi_1 l_1, \xi_2 l_2}$ in eq. (4.4.16). (In fact, their method was developed as a generalization of the method of Barisic et al. (1970) to the case of non-orthogonalized orbitals.) These authors obtain the tight-binding and overlap integrals by performing a Slater–Koster fit involving s-, p- and d-orbitals to a self-consistent (APW) energy-band calculation at two different densities, and obtain the gradients by numerical differentiation. (They also obtain these same quantities by differentiating overlap integrals involving Herman–Skillman atomic wavefunctions and find agreement with the first set of values within about 20%.) They find that, for Nb, the ∇S terms have a large effect in reducing the matrix elements of the electron–phonon interaction and the net result is a λ which comes out to be comparable to the values obtained using the RMTA (Butler et al. 1976, Klein and Papaconstantopoulos, 1974).

In attempting to compare the merits of the tight-binding RAA and the APW- or KKR-based RMTA methods of calculating the screened electron–phonon matrix element, one sees that one has something of a trade-off between competing advantages and disadvantages. The tight-binding RAA method does partially include the effect of the tails of the potentials $V_{\mathrm{a}}(\boldsymbol{r} - \boldsymbol{r}_l)$ in neighboring atomic cells (although it neglects the "crystal field" terms, as discussed after eq. (4.44)), while the cellular methods effectively include the three-center integrals arising from wavefunctions of two neighboring "orbitals" overlapping on a particular site (since these are included in an angular momentum expansion of the total wavefunction about that site). Neither method calculates the potential due to the electron response exactly, the RMTA moving all the charge inside the muffin-tin spheres rigidly with the ions, while the RAA moves the overlapping "atomic" densities rigidly through each other (and neglects also "interference" charge densities caused by the overlap of the rigidly displaced part of $\delta\psi_{k\lambda}$ (eq. (4.4.19) with the unperturbed $\psi_{k\lambda}$). Both methods start from "best" self-consistent calculation of the electronic band structure. However, the RMTA directly uses quantities which are calculated (or readily available) in the course of the band structure calculation, such as phase shifts and partial densities of states, and so is perhaps computationally simpler. The tight-binding methods usually involve first some Slater–Koster fit to the band-structure (often at two different densities of the crystal) and some process of differentiation to obtain the relevant quantities to be used for the matrix-element calculations. (The fact that a *self-consistent* band structure may be fitted to at two different densities does not imply that the derivatives yield a properly screened matrix element, since the nature of the electron response to a phonon and to a uniform compression of the lattice are quite different in principle.)

We turn once more to the calculation of the phonon spectrum. To date there have been very few attempts to calculate phonon spectra using the tight-binding form for the electron–phonon matrix elements. The most successful of these has been that of Varma and Weber (1977) using the non-orthogonal tight-binding representation discussed above. We shall discuss these and other calculations in more detail in §5. We outline the formalism here. Let us start with the general expression for the $\boldsymbol{q}$-dependent part of the electronic contribution to the dynamical matrix, $E'_{\alpha\beta}(\boldsymbol{q})$, as given by eqs. (2.23), and (2.24). In these expressions $\langle\xi_3|\delta V^{\beta}_{\boldsymbol{q}}|\xi_4\rangle$ represents the matrix element of the first-order change in the *bare* ionic potential. Note that the corresponding *screened* matrix element is given by

$$\langle\xi_3|\delta V^{\beta}_{\boldsymbol{q},\mathrm{sc}}|\xi_4\rangle = \sum_{\xi_5\xi_6} \left[\mathbf{1}-\mathbf{V}_{\mathrm{eff}}(\boldsymbol{q})\boldsymbol{\chi}^0(\boldsymbol{q})\right]^{-1}_{\xi_3\xi_5,\xi_4\xi_6}\langle\xi_5|\delta V^{\beta}_{\boldsymbol{q}}|\xi_6\rangle. \tag{4.4.24}$$

We also note that $\boldsymbol{\chi}=\boldsymbol{\chi}^0(\mathbf{1}-\mathbf{V}_{\mathrm{eff}}\boldsymbol{\chi}^0)^{-1}$ (eq. (4.1.19)).
Thus

$$\begin{aligned} E'_{\alpha\beta}(\boldsymbol{q}) &= \sum_{\xi_1\ldots\xi_6} \langle\xi_1|\delta V^{\alpha}_{-\boldsymbol{q},\mathrm{sc}}|\xi_2\rangle\left[\mathbf{1}-\mathbf{V}_{\mathrm{eff}}(\boldsymbol{q})\boldsymbol{\chi}^0(\boldsymbol{q})\right]_{\xi_1\xi_3,\xi_2\xi_4}\chi^{(0)}_{\xi_3\xi_5,\xi_4\xi_6}\langle\xi_5|\delta V^{\beta}_{\boldsymbol{q},\mathrm{sc}}|\xi_6\rangle \\ &= \sum_{\xi_1\ldots\xi_4} \langle\xi_1|\delta V^{\alpha}_{-\boldsymbol{q},\mathrm{sc}}|\xi_2\rangle\chi^0_{\xi_1\xi_3,\xi_2\xi_4}\langle\xi_3|\delta V^{\beta}_{\boldsymbol{q},\mathrm{sc}}|\xi_4\rangle \\ &\quad -\int\!\!\int \mathrm{d}\boldsymbol{r}\,\mathrm{d}\boldsymbol{r}'\,v_{\mathrm{eff}}(\boldsymbol{r},\boldsymbol{r}')\,\delta n^{\alpha}(\boldsymbol{r})\,\delta n^{\beta}(\boldsymbol{r}'), \end{aligned} \tag{4.4.25}$$

where $v_{\text{eff}}(\boldsymbol{r},\boldsymbol{r}')$ is the non-local effective electron–electron interaction including exchange and correlation. In deriving the last term in eq. (4.4.25) we have used the fact that the first-order change in electron density in response to the total self-consistent perturbation δV^{α} is given by

$$\delta \mathbf{n}^{\alpha} = \boldsymbol{\chi}^{0} \delta \mathbf{V}^{\alpha} \tag{4.4.26}$$

The first term in eq. (4.4.25) now involves the *screened* matrix elements and the *non-interacting* density response function $\boldsymbol{\chi}^0$. It is in fact just the change in the total band structure energy of the electrons due to the phonon perturbation. The second term subtracts off the electron–electron interaction energy which has been double-counted in the first term. This term in fact can be shown to cancel the bare ion–ion interaction $C_{\alpha\beta}(\boldsymbol{q})$ to a large extent, thus providing an *alternative* method of achieving this desirable cancellation to the pseudo-atom method discussed earlier. This idea was first proposed by Pickett and Gyorffy (1976). That this cancellation must take place may be seen by noting that $\delta n^{\alpha}(\boldsymbol{r})$ consists of a part which corresponds to the rigid displacement of the valence electron density with the ion cores as we have already seen earlier. If we choose the particular representation, where the ξ's are the Bloch states $\boldsymbol{k}\lambda$ themselves, we obtain immediately

$$D_{\alpha\beta}(\boldsymbol{q}) = M^{-1}\Bigg[\sum_{\boldsymbol{k}\lambda\lambda'} \langle \boldsymbol{k}\lambda | \delta V^{\alpha}_{-\boldsymbol{q},\text{sc}} | \boldsymbol{k}+\boldsymbol{q}\lambda' \rangle \langle \boldsymbol{k}+\boldsymbol{q}\lambda' | \xi V^{\beta}_{\boldsymbol{q},\text{sc}} | \boldsymbol{k}\lambda \rangle \times \frac{\langle n_{\boldsymbol{k}\lambda}\rangle - \langle n_{\boldsymbol{k}+\boldsymbol{q}\lambda'}\rangle}{E_{\boldsymbol{k}\lambda} - E_{\boldsymbol{k}+\boldsymbol{q}\lambda'}} + \left(C_{\alpha\beta}(\boldsymbol{q}) + E''_{\alpha\beta}(\boldsymbol{q}) - E'_{\alpha\beta}(0)\right) \Bigg], \tag{4.4.27}$$

where $E''_{\alpha\beta}(\boldsymbol{q})$ is the contribution coming from the second term in eq. (4.4.25). Varma and Weber (1977) parameterize the last term in terms of short-range force constants and use the RAA approximation for the screened matrix elements (as given by eq. (4.4.9)) to evaluate the first term.

5. *Review of calculations of phonon spectra, phonon anomalies, and lattice instabilities in d-band metals*

We have seen in the last section how recent theoretical developments have attacked the problems of screening and density response in non-free-electron-like solids, the large cancellation of the (ionic) Coulomb and electronic contributions to the dynamical matrix and the elimination of the large potential well near the ion cores. Using eqs. (4.1.19) and (4.2.28), we may write

$$D_{\alpha\beta}(\boldsymbol{q}) = \frac{1}{M}\left[C'_{\alpha\beta}(\boldsymbol{q}) - \sum_{\mu_1\mu_2\mu_3} I^{\alpha}_{\mu_1}(-\boldsymbol{q}) \chi^{0}_{\mu_1\mu_2}(\boldsymbol{q}) \left[\mathbf{1} - \mathbf{V}_{\text{eff}}(\boldsymbol{q})\boldsymbol{\chi}^{0}(\boldsymbol{q})\right]^{-1}_{\mu_1\mu_3} I^{\beta}_{\mu_3}(\boldsymbol{q}) \right] \tag{5.1}$$

which represents a rigorous expression for the dynamical matrix. The second (electronic) contribution is seen to depend on $I^{\alpha}_{\mu_1}(-\boldsymbol{q})$ that represent the scattering properties of the displaced ions, and are most conveniently calculated in terms of the angular momentum representation using eqs. (4.3.6) and (4.3.9), and on χ^0 (given by eq. (4.3.12)) which contains all the band structure and wavefunction information on the solid (including Fermi surface effects, bonding and overlap effects, etc.). The real problem is usually the size of the matrix $(\mathbf{1}-\mathbf{V}_{\text{eff}}\boldsymbol{\chi}^0)$ which is to be inverted. Thus for instance if we include the 14 orbitals corresponding to $l=0,1$, and 2 (including two orbitals per d-function to allow for the energy-dependence of the $l=2$ radial function), we have 196 possible virtual transitions μ, and a 196×196 matrix to invert. The problem becomes worse if one includes $l=3$ orbitals. The size may be restricted somewhat by keeping only elements in $\mathbf{V}_{\text{eff}}(\boldsymbol{q})$ which correspond to interactions between monopolar or dipolar charge fluctuations, and by neglecting the smaller elements of χ^0 corresponding to cross-product terms with different (l,m) values. (It may be seen from eq. (4.3.12) that destructive interference will occur between the mixing coefficients for such terms.) As a result, most of the calculations using this formulation (which we may term the density response formulation) have made severe approximations and concentrated on emphasizing the basic *physical* effects which lead to the observed structure in the phonon spectra. They may be said to fall into the category of model calculations *based* on the microscopic theory. We shall review such calculations in this section. (There has also been a great deal of work on fitting phonon dispersion curves in transition metals using purely phenomenological models which are not directly based on microscopic theories, for example, Sharma and Upadhyaya (1977). We shall not discuss such models here.) We shall also review calculations based on the formulation given in eq. (4.4.27) (which we may term the rigid atom formulation), which have concentrated mainly on the "band structure term" (the first term in eq. (4.4.27)) and have parametrized or ignored the other terms, again concentrating mainly on getting the correct structure to the dispersion curves.

Let us first review some basic facts regarding instabilities in these systems. The poles of the frequency-dependent electron density response function $\chi_{\mu_1\mu_2}(\boldsymbol{q},\omega)$ determine the *plasmon* frequencies of the electron system. In the absence of single particle excitations and other damping effects, these would be given (from eq. (4.1.19)) by

$$\|\mathbf{1}-\mathbf{V}_{\text{eff}}(\boldsymbol{q})\boldsymbol{\chi}^0(\boldsymbol{q},\omega)\|=0. \tag{5.2}$$

If one of these plasmon modes goes "soft", i.e. if eq. (5.2) is satisfied for $\omega=0$ at some $\boldsymbol{q}$, this heralds an instability of the electron system itself. If this results in the divergence of elements of $\chi_{\mu_1\mu_2}$ involving monopolar charge fluctuations, we have an *ionic*-type instability, if it involves dipolar charge fluctuations we have a dipolar instability, and if it involves charge fluctuations between overlapping orbitals on neighboring sites we have a covalent-type instability. However, as may be seen from eq. (5.1), such electronic instabilities cannot occur without first driving the lattice into a structural transformation, since the second term will certainly drive a phonon frequency at that $\boldsymbol{q}$ "soft" because of the divergence of $[\mathbf{1}-\mathbf{V}_{\text{eff}}\boldsymbol{\chi}^0]^{-1}$. We thus see that in the density response formalism, a sufficient condition for a lattice instability is a tendency for the electron system to manifest an instability (heralded by some large elements of $\chi_{\mu_1\mu_2}(\boldsymbol{q})$). In the "rigid ion" picture, it is presumably the "band structure" term which manifests anomalous softening at certain $\boldsymbol{q}$-values, and this has been attributed to structure in the factor involving the energy denominators (Gupta and Freeman 1976a, b, c, Myron et al. 1977), and also to $\boldsymbol{q}$-dependence of the screned electron–phonon matrix elements (Varma and Weber 1977). It should be noted that a tendency for elements of $(\mathbf{1}-\mathbf{V}_{\text{eff}}\boldsymbol{\chi}^0)^{-1}$ to become large at certain $\boldsymbol{q}$-values will also result in large screened electron–phonon matrix elements at those wavevectors so that the two approaches are not inconsistent, although there are differences in detail as will be discussed later.

One of the earliest attempts to explain the observed anomalies in Nb was to invoke the idea of acoustic plasmons (Ganguly and Wood 1972). According to Frölich (1968), the adiabatic screening by light s-electrons of the collective charge fluctuations of d-electrons in a transition metal can lead to a plasmon mode which possesses a linear dispersion law at small wavevectors. Ganguly and Wood (1972) postulated such an acoustic plasmon in Nb and were able to explain the anomaly in the LA[100] branch (see fig. 1) in terms of the coupling of the phonon and plasmon branches (the latter, of course, being a "silent" mode as far as neutron spectroscopy is concerned). The treatment of the acoustic plasmon given by Frölich is based on the effective mass approximation for the s- and d-electrons, and is probably not valid for large wavevectors. If such modes exist, they should occur as the roots of eq. (5.2) which is valid for a general electron system. A treatment based on the localized orbital representation is given by Sinha and Varma (to be published). However, it is generally believed that the condition for these modes to exist, namely,

$$v_{\text{s}} \gg v_{\text{a.p.}} \gg v_{\text{d}}, \tag{5.3}$$

where v_{s}, $v_{\text{a.p.}}$ and v_{d} are the s-electron Fermi velocity, the acoustic plasmon velocity and the d-electron Fermi velocity respectively, is not satisfied in transition metals. Further, strong s–d hybridization effects and Landau damping effects appear to make it difficult for a normal transition metal to sustain such modes. It is possible, however, that such modes might exist in systems which have f-electrons with very large band masses at the Fermi level, e.g. in the so-called "mixed valence" or actinide materials.

5.1. Charge fluctuation models

Let us take the rigorous expression given in eq. (4.2.27) and discuss the physical consequences if we had only *one* type of orbital ξ of importance involved in the conduction band around the Fermi level. Let us further assume this orbital is sufficiently well localized so that we may neglect overlap effects. Then there is only one type of charge fluctuation μ involved, namely the monopole type illustrated in fig. 10a. We may thus drop the subscripts $\mu_1 \mu_2$, and write

$$\omega_{\boldsymbol{q}j}^2 = \tilde{\omega}_{\boldsymbol{q}j}^2 - \frac{1}{M}\sum_{\alpha\beta} e_\alpha^*(\boldsymbol{q}j) I^\alpha(-\boldsymbol{q}) \frac{1}{V_{\mathrm{eff}}(\boldsymbol{q}) - 1/\chi^0(\boldsymbol{q})} I^\beta(\boldsymbol{q}) e_\beta(\boldsymbol{q}j) \qquad (5.1.1)$$

where $\tilde{\omega}_{\boldsymbol{q}j}$ is the renormalized "pseudo-atom" frequency, and we have substituted the explicit expression for $\chi(\boldsymbol{q})$ (which is now a scalar quantity) from eq. (4.1.19). From eq (4.3.9) we see that

$$I^\beta(\boldsymbol{q}) = -\mathrm{i}\sum_{\boldsymbol{G}} (\boldsymbol{q}+\boldsymbol{G})_\beta W(\boldsymbol{q}+\boldsymbol{G})(\xi|\exp[-i(\boldsymbol{q}+\boldsymbol{G})\cdot\boldsymbol{r}]|\xi), \qquad (5.1.2)$$

since the $N_{\xi\xi}^\alpha$ vanish because of the $l \to l \pm 1$ selection rule (eq. (4.3.6)). Also from eq. (4.3.12),

$$\chi^0(\boldsymbol{q}) = \sum_{\boldsymbol{k}\lambda_1\lambda_2} \frac{\langle n_{\boldsymbol{k}\lambda_1}\rangle - \langle n_{\boldsymbol{k}+\boldsymbol{q}\lambda_2}\rangle}{E_{\boldsymbol{k}\lambda_1} - E_{\boldsymbol{k}+\boldsymbol{q}\lambda_2}}. \qquad (5.1.3)$$

As defined, this is a negative quantity. $[-\chi^0(\boldsymbol{q})]$ is often referred to as the generalized susceptibility function in the constant matrix element approximation (Gupta and Freeman 1976a). We may also write the expression for $V_{\mathrm{eff}}(\boldsymbol{q})$ (eq. (4.1.32)) as a direct lattice sum

$$V_{\mathrm{eff}}(\boldsymbol{q}) = \sum_j (v_{\mathrm{c},j} + v_{\mathrm{xc},j}) \exp(\mathrm{i}\boldsymbol{q}\cdot\boldsymbol{r}_j), \qquad (5.1.4)$$

where $v_{\mathrm{c},j}$ represents the Coulomb interaction between charge distributions $\phi_\xi^*\phi_\xi$ centered on the origin site and site j, and $v_{\mathrm{xc},j}$ the exchange-correlation interaction. Let us also introduce, in addition to the tight-binding band involving the orbital ξ, a free-electron-like band, where we neglect hybridization with the ξ-band and also neglect electron–phonon matrix elements connecting the ξ-band with the free-electron band. The effect of the free-electron band is then (a) to further renormalize the frequencies $\tilde{\omega}_{qj}$, and (b) to screen the potentials $W(\boldsymbol{q}+\boldsymbol{G})$ and $v(\boldsymbol{q}+\boldsymbol{G})$ (see eq. (4.1.32)) by a free-electron–dielectric function $\varepsilon(\boldsymbol{q}+\boldsymbol{G})$. In eq. (5.1.4), this will result in a $v_{c,j}$ which is short-ranged (except for possible Friedel oscillations). Note that $v_{xc,j}$ which is short-ranged in any case, tends to cancel $v_{c,j}$ for $j=0$, i.e. to remove the "self-interaction" of the ξ-orbital charge distribution on a particular site. Thus we may write

$$V_{\text{eff}}(\boldsymbol{q}) = U + \sum_j{}' V_j \exp(i\boldsymbol{q}\cdot\boldsymbol{r}_j), \tag{5.1.5}$$

where U is the effective intra-atomic Coulomb integral reduced by exchange, correlation and screening (to values typically of the order of 2–3 eV), while V_j are the screened inter-atomic Coulomb integrals, also reduced by exchange. Eq. (5.1.2) may also be written in real-space form as

$$I^{\beta}(\boldsymbol{q}) = -\sum_j \int \mathrm{d}\boldsymbol{r}\phi_\xi^*(\boldsymbol{r})\phi_\xi(\boldsymbol{r})\nabla_\beta W_{\text{sc}}(\boldsymbol{r}-\boldsymbol{r}_j)\exp(\mathrm{i}\boldsymbol{q}\cdot\boldsymbol{r}_j), \tag{5.1.6}$$

where $W_{\text{sc}}(\boldsymbol{r})$ is the Fourier transform of $W(\boldsymbol{K})/\varepsilon(\boldsymbol{K})$. $I^{\beta}(\boldsymbol{q})$ measures the $\boldsymbol{q}$-dependent coupling coefficient between the charge fluctuation

$$\delta n(\boldsymbol{r}) = A_{qj}\sum_l \left[\phi_\xi^*(\boldsymbol{r}-\boldsymbol{r}_l)\phi_\xi(\boldsymbol{r}-\boldsymbol{r}_l)\right]\exp(i\boldsymbol{q}\cdot\boldsymbol{r}_l) \tag{5.1.7}$$

(i.e. a periodic redistribution of electron density amongst the ξ-orbitals on the sites $\boldsymbol{r}_l$) and the displaced "pseudo-atoms". For α along the direction of $\boldsymbol{q}$, the function $I^{\alpha}(\boldsymbol{q})$ is denoted schematically in fig. 11 where it is seen to go to zero at $\boldsymbol{q}=0$ (where all the pseudo-atoms move together, thus cancelling their effect on the charge fluctuation on the origin atom), and at the zone boundary (where again pseudo-atoms equidistant from the origin atom move in pairs cancelling their effect on the charge fluctation on the

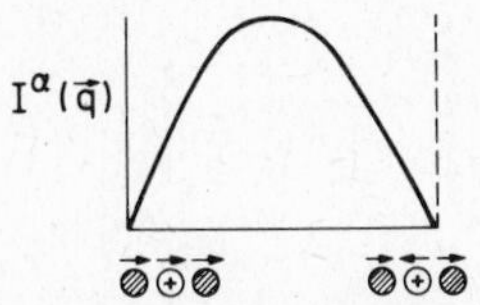

Fig. 11. Schematic of the function $I^{\alpha}(\boldsymbol{q})$ plotted versus $\boldsymbol{q}$ for a simple nearest-neighbor coupling between charge fluctuations and ionic displacements. Also shown are the configurations of the neighboring displaced ions (shaded circles) relative to the origin ion with the charge fluctuation (marked with a +) for the particular cases $\boldsymbol{q}=0$ and $\boldsymbol{q}=$zone boundary wavevector, illustrating why symmetry requires $I^{\alpha}(\boldsymbol{q})$ to vanish at those points.

origin atom). The function $V_{\text{eff}}(\boldsymbol{q})$ measures the increase in potential energy due to the mutual interaction of unit amplitude charge fluctuations on the atoms of the form given by eq. (5.1.6). It is indicated schematically in fig. 12, where it is seen to be a maximum at $\boldsymbol{q}=0$ (since the V_j are in general positive) and decreases to a negative minimum at the zone boundary where the charge fluctuation takes on an ionic character (assuming that the sum of the nearest-neighbor V_j's is sufficient to overwhelm U, as seems likely from order of magnitude considerations in most crystal structures). The denominator $[V_{\text{eff}}(\boldsymbol{q})-1/\chi^0(\boldsymbol{q})]$ is thus kept from vanishing by $-1/\chi^0(\boldsymbol{q})$, which by eq. (5.1.3) is approximately $n_\xi(E_F)^{-1}$, where $n_\xi(E_F)$ is the density of ξ-orbital states at the Fermi level. (We are neglecting here fine structure in the function $\chi^0(\boldsymbol{q})$ due to Fermi surface effects, etc.) Note that as $n_\xi(E_F)$ increases, the denominator becomes smaller, particularly for large $\boldsymbol{q}$'s (where $V_{\text{eff}}(\boldsymbol{q})$ is a minimum) and there is a corresponding enhancement of structure in the $I^\alpha(\boldsymbol{q})$ which produces mode softening at particular $\boldsymbol{q}$-values. The tendency of the denominator to become small at a certain $\boldsymbol{q}$ is an indication here of a tendency of the electron system to become unstable against a charge density wave of the type given in eq. (5.1.6) for that wavevector, since the denominator measures the total energy cost of creating unit amplitude of such a fluctuation. Note that in this simple model it is the more rapidly varying function $I^\alpha(\boldsymbol{q})$ however, which predominantly determines the $\boldsymbol{q}$ of the maximum phonon softening, although the denominator controls its magnitude.

The screened electron phonon matrix element between states $\boldsymbol{k}'\lambda'$ and $\boldsymbol{k}\lambda$ separated by a wavevector $\boldsymbol{q}$ is given in this simple model by

$$\langle \boldsymbol{k}'\lambda'|\delta V_s|\boldsymbol{k}\lambda\rangle = A_{\boldsymbol{q}j}\sum_\alpha e_\alpha(\boldsymbol{q}j)\frac{I^\alpha(\boldsymbol{q})A^{*\xi}_{\boldsymbol{k}'\lambda'}A^\xi_{\boldsymbol{k}\lambda}}{\varepsilon(\boldsymbol{q})\left[1-V_{\text{eff}}(\boldsymbol{q})\chi^0(\boldsymbol{q})\right]}. \tag{5.1.8}$$

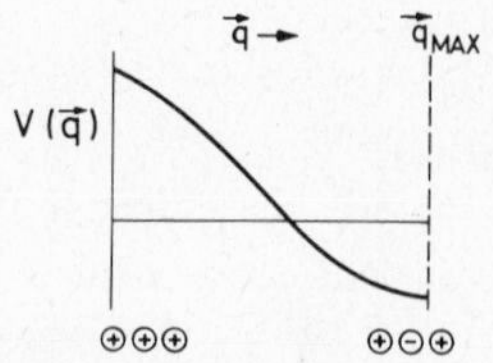

Fig. 12. Schematic of the function $V_{\text{eff}}(\boldsymbol{q})$ plotted versus $\boldsymbol{q}$. Also shown are the configurations of the monopole-type charge fluctuations on the ions for the particular cases $\boldsymbol{q}=0$ and $\boldsymbol{q}$ = zone boundary wavevector.

Note that in regimes of $\boldsymbol{q}$-space where the denominator is small, it tends to offset the normal screening of the matrix element, thus enhancing the electron-phonon interaction.

We thus see that the above simple model qualitatively accounts for the empirical correlations discussed in §1, namely that between the presence of anomalies or instabilities, $n(E_F)$ and high superconducting transition temperatures. It also explains why the non-magnetic 4d transition metals tend more to show lattice instabilities than the magnetic 3d metals, since the occurrence of a large U (eq. (5.1.5)) will tend to *increase* $V_{\text{eff}}(\boldsymbol{q})$ and increase the denominator, while on the other hand it will tend to promote a *spin* instability. However, it should be emphasized that apart from the enhancement effect mentioned above, there is a direct correlation between λ and $n(E_F)$ from the Fermi surface averaging above (see eqs. (4.3.11), (4.3.14) and (4.3.15)). Such correlations between λ and $n(E_F)$ or some symmetry projected components of $n(E_F)$ have been noted in the literature (Varma and Dynes 1976, Mueller and Myron 1977).

It is easy enough to generalize the above case of only one localized orbital to that of several orbitals in the conduction bands at the Fermi level. Thus in the initial calculations for Nb using this model (Sinha and Harmon 1975, 1976), the three t_{2g}-like d-orbitals were chosen to represent the lower sub-bands at the Fermi level, resulting in 9×9 matrices for the quantities $\mathbf{V}$, $\boldsymbol{\chi}^0$, etc. (This is not quite accurate since this sub-band is not pure t_{2g} in character.) The s-bands were taken to be free-electron-like and unhybridized with the d-bands, overlap effects were neglected, and the $\boldsymbol{\chi}^0$ matrix was taken to be of the form

$$\chi^0_{\mu_1\mu_2} \simeq \delta_{\mu_1\mu_2}\tfrac{1}{9}\chi^0(\boldsymbol{q}), \tag{5.1.9a}$$

where $\chi^0(\boldsymbol{q})$ is given by eq. (5.1.3). This function was calculated from the APW energy bands for Nb and found to have only minor structure as a function of $\boldsymbol{q}$. Eq. (5.1.8) is based on the approximation that each of the t_{2g} orbitals is equally mixed into the bands at the Fermi level, so that $A_{\boldsymbol{k}}^{\xi_1} \simeq 1/\sqrt{3}$, and that there is destructive interference among these coefficients for $\mu_1 \neq \mu_2$ (see eq. (4.3.12)). Fig. 13 shows the deviation of $\tilde{\omega}^2_{qj}$ for the LA branches of Nb below the "smooth" frequencies $\tilde{\omega}^2_{qj}$, as given by the second term in eq. (5.2), for different scaling factors applied to the function $\chi^0(\boldsymbol{q})$ to simulate the effect of changing the density of states at the Fermi level. One sees that there is a strong tendency, with increasing $n(E_F)$, to produce dips in the regimes of $\boldsymbol{q}$-space where they are observed experimentally. The anomaly along the [111] axis is particularly interesting since it is very sensitive to the magnitude of $\chi^0(\boldsymbol{q})$ and occurs close to the value $(\frac{2}{3},\frac{2}{3},\frac{2}{3})\ 2\pi/\alpha$ which corresponds to the unstable mode for the formation of the ω-phase (Moss et al. 1973). Measurements for $Nb_{0.8}Zr_{0.2}$

alloys (for which $n(E_F)$ should be larger then in pure Nb on the rigid band model) by Traylor and Wakabayashi (private communication) show however that there is a general softening of LA frequencies relative to those in pure Nb rather than an enhancement of the anomalies themselves. However, recent measurements by Stassis et al. (1978a) show a dramatic anomaly of this mode in bcc Zr. On the other hand, as we have seen, on alloying Mo into Nb (Powell et al. 1968), the anomalies continuously disappear as expected on the basis of this model. The same is true for $NbH(D)_x$ non-stoichiometric compounds (Lottner et al. 1978). It is reasonably well established that the addition of interstitial hydrogen to Nb pulls some of the d-band states below the Fermi level, resulting in a lower $n_d(E_F)$ (Swittendick 19XX).

We can see that the above model is grossly oversimplified, primarily because of its neglect of overlap between the d-orbitals, and the neglect of hybridization with the s-band and the e_g-type orbitals. One immediate consequence of the neglect of overlap is the fact that $I^{\beta}(\boldsymbol{q})$ vanishes for *transverse* modes, as may be seen from the symmetry of the expression (5.1.2). Thus the "monopole" form of the model is unable to explain the anomalous features in the *transverse* branches in Nb. Note however that including d–d overlap implies, in the (l,m) representation, the inclusion of $l=1$ and $l=3$ orbitals as well, so that the $N^{\alpha}_{\xi_1\xi_2}$ terms in eq. (4.3.9) no longer vanish. The monopole term of the model is also unable to account for the anomaly at the point H in Mo (which has larger $l=1$ and $l=3$ character at the Fermi level although a much smaller overall density of states), since $I^{\alpha}(\boldsymbol{q})$ vanishes at the zone boundary in this model (see fig. 11). Recently,

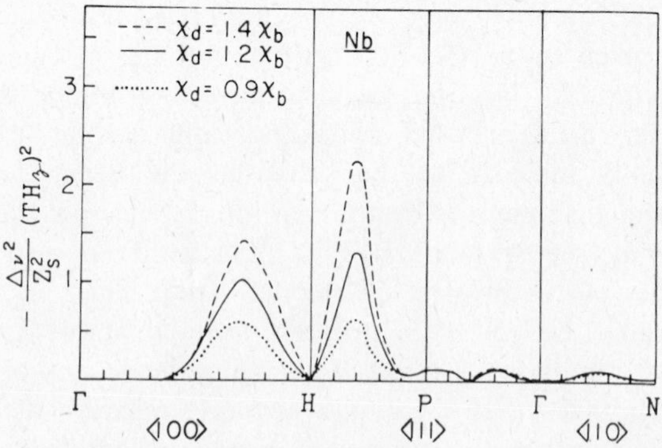

Fig. 13. The decrease in ν^2 for the longitudinal branches of Nb (in units of the pseudo-atom charge Z_s) as calculated from the monopole-type charge fluctuation model of Sinha and Harmon (1975, 1976). The different curves represent the results obtained by scaling the band-calculated $\chi^0(\boldsymbol{q})$ function to simulate the effect of changing the density of states at E_F.

Stassis et al. (1978b) observed an anomalous softening with decreasing temperatures of the zone-center longitudinal optic phonon mode for hcp Zr. (This really corresponds to a zone boundary phonon along the c-axis in the double-zone scheme.) They were able to explain this qualitatively in terms of mixed $l=1$ and $l=2$ character and its temperature dependence in terms of the loss of $l=1$ character at the Fermi level due to thermal smearing with increasing temperature. The same effect is also seen in h.c.p. Ti (Stassis et al. 1979). The most dramatic anomaly of this type (causing the LO phonon frequency at Γ to become almost degenerate with the TO phonon frequency) is exhibited by the metal Tc (Smith et al. 1976).

Basically, the microscopic charge fluctuation model described above simply emphasizes the physics connected with the subset of terms involving only single-site d–d transitions in the full set of terms yielding the electronic contribution to the phonon frequencies. This corresponds to the monopolar-type charge fluctuations, and the model then identifies a subset of the observed anomalies as being associated with an enhanced electronic response (in terms of such charge fluctuations) to the lattice displacements.

Wakabayashi (1977) has developed a model which introduces a monopolar "charge fluctuation" degree of freedom which is coupled to the lattice displacements and to other such charge fluctuations by short-range force constants of the appropriate symmetry while the rest of the interactions between atoms are parametrized with the usual tensor force constants. This exactly reproduces the structure of the equations of motion from the microscopic model which emphasizes only the d–d single site transitions (see eqs. (4.2.29)). Wakabayashi finds a much better fit to the observed anomalous structure in the longitudinal branches for Nb using this model than the conventional force-constant analysis. Interestingly, he finds that the charge fluctuation degree of freedom does not help appreciably in improving the fit for Mo, which has a much lower $n(E_F)$, in accordance with our expectations from the microscopic theory. He also finds that the charge fluctuation model (with second-neighbor interplanar force constants along the c-axis) can be used to explain the flat behavior of the LO phonon modes in Ti and Zr much better than a normal Born–van Karman model.

Allen (1977) has developed a very similar phenomenological model for Nb, starting from general symmetry considerations regarding the types of charge fluctuation degrees of freedom. He finds that a monopole-type charge fluctuation is important for determining the anomalies in the LA branches, in agreement with Wakabayashi, and also finds that a quadrupolar ($\Gamma_{25'}$-type symmetry) type of charge fluctuation yields the anomalous upward curvature found in the TA branches. Allen also makes the interesting conjecture that the $\frac{1}{2}(C_{11}-C_{12})$ mode shear instability in the A-15 compounds can be explained in terms of coupling of monopole-type charge fluctuations on *different* metal atom chains.

Hafstrom et al. (1978) have used the simplified scalar form of the charge fluctuation model outlined at the beginning of this subsection to explain the temperature dependence of the geometric mean squared frequency of $V_2Hf_{1-x}Ta_xC$-15 compounds as obtained from heat capacity experiments. They assume $\chi^0(\boldsymbol{q},T)$ may be written as $\chi(T)A(\boldsymbol{q})$, where $A(\boldsymbol{q})$ is a universal function which is independent of temperature and composition, and $\chi(T)$ is the (temperature-dependent) susceptibility of the material. They then show that the square of its geometric mean frequency may be written as

$$\omega_g^2(T)=\Omega_g^2(T)[1-F\{\chi(T)\}], \tag{5.1.9b}$$

where $\Omega_g(T)$ is the "bare" (pseudo-atom) mean frequency and $F\{\chi(T)\}$ is a function which increases monotonically with $\chi(T)$. $\Omega_g(T)$ is independent of the alloy concentration x because the masses and unit cell volumes are similar. However, since $n_d(E_F)$ varies across the series so will $\chi(T)$. As we have seen in fig. 7 eq. (5.1.9) seems to provide a remarkable universal correlation between $\Omega_g^2(T)$ and $\chi(T)$ across the whole range of concentrations and temperatures. Similar correlations appear to exist between $\omega_g^2(T)$ and $\chi(T)$ in the A-15 compounds (Knapp et al. 1976).

An interesting application of a scalar (monopolar) charge fluctuation model was made some years ago by Yamada (1975) in connection with the so-called "Verwey transition" which occurs in magnetite at 119 K, as discussed in §1. Yamada wrote the charge density response of the conduction electrons in the form of eq. (5.1.6) generalized to the case of the four Fe atoms in the unit cell. He then decomposed these according to symmetry types and considered only the charge fluctuations which have symmetry Δ_4 and Δ_5 in the unit cell. Such charge fluctuations only couple to phonons of the same symmetry, and the coupling coefficients (analogous to the $I^{\beta}(\boldsymbol{q})$ in eq. (5.1.5)) were taken as parameterized functions, which were *postulated* to have a maximum at the point $(0,0,\frac{1}{2})$ in the reduced zone. Interactions between the charge fluctuations on the Fe sites were not considered. By casting the "bare" electron–phonon coupling hamiltonian into a coupled pseudospin–phonon form, and considering the phonon response function, Yamada was able to account for the nature and symmetry of the observed critical scattering at the phase transition by choosing the charge fluctuation to be of symmetry type $\Delta_5^{(1)}$.

We turn our attention next to the transition metal carbides for which materials the neutron scattering measurements of the phonon spectra have aroused a great deal of interest for some years (Smith and Glaser 1970, 1971, Smith 1972, Smith et al. 1976). The microscopic charge fluctuation model has also been applied to NbC (Sinha and Harmon 1975, 1976). Band structure calculations (Gupta and Freeman 1976c, Klein and Papaconstantopoulos 1976) have shown that the conduction bands consist of d-orbitals on the Nb sites hybridized with C p-orbitals. The Fermi level lies in a region where the d-orbitals are mainly t_{2g}-like in character (see fig. 6). In the phonon calculation the effect of the interband transitions was explicitly included by introducing dipolar-type charge fluctuations on the Nb sites as well as charge fluctuations on the Nb and C sites corresponding to the charge fluctuations on the Nb and C sites arising from the intraband transitions around the Fermi level. Electron–phonon matrix elements coupling the C p-orbitals to the Nb d-orbitals were, however, neglected, so that there was no coupling of the transverse modes to the charge fluctuations. The equations of motion for the lattice then are given by (see eq. (4.2.29))

$$\omega^2 e_\alpha(\kappa) = \sum_{\beta\kappa'} R^{\kappa\kappa'}_{\alpha\beta} e_\beta(\kappa') + \sum_{\beta} T^{\kappa 1}_{\alpha\beta} p_\beta(1) + \sum_{\kappa'\xi'} I^{\alpha,\kappa\kappa'}_{\xi'}(\boldsymbol{q}) Q_{\mu'}(\kappa'),$$

$$0 = \sum_{\beta\kappa'} (I^+)^{1\kappa'}_{\alpha\beta} e_\beta(\kappa') + \sum_{\beta} \left(V^{\mathrm{dd}}_{\alpha\beta}(\boldsymbol{q}) + \alpha^{-1}_{\alpha\beta}(\boldsymbol{q}) \right) p_\beta(1) + \sum_{\kappa'\mu'} J^{\alpha,1\kappa'}_{\mu'}(\boldsymbol{q}) Q_{\mu'}(\kappa'),$$

$$0 = \sum_{\beta\kappa'} (I^+)^{\beta,\kappa\kappa'}_{\mu}(\boldsymbol{q}) e_\beta(\kappa') + \sum_{\beta} (J^+)^{\beta,\kappa 1}_{\mu} p_\beta(1)$$

$$+ \sum_{\kappa'\mu'} \left[\mathbf{V}_{\mathrm{eff}} - (\boldsymbol{\chi}^0)^{-1} \right]^{-1}{}_{\kappa\kappa'}{}_{\mu\mu'} Q_{\mu'}(\kappa'), \qquad (5.1.10)$$

where $\kappa = 1$ labels the metal site, and $\kappa = 2$ the carbon site. $p_\beta(1)$ is the β-component of the dipolar amplitude introduced on the Nb sites, and $Q_\mu(\kappa)$ the amplitude of the μth type of $t_{2g} - t_{2g}$ (e.g. $xy \to xy$) or p→p charge fluctuation induced on the site κ. **R**, **I** and **J** describe $\boldsymbol{q}$-dependent coupling coefficients between the pseudo-atoms themselves, between the pseudo-atoms and the dipolar fluctuations, and between the dipolar and charge fluctuations. $V^{\mathrm{dd}}_{\alpha\beta}(\boldsymbol{q})$ is the coupling coefficient between dipolar fluctuations, and $\alpha_{\alpha\beta}$ is the dipolar polarizability (given by interband transitions) on the Nb sites. $I^{\alpha,\kappa\kappa'}_{\mu}(\boldsymbol{q})$, $V^{\kappa\kappa'}_{\mathrm{eff},\mu\mu'}(\boldsymbol{q})$ and $\chi^{0,\kappa\kappa'}_{\mu\mu'}(\boldsymbol{q})$ have the same meanings as discussed previously, but generalized to the case of more than one atom per cell. For the calculations, $\chi^{0,\kappa\kappa'}_{\mu\mu'}(\boldsymbol{q})$ was taken to be roughly given by

$$\chi^{0,\kappa\kappa'}_{\mu\mu'}(\boldsymbol{q}) \simeq \delta_{\mu\mu'} \delta_{\kappa\kappa'} n_\kappa(E_\mathrm{F}), \qquad (5.1.11)$$

where $n_\kappa(E_\mathrm{F})$ is the partial density of states for d-orbitals on the metal site ($\kappa = 1$) or p-orbitals on the C site ($\kappa = 2$). The p-orbitals were approximated by a single orbital on the C site instead of three so that the **V** and $\boldsymbol{\chi}^0$ were

10×10 matrices (corresponding to 9 d–d transitions and one p–p transition). The **I** and **J** matrices were written in the form of eq. (5.1.2) with the appropriate phase factors for the basis sites, and the **R**, **I** and $(\mathbf{V}+\boldsymbol{\alpha}^{-1})$ matrices were taken to be of the simple shell-model form.

Fig. 14 shows the results of the calculations for the longitudinal modes of NbC. One sees that the positions of the anomalies are very well determined by introduction of the charge fluctuation terms. Also shown are the results of the same model on setting $n_\kappa(E_F)=0$ (κ-1, 2), i.e. "freezing out" the charge fluctuations by giving them zero polarizability, and one can see that the dispersion curves then are reminiscent of the curves for ZrC. Thus, although the above calculation is not truly microscopic in the sense that the "dipolar" part of the response and the short-range contributions to the phonon frequencies are parametrized, it emphasizes that the phonon anomalies are driven by the coupling between the ionic displacements and the charge fluctuations on the ion sites.

One of the most interesting purely phenomenological models which was developed to account for the anomalies in the dispersion curves of the transition metal carbides was the "double-shell" model of Weber (1973). In this model an extra set of dipolar degrees of freedom was introduced on

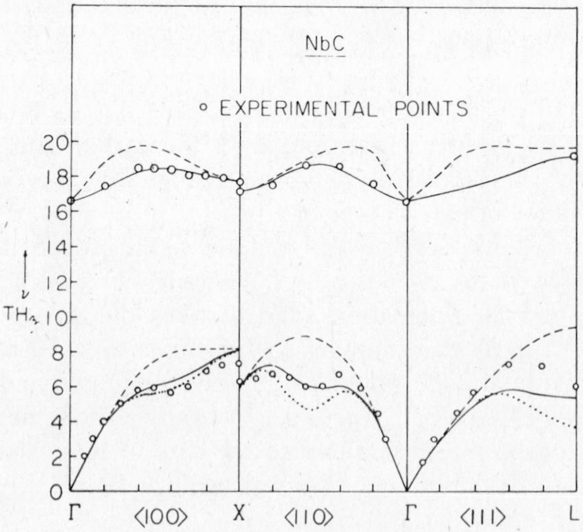

Fig. 14. Dispersion curves for the longitudinal branches of NbC as calculated by Sinha and Harmon (1975, 1976) from a microscopic charge fluctuation model as discussed in the text. The experiment results are those of Smith and Gläser (1971). The smooth curve corresponds to a best fit within the model for the partial densities of states at the Nb and C sites at E_F. The dotted curve corresponds to a 10% increase in these parameters. The dashed curve corresponds to setting these to zero (i.e. removing the charge fluctuation effects entirely).

the metal atoms. The equations of motion look formally like those in eq. (5.1.10) but with the $Q_\mu(\kappa)$ replaced by $p'_\beta(1)$. The anomalies are accounted for in this model by the fact that the coupling coefficient between these "secondary dipoles", i.e. the coefficient analogous to $[\mathbf{V}_{\text{eff}} - (\chi^0)^{-1}]$ in eq. (5.1.10) becomes small in certain regions of $\boldsymbol{q}$-space corresponding to the positions of the anomalies. Since this coupling coefficient appears in the denominator when the secondary dipolar degrees of freedom are eliminated from the equations of motion, this was referred to as "resonance-like" behavior, leading to the anomalies. This model in fact played an important role in stimulating the development of the microscopic theory. There are, however, some differences between the microscopically based model as given by eq. (5.1.10), and the double-shell model. In the latter, the smallness of the secondary dipole–dipole coupling coefficients is achieved by introducing a predominantly second-neighbor (and weaker first-neighbor) attractive interaction between the "supershells" whose displacements relative to the cores yield the secondary dipoles. In eqs. (5.1.10), the $Q_\mu(\kappa)$ degrees of freedom are predominantly monopolar in nature, and while it is true that it is the smallness of $[\mathbf{V} - (\chi^0)^{-1}]$ which enhances the effect, it is the $\boldsymbol{q}$-dependence of the **I** (and **J**) coefficients which determines the positions of the anomalies. An earlier version of the "double-shell" model where the extra degree of freedom was monopolar (i.e. a scalar) rather than dipolar had in fact been given by Weber et al. (1972).

A point of interest in connection with such models has to do with the L point, i.e., the [111] zone boundary point. One may see that in NbC (relative to ZrC) there is considerable softening of the acoustic mode frequencies at this point. From the structure of the dynamical matrix at this high-symmetry point, it may be shown that the pattern of ionic displacements in the longitudinal acoustic modes corresponds to stationary (111) sheets of C atoms and alternately positive and negative displacements of (111) sheets of Nb atoms along the [111] axis, while for longitudinal optic modes the pattern is interchanged between Nb and C atoms. Thus it may be seen that lattice coupling to charge fluctuations on the Nb sites vanishes for the acoustic modes (see fig. 11b), and to charge fluctuations on the C sites vanishes for the optic modes at the point L. Thus the softening observed in the LA mode in NbC at this point implies charge fluctuations on the C sites. In materials for which there is only one predominant type of charge fluctuation associated with the *metal* site (as might be expected for the "mixed valence" or "actinide" compounds of the rock-salt structure such as SmS or UN), one might expect that there would instead be anomalous behavior of the *LO* mode, and indeed this appears to be the case experimentally (Dolling et al. 1978), although the anomalies along the [100] and [110] directions do not appear.

As has been mentioned already, the most noticeable deficiency in the above microscopically based model is the absence of coupling of the transverse modes to the charge fluctuations, whereas experimentally the phonon anomalies in NbC (and TaC as well) exhibit anomalies in the TA[110] and TA [111] modes as well. Allowing explicitly for overlapping orbitals, so that the electron–phonon interaction can couple d-orbitals on the metal site to p-orbitals on the C site, will change the symmetry of the $I_{\mu}^{\alpha,\kappa\kappa'}(\boldsymbol{q})$ sufficiently to allow transverse mode coupling. A microscopically based model which explicitly introduces such coupling was introduced by Hanke et al. (1976) for NbC. Instead of choosing as basis orbitals the Nb-based t_{2g}-orbitals and the C-based p-orbitals, however, these authors choose to work with linear (bonding and antibonding) combinations of these orbitals; associating the states below E_F with the pure bonding combination and the states above E_F with the pure antibonding combination. Although this approximation seems somewhat crude, it has some vindication from a more detailed recent analysis by Terakura (1978). For this calculation, the only charge fluctuations considered were between these bonding and antibonding states, the other charge fluctuations being treated as giving rise to free-electron screening as in the Nb calculations of Sinha and Harmon (1975). For the calculation of $\chi^0(\boldsymbol{q})$ the band structure energies $E_{\boldsymbol{k}\lambda}$ were treated in the effective mass approximation as parabolic bands. Finally the electron–phonon interaction was calculated using a local pseudopotential based on the Nb site and the interaction between the electrons and the C sublattice was neglected. The main point of the paper is to emphasize that the phase relations of the bonding and antibonding wavefunctions are such that phase factors appear in $\boldsymbol{\chi}^0(\boldsymbol{q})$ itself, which cause it to reach a peak in its absolute value at the positions of the observed anomalies, which are close to $(2\pi/a)(\frac{1}{2},0,0)$, $(2\pi/a)(\frac{1}{2},\frac{1}{2},0)$ and $(2\pi/a)(\frac{1}{2},\frac{1}{2},\frac{1}{2})$, the magnitude of these being controlled by the energy difference between the bonding and antibonding sets of bands in the denominator of the expression for $\boldsymbol{\chi}^0(\boldsymbol{q})$. Their calculations for the acoustic branches of the phonon spectrum of NbC are illustrated in fig. 15. Although the Hanke–Hafner–Bilz model, like the earlier model (Sinha and Harmon 1975) suffers from a number of drastic oversimplifications, it emphasizes another piece of the physics, namely the role of p–d hybridization in coupling the conduction electrons in these compounds to the transverse modes, and in giving rise to a $\boldsymbol{q}$-dependence of $\boldsymbol{\chi}^0(\boldsymbol{q})$ which can result in "resonant-like" screening at certain $\boldsymbol{q}$-values. In the latter sense it is similar in spirit to the models of Weber et al. (1972) and the double-shell model of Weber (1973). Note that in the Sinha–Harmon model for NbC, the anomalous structure, which arises from maxima in $I_{\mu}^{\alpha}(\boldsymbol{q})$ (the electron–phonon coupling) is due to a general "structure factor" which peaks at those particular $\boldsymbol{q}$-values and for similar reasons as the peaks in the Hanke–Hafner–Bilz model, namely phase relations imposed by the geometrical structure of the NaCl lattice. One may conclude that both effects must contribute to the anomalies observed in NbC and TaC. A possible way to isolate the "d-shell charge fluctuation effect" from the "p–d hybridization effect" would be to examine a compound with the same structure and with strong d-like character at the Fermi level but with little or no p–d hybridization. From what has been said above, such a material should show anomalies in the LA modes at roughly the same positions as those in NbC and TaC, but no TA mode anomalies. It appears that YS is exactly such a material since this is precisely the behavior which is observed (Roedhammer et al. 1978) although the authors choose to analyze the result in terms of a double-shell model. In TiN (Kress et al. 1978), anomalies similar to those in NbC are observed, but the TA anomalies are much weaker. Another possible candidate which should be interesting to study from this point of view is NbO.

We note that the first applications of the simple scalar form of the monopolar charge fluctuation model for the d-electrons (with free-electron screening to account for the s-electrons) to a transition metal were the calculations of Hanke and Bilz (1972) and Hanke (1973), who used them (with a parametrized local pseudopotential) to calculate the phonon spectrum of Ni and Pd. Nickel shows no anomalies in its phonon spectra and the role of the charge fluctuations in promoting anomalies was not emphasized at that time. Prakash (1978) has also used a method of

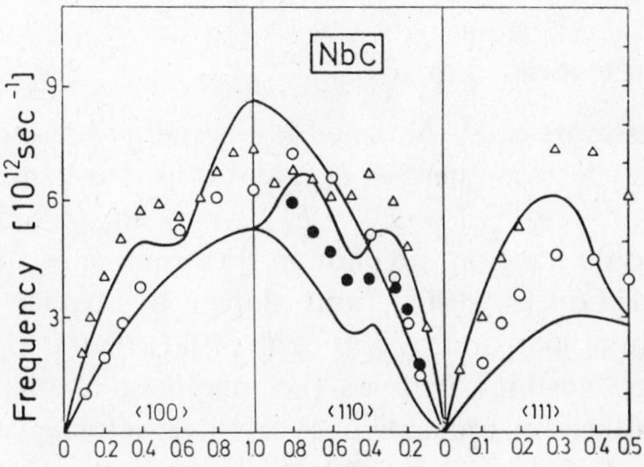

Fig. 15. Acoustic branches of dispersion curves for NbC for the model of Hanke et al. (1976) (as discussed in the text) compared with the experimental results of Smith and Glässer (1971).

inverting the dielectric matrix based on a factorization Ansatz for $\chi^0(\boldsymbol{q}+\boldsymbol{G},\boldsymbol{q}+\boldsymbol{G}')$ (Sinha et al. 1974, Price et al. 1974) which effectively amounts to a combined charge and dipole fluctuation model, together with a local pseudopotential, to calculate the phonon spectra of Pd, V and Sc, obtaining fair agreement with experiment.

In one of the most recent applications of the density response function method, Cooke (1978) has performed a calculation for Nb, where instead of transforming the electron–phonon matrix elements to reduce the cancellation between large $C_{\alpha\beta}(\boldsymbol{q})$ and $E_{\alpha\beta}(\boldsymbol{q})$ terms, he works directly with the potential due to the bare Nb^{5+} ion cores and uses the basic eqs. (2.12)–(2.14) to calculate the phonon spectrum. The density response matrix $\chi(\boldsymbol{q}+\boldsymbol{G},\boldsymbol{q}+\boldsymbol{G}')$ is evaluated by making a representation of the wavefunctions in terms of angular momentum orbitals as discussed in §§4.2 and 4.3. He finds that the LA[100] anomaly in Nb in this picture arises owing to the way in which $C_{\alpha\beta}(\boldsymbol{q})$ and $E_{\alpha\beta}(\boldsymbol{q})$, each of which has no anomalies, cancel each other. This is not necessarily inconsistent with our previous picture of the anomalies arising from the $E_{\alpha\beta}(\boldsymbol{q})$ in the pseudo-atom representations, since the above-mentioned cancellation is already built into such a picture. Cooke omits $l=3$ orbitals and also neglects the energy dependence of the $l=2$ radial wavefunction in obtaining the density response function. He also neglects all exchange and correlation effects. We must await results of further such calculations for other symmetry directions of Nb and for other materials before we can be sure that such a procedure is justified. If so, the implications for the use of the density response formalism in calculating phonon spectra are quite encouraging.

5.2. Rigid atom models

The quantities most easily obtained in an energy band calculation for a solid are the one-electron energies $E_{\boldsymbol{k}\lambda}$, and thus it is tempting to focus on the factor $(\langle n_{\boldsymbol{k}\lambda}\rangle-\langle n_{\boldsymbol{k}+\boldsymbol{q}\lambda'}\rangle)/(E_{\boldsymbol{k}\lambda}-E_{\boldsymbol{k}+\boldsymbol{q}\lambda'})$ in eq. (4.4.27) and examine what effect it would have on the phonon spectrum. It was first pointed out by Keeton and Loucks (1968), and shown by explicit calculation by Evenson and Liu (1968), and by Liu et al. (1971), that if the Fermi surface had appreciable "nesting" features (i.e. mapping of one portion of the surface into another on translation by a wavevector $\boldsymbol{q}_0$) then one would obtain a peak in the absolute magnitude of the "generalized susceptibility function" $\tilde{\chi}^0(\boldsymbol{q})$ defined by

$$\tilde{\chi}^0(\boldsymbol{q})=\sum_{\boldsymbol{k}\lambda\lambda'}\frac{\langle n_{\boldsymbol{k}\lambda}\rangle-\langle n_{\boldsymbol{k}+\boldsymbol{q}\lambda'}\rangle}{E_{\boldsymbol{k}\lambda}-E_{\boldsymbol{k}+\boldsymbol{q}\lambda'}}. \tag{5.2.1}$$

This fact has been used to account for the anomaly in Nb along the [111] axis (Myron et al. 1975), the "charge density wave" instability in the layered chalcogenides (Myron and Freeman 1975, Ricco 1977, Myron et al. 1977) the ever-popular anomalies in the transition metal carbides (Gupta and Freeman 1976a, b, c, Klein and Papaconstantopoulos 1976) and the instability causing the metal–insulator transition in VO_2 (Gupta et al. 1977). The most pronounced nesting occurs in the layered dichalogenides (enhanced by the essentially two-dimensional nature of the energy bands. Thus fig. 16a shows the Fermi surface for 1T-TaS_2 calculated by Myron and Freeman (1975). It is seen that there is indeed nesting along the a^*-axis at a $\boldsymbol{q}_0$ close to that observed for the periodic lattice distortion which this material undergoes at low temperatures, and this reflects itself in a corresponding peak in $\tilde{\chi}^0(\boldsymbol{q})$ (Myron et al. 1977) (see fig. 16b). There is, however, also nesting along the b^*-axis, and in fact there is also a peak in $\tilde{\chi}^0(\boldsymbol{q})$ along this direction. As we shall see later, it is likely that the electron–phonon matrix elements may be the determining factor in the $\boldsymbol{q}_0$ the lattice actually "chooses". Fig. 17 shows the Fermi surface for NbC as calculated by Gupta and Freeman (1976b, c). It is seen that the flat portions of this surface are connected by wavevectors, which in the [100], [110] and [111] directions, respectively, are strikingly close to the positions of the observed anomalies. The corresponding $\tilde{\chi}^0(\boldsymbol{q})$ (Gupta and Freeman 1976b, c) is shown in fig. 18.

One of the problems with explanations of the phonon anomalies in terms of structure in $\tilde{\chi}^0(\boldsymbol{q})$ is that they ignore the $\boldsymbol{q}$-dependence to the matrix elements of the electron–phonon interaction in eq. (4.4.28), and that the peaks are numerically rather small riding on a rather large smooth background, especially if interband transitions are included. Thus Varma and Weber (1979) show that for NbC, assuming constant matrix elements, the first term in eq. (4.4.28) has rather little structure, notwithstanding the large peaks in fig. 18 arising from the intra-band transitions alone. In view of the strong $\boldsymbol{q}$-dependence of the electron–phonon matrix elements as shown in several other calculations discussed already, there does not appear to be a case where an anomaly or instability can be unambiguously assigned *solely* to an absolute maximum in $\tilde{\chi}^0(\boldsymbol{q})$.

Gale and Pettifor (1977) and Terakura (1978) have calculated the "band structure" term in eq. (4.4.28) in the rigid atom approximation for NbC by actually giving the Nb atoms a periodic pattern of (longitudinal or transverse) displacements corresponding successively to different wavevectors (the C atom displacements were ignored) for a large cluster (~1500) of atoms. They then calculate the total density of states (and hence the total one-electron energy) by using a recursion-relation method of Haydock et al. (1975), starting from a tight-binding representation which includes only d-orbital overlaps. By calculating the coefficient of the square of the

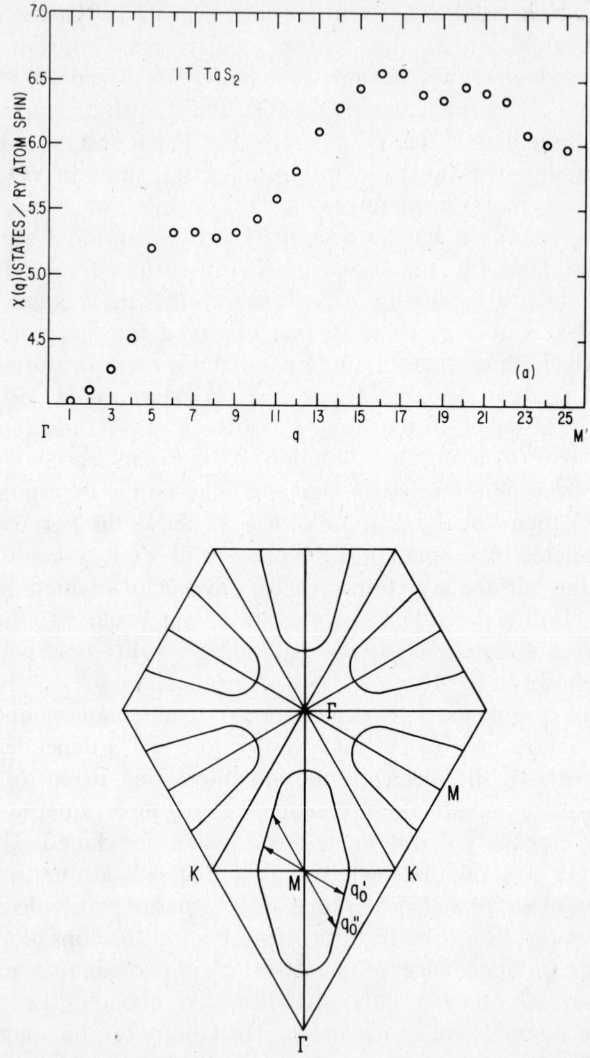

Fig. 16. (a) Fermi surface cross sections for 1T-TaS_2 in the basal plane, showing the "nesting" wave vectors $\boldsymbol{q}_0'$ (in the ΓM direction) and $\boldsymbol{q}_0''$ (in the K direction). From Myron and Freeman (1975). (b) The function $\tilde{\chi}^0(\boldsymbol{q})$ for 1T-TaS_2 for the direction ΓM. From Myron et al. (1977).

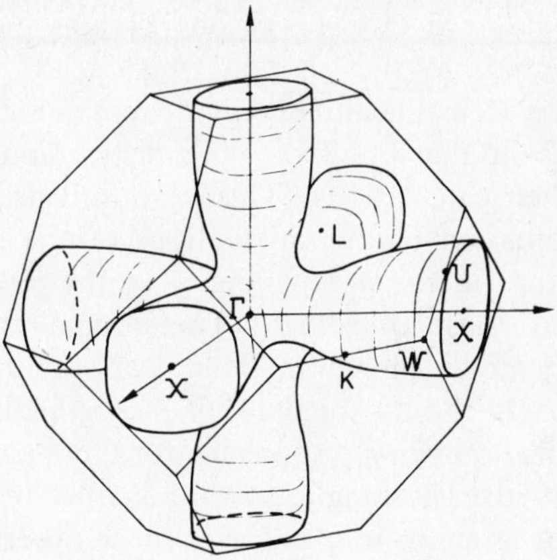

Fig. 17. The band 4 Fermi surface for NbC in the first Brillouin zone, as calculated by Gupta and Freeman (1976b, c).

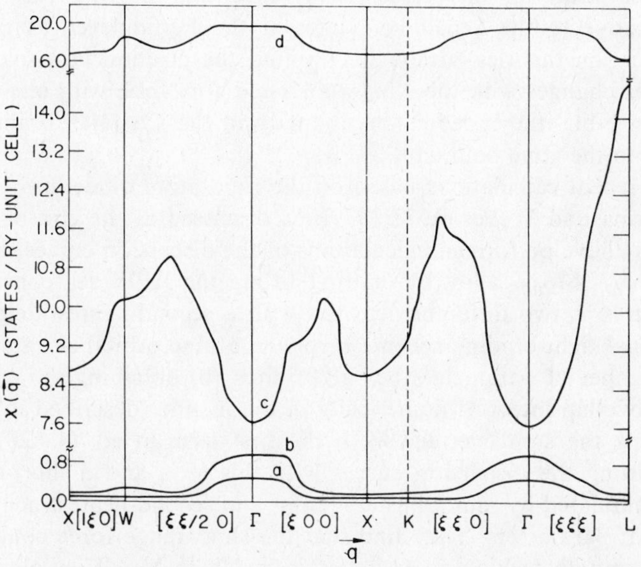

Fig. 18. The function $\tilde{\chi}^0(\boldsymbol{q})$ for NbC as calculated by Gupta and Freeman (1976b, c) along several symmetry and off-symmetry directions: (a) the intraband contribution from band 6; (b) the intraband contribution from band 5; (c) the intraband contribution from band 4, and (d) the total interband contribution from bands 4, 5, 6.

displacement amplitude in this energy one obtains the "rigid atom band structure" contribution to the phonon frequencies, which includes the effect of both matrix elements and energy denominator. These authors find that the anomalies in the longitudinal acoustic modes of NbC can be accounted for in their calculations. (This is not inconsistent with the result of the charge fluctuation model calculations (Sinha and Harmon 1975) that there is structure in the $\boldsymbol{q}$-dependence of the electron–phonon coupling coefficient at those $\boldsymbol{q}$-vectors.) The anomalies in the transverse $T_1[110]$ and T[111] modes, however, do not appear in their calculations, which considered only the d-orbitals, thus strengthening the case for the role of p–d hybridization being responsible for these anomalies, as claimed by Hanke et al. (1976). The longitudinal anomalies in these "rigid atom" calculations are not as sharp in $\boldsymbol{q}$-space as those observed experimentally, however.

Gale and Pettifor and Terakura leave open the possibility that screening effects can enhance the magnitude of these anomalies, as in the response-function-based models. These authors have also examined the effect of the electron/atom ratio on this result. They find that the anomalies are strongly sensitive to the density of states at the Fermi level, as in the charge fluctuation theories, although they find the position in $\boldsymbol{q}$-space of the anomalies changes with the e/a ratio. Thus, for NbN, with one more electron than NbC, they predict the anomaly in the LA [100] branch to have moved to the zone boundary.

The most recent calculations based on the rigid atom model have been those of Varma and Weber (1977, 1979), as discussed at the end of §4.4. These authors have performed calculations of the dispersion curves of Nb, Mo and a $Nb_{0.25}Mo_{0.75}$ alloy by (a) first fitting the APW self-consistent band structures at two different densities with a second-nearest-neighbor non-orthogonal tight-binding scheme involving a nine orbital s–p–d basis (the total number of parameters being 20), then (b) obtaining the derivatives of the overlap integrals numerically in the manner described in §4.4, (c) performing the sum over $\boldsymbol{k},\lambda,\lambda'$ in the first term in eq. (4.4.28) and finally (d) fitting the dispersion curves with this term and a short-range interaction simulated by four adjustable first- and second-nearest-neighbor force-constant parameters. They find that the short-range force constants behave in a smooth fashion as one goes from Nb to Mo. They also find that varying the quantities representing the derivatives of the overlap integrals by about 20–30% or omitting the ∇S terms in the matrix elements (eq. (4.4.11)) changes the magnitude of the "band structure" term but not significantly its $\boldsymbol{q}$-dependence. The results of their fits to the phonon spectra of Nb, Mo and $Nb_{0.25}Mo_{0.75}$ are shown in fig. 19. It may be seen that the results are very encouraging, reproducing all the observed anomalous features in the dispersion curves, including those in the transverse modes. These authors have recently also applied their model to NbC (Varma et al. 1979) with results which are about as good in agreement with experiment as the density response model calculations discussed previously.

Varma and Weber point out that the screened electron–phonon matrix elements in their representation between states $\boldsymbol{k}'\lambda'$ and $\boldsymbol{k}\lambda$ are roughly proportional to $(v^{\alpha}_{\boldsymbol{k}'\lambda'} - v^{\alpha}_{\boldsymbol{k}\lambda})$, where $v^{\alpha}_{\boldsymbol{k}\lambda}$ is the α-component of the velocity $\nabla E_{\boldsymbol{k}\lambda}$. The occurrence of the anomalies in their theory, which is strongly influenced by the $\boldsymbol{q}$-dependence of the electron–phonon matrix elements, is therefore correlated with regions around the Fermi surface which are nested in two directions, but not in the third, so that a large *mismatch*

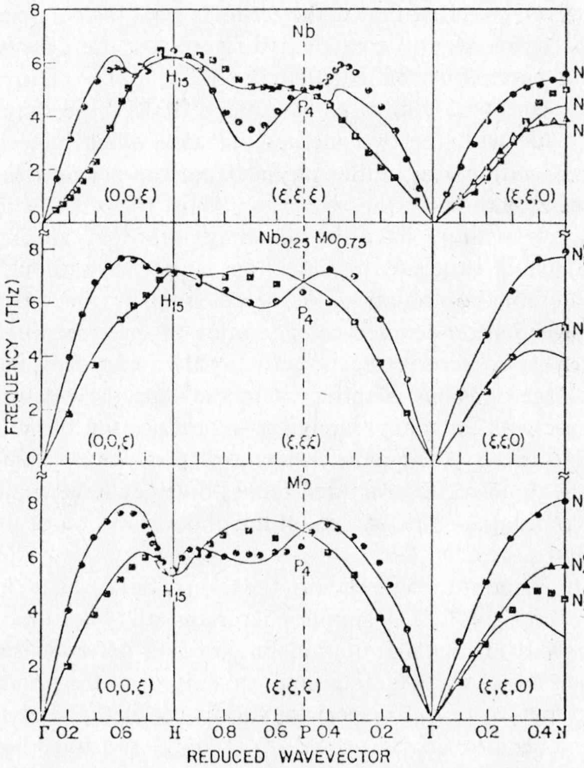

Fig. 19. Dispersion curves for Nb, $Nb_{0.25}$ $Mo_{0.75}$ and Mo calculated in the rigid atom approximation by Varma and Weber (1977) as discussed in the text. The experimental points are those of Powell et al. (1968).

between $\nabla_\alpha E_{k\lambda}$ and $\nabla_\alpha E_{k+q\lambda'}$ is realized. This is not quite the same requirement as that for producing a peak in $\tilde{\chi}^0(\boldsymbol{q})$ although the two are not necessarily mutually exclusive. It should be noted, however, that a large density of states at the Fermi level generally implies small $|\nabla E_{k\lambda}|$ so that their model should imply that the anomalies *decrease* with increasing density of states at the Fermi level, unless compensatory one-dimensional nesting features appear simultaneously.

5.3. Discussion

Both the density response method as represented by eq. (4.2.27) and the formulation represented by eq. (4.4.27) would give the same result if *all* terms were calculated properly and one would then have solved the problem of a first-principles understanding of the phonon spectra of solids. It should however be noted that if the terms in eq. (4.4.27) other than the band structure terms are to be calculated rigorously, the density response function must necessarily be calculated, and the regrouping of terms represented by eq. (4.4.27) loses its advantage. Thus the difference in the applications made with the two methods so far (which have necessarily implied approximations) is in the physical processes emphasized in the respective approximations.

The rigid atom methods have the advantage that they go directly from the electronic band structure to the phonon spectra, although they do involve considerable computation. As represented by the Varma–Weber theory, they sweep aside detailed consideration of the screening (or equivalently the density response) effects, claiming that a rigid displacement of the atomic charge densities is sufficient to calculate the electron–phonon matrix elements with accuracy enough to reproduce the basic features of the phonon spectra. By their nature, they are necessarily parametrized (to account for the "non-band structure" terms) but they have achieved some success in accounting for the features of the phonon spectra of the 4d b.c.c. transition metal series.

The density response methods on the other hand, as actually used hitherto, have been applied as simple microscopically-based *models* which have not involved much computation but used in parametrized form to emphasize the physics of certain anomalous features of the phonon spectra of the d-band metals. In this respect they too have met with some success. They have not been taken far enough in attempting a fully-fledged first-principles calculation of phonon spectra (except for the recent calculation of Cooke referred to in §5.1). This is because such a program involves quite severe computational difficulties, although in principle capable of yielding quite rigorous expressions for the phonon frequencies.

The question then arises – how important for instance are charge fluctuation effects in determining the anomalies in the phonon spectra of certain transition metals? Both the charge fluctuation models and the rigid atom tight-binding models show peaked electron–phonon coupling at the $\boldsymbol{q}$-vectors corresponding to the anomalies, which is not really surprising since the basic electron–phonon matrix element must be invariant with respect to the representation used for it. In the charge fluctuation models, however, the implication is that charge fluctuation effects in the screening actually enhance this structure. Further if such charge fluctuation effects are appreciable, then their contribution to $E''_{\alpha\beta}(\boldsymbol{q})$ (eqs. (4.4.25) and (4.4.27)) must also contribute to the anomalies, unlike the assumption of the Varma–Weber theory that $E''_{\alpha\beta}(\boldsymbol{q})$ simply smoothly cancels $C_{\alpha\beta}(\boldsymbol{q})$ to yield short-range forces. It should be noted that according to the pseudo-atom treatment given in §§2 and 4.2, the "rigid pseudo-atom" part of $\delta n^{\alpha}(\boldsymbol{r})$ in the expression for $E''_{\alpha\beta}(\boldsymbol{q})$ exactly cancels $C_{\alpha\beta}(\boldsymbol{q})$, and we are left with interactions between the further *distortions* in $\delta n^{\alpha}(\boldsymbol{r})$. The longest-range interactions of this type are the monopole–monopole interactions, which would exist if appreciable monopole large fluctuations existed. It is true, however, that the simple non-overlapping d-orbital representation as used for instance by Sinha and Harmon (1975, 1976) probably overestimates the charge fluctuation effect. This is because the density response *outside* the muffin-tin spheres to $\nabla_{\alpha} W(\boldsymbol{r}-\boldsymbol{r}_l)$ in that theory is expanded only in d-orbitals centered on that site, thus missing that portion of the response in the interstitial region which corresponds to *gradients* of these d-orbital densities as they move with the atoms. This part of the response would in fact tend to screen and reduce the effect of $\nabla_{\alpha} W(\boldsymbol{r}-\boldsymbol{r}_l)$ and hence the charge fluctuations in the d-orbitals, although it is partly compensated for in the models by the introduction of the "s-electron" screening effect. It is a pity that there appears at present no experimental method of checking the nature of the true electron response to a particular phonon to see to what extent any charge transfer effects are involved. Sham (1978) has suggested that a true self-consistent electronic band structure calculation be done for a crystal with a reasonably symmetric phonon wave built into it and the charge density calculated to test this point. It is hoped that such a calculation will indeed be forthcoming. Another test would be an actual calculation of the full matrix $(\mathbf{1}-\mathbf{V}_{\text{eff}}\boldsymbol{\chi}^{0})^{-1}$ so that the rigorous electron response to a phonon may be calculated and compared with the predictions of the charge fluctuation models and the rigid atom models.

It is possible that the models can also be differentiated on the basis of their experimental predictions. Thus, the charge fluctuation models predict an increase in the electron–phonon interaction with increasing $n_{\text{d}}(E_{\text{F}})$ (although not a change in its *structure* in $\boldsymbol{q}$-space *unless* new orbitals are involved around the new E_F), and thus a softening of phonon frequencies (but not necessarily a shift in the *positions* of the anomalies) with increasing $n(E_F)$. This seems to be borne out by experiments on the A-15 and C-15 materials, where we have seen the remarkable correlation between temperature and concentration dependence of the frequencies and the susceptibility (a measure of $n_d(E_F)$). In the rigid atom tight-binding method, the magintude and positions of the anomalies depend not so much on $n(E_F)$ but on certain nesting features of the bands around the Fermi surface, and thus the two theories might give rise to differing predictions on, say, the effect of changing the electron/atom ratio in the carbides. Also the temperature dependence of the phonon frequencies in the rigid atom tight-binding models would presumably be controlled more strongly by the effect of lattice thermal expansion on the overlap integrals rather than the smearing out of the density of states.

With regard to actual lattice instabilities, such as those which occur is the so-called "charge density wave" group of layered chalcogenides, as the name itself implies, the generally accepted picture is one which is in closer accord with the charge fluctuation models than the rigid atom models. In other words, one envisages an incipient instability with regard to the appearance of an order parameter which is phonon displacement *coupled to* a charge fluctuation on the atom. The fact that such an order parameter appears below the transition (Wilson et al. 1975) necessarily implies *fluctuations* in this parameter above the transition temperature, and consequently implies the presence of actual charge fluctuations. It seems at least reasonable that this should qualitatively be the case for all lattice phase transitions driven (or nearly driven) by the electron system, although the amount of charge fluctuation coupled in to the lattice motion may of course vary from system to system. As we shall see in the next section, another physical effect which such dynamical charge fluctuations may on occasion give rise to is the central peak phenomenon.

5.4. Lattice instabilities and the central peak

We have so far discussed lattice instabilities in terms of mode softening effects and concentrated therefore on microscopic theories of the phonon spectrum. However, we must take into account the fact that for certain materials, e.g. NbO_2 (Pynn et al. 1978), Nb_3Sn (Shirane and Axe 1971b), $NbSe_2$ or $TaSe_2$ (Moncton et al. 1977) or Fe_3O_4 (Fujii et al. 1975), the neutron scattering experiments reveal that the phonon at the wavevector of the instability does not actually soften to zero frequency, but that instead a central component (centered around $\omega = 0$) appears in the neutron scattering and grows as the instability temperature is approached, behaving like critical scattering in the actual vicinity of the transition. This is the well-known "central peak phenomenon" which has aroused lively interest amongst solid state theorists over the past few years. (For a recent collection of articles on the subject, the reader is referred to the Proceedings of the International Conference on Lattice Dynamics, Paris, 1977 (Balkanski 1978).) Much of this theoretical work is concerned with the appearance of the central peak near phase transitions in insulators, such as $SrTiO_3$, and we shall not discuss these approaches here, since we are primarily concerned with crystallographic phase transformations driven by the electron–phonon interaction.

Bhatt & McMillan (1975) have constructed a phenomenological Landau theory of the central peak phenomenon considering the dynamics of a coupled phonon and charge density wave (CDW) system. By introducing a dissipative damping mechanism for the charge density wave, they are able to show from their equations of motion that for the regime $\omega_0\tau \ll 1$ (where ω_0 is the phonon frequency and τ is the relaxation time for the damping of the CDW) the CDW is able to follow the lattice motion adiabatically and phonon softening occurs, whereas for $\omega_0\tau \gg 1$, there occurs broadening of the phonon line and the appearance of a central peak.

Let us see how this phenomenon may be incorporated within the framework of the microscopic charge fluctuation model discussed in §5.1. For a simple semi-quantitative treatment, we consider here only the scalar form of the theory (i.e. only one type of charge fluctuation) as done in §5.1. Instead of taking $\chi^0(\boldsymbol{q})$ to be given by the perfect crystal set of energy bands of the Bloch states, we allow for electron scattering effects by giving it an imaginary component. Let us for the moment assume that we have the extreme limit where the electrons exhibit "hopping" conductivity, or small polaron behavior. This may in fact be a reasonable assumption for the case of the transition metal oxides which undergo metal–insulator transitions (Honig and Van Zandt 1975). In this case we may introduce a Debye-like relaxation for the non-interacting density response function and write (including now the frequency dependence)

$$\chi^0(\boldsymbol{q},\omega)=\chi^0(\boldsymbol{q})/(1-i\omega\tau), \tag{5.4.1}$$

where $\chi^0(\boldsymbol{q})$ is the "static" response function. (For Bloch electrons subject to scattering effects, the $\boldsymbol{q}$ and ω-dependence of the full complex $\chi^0(\boldsymbol{q},\omega)$ function are more complicated, but the qualitative results are probably not much altered hereby.) Let us consider a simple pure longitudinal mode and examine the expression for the lattice Green's function given by eq.

(4.2.25)

$$G_{qj}(\omega)=-\frac{\tilde{\omega}_{qj}}{\pi}\left[\omega^2-\tilde{\omega}_{qj}^2+\frac{1}{M}|I^{\alpha}(q)|^2\left(V_{\text{eff}}(q)-\frac{1-i\omega\tau}{\chi^0(q)}\right)^{-1}\right]^{-1}. \quad (5.4.2)$$

The neutron scattering cross section, which is proportional to the imaginary part of this Green's function (multiplied by the detailed balance factor) is then given by

$$I(q,\omega)\propto\frac{1}{e^{\beta\omega}-1}\,\mathrm{Im}\left\{\omega^2-\tilde{\omega}_{qj}^2+\frac{1}{M}|I^{\alpha}(q)|^2\left(V_{\text{eff}}(q)-\frac{1-i\omega\tau}{\chi^0(q)}\right)^{-1}\right\}^{-1}, \quad (5.4.3)$$

where $\beta=\hbar/k_{\mathrm{B}}T$, which for $\beta\omega\ll 1$ gives

$$I(q,\omega)\propto|I^{\alpha}(q)|^2k_{\mathrm{B}}T\tau\chi^0(q)/\left[1-V_{\text{eff}}(q)\chi^0(q)\right]^2$$
$$\times\left\{\left(\omega^2-\omega_{qj}^2\right)^2+\omega^2\tau'^2\left(\omega^2-\tilde{\omega}_{qj}^2\right)\right\}^{-1}, \quad (5.4.4)$$

where

$$\omega_{qj}^2=\tilde{\omega}_{qj}^2-\frac{1}{M}|I^{\alpha}(q)|^2\chi^0(q)\left[1-V_{\text{eff}}(q)\chi^0(q)\right]^{-1} \quad (5.4.5)$$

and represents the square of the "virtual soft mode" frequency, i.e. the renormalized frequency in the absence of relaxation effects, and

$$\tau'=\tau/\left(1-V_{\text{eff}}(q)\chi^0(q)\right) \quad (5.4.6)$$

represents a *collective* electron relaxation time. Note that τ' can be quite large compared to τ, the single electron relaxation time in the vicinity of an electronic charge fluctuation instability, as the denominator in eq. (5.4.6) can become quite small.

Eq. (5.4.4) yields a three-peaked structure with "phonon sidebands" and a central component. These may or may not merge into each other. If we assume they are well separated, the width of the central component in ω is given by

$$\Gamma_{\text{ep}}=\frac{1}{\tau'}\left(\omega_{qj}^2/\tilde{\omega}_{qj}\right). \quad (5.4.7)$$

Note that it becomes infinitely narrow in this approximation as the lattice instability is approached, which is when $\omega_{qj} \to 0$. The intensity of this central component is given by

$$I = \frac{k_B T}{\tilde{\omega}_{qj}^2} \frac{\left(\tilde{\omega}_{qj}^2 - \omega_{qj}^2\right)}{\omega_{qj}^2} \tag{5.4.8}$$

and hence also diverges at the transition temperature. Eq. (5.4.4) is identical in form to the expressions derived by Bhatt and McMillan (1975) and by Yamada et al. (1974). The latter authors developed a phenomenological theory for a lattice coupled to a spin system which exhibited some dissipative damping effects, and obtained a theory which has been used, among other things, to analyze the central peak observed in the scattering from magnetite (Fe_3O_4) in the vicinity of the Verwey transition (Yamada 1975). It was found that the relaxation time for the "spin" system (interpreted as the electron hopping relaxation time) was greater than 10^{-10} sec. This was thought to be surprisingly long, but eq. (5.4.6) in the microscopic theory yields a possible explanation, since if the electronic system itself is close to a charge density wave instability (in this case with regard to Fe^{2+} to Fe^{3+} fluctuations on the Fe sites), τ' can be quite large due to "critical slowing down" effects.

The charge fluctuation model, as applied by Sinha and Harmon (1975) to the calculation of the anomalies in Nb and NbC, has been used with the modification in eq. (5.4.1) for the diagonal components of the $\chi^0_{\mu\mu'}$ matrix to discuss the central peak in 2H-$NbSe_2$ by Das et al. (1978). The d-bands at the Fermi level were taken to be composed of (xy), $(x^2 - y^2)$ and $(3z^2 - r^2)$ orbitals (Mattheiss 1973) and the effect of the other charge fluctuations was simulated by a Lindhard-type screening function. The squared phonon frequencies (including the imaginary part of the self-energy) are given by (eq. (4.2.27))

$$\omega_{qj}^2 = \tilde{\omega}_{qj}^2 - \sum_{\alpha\beta} \sum_{\kappa_1\kappa_2\kappa_3\kappa_4} \sum_{\mu_1\mu_2} \left(M_{\kappa_1} M_{\kappa_4}\right)^{-1/2} e_\alpha^*(\kappa_1|qj)$$
$$\times I_{\mu_1}^{\alpha_1,\kappa_1\kappa_2}(-q) \chi_{\mu_1\mu_2}^{\kappa_2\kappa_3} I_{\mu_2}^{\beta,\kappa_4\kappa_3}(q) e_\beta(\kappa_4|qj), \tag{5.4.9}$$

where κ denotes the basis atom in the unit cell (of mass M_κ) and $e_\alpha(\kappa|qj)$ denotes the component of the eigenvector of a particular mode associated with the κth atom. $I_{\mu_2}^{\beta,\kappa_4\kappa_3}$ denotes the coupling between displacement of the κ_4th atom in the unit cell in the β-direction and a charge fluctuation on the κ_3 atom of type μ_2 (which was one of the 9 types $(xy) \to (x^2 - y^2)$ etc.). Note that there are 6 atoms in the unit cell for the 2H-polytype of $NbSe_2$

(consisting of two formula units). Overlap between the orbitals was neglected so that the size of the $\mathbf{V}_{\text{eff}}$ and $\boldsymbol{\chi}^0$ matrices was 18×18 (corresponding to 9 possible types of transitions on each of the two Nb sites in the unit cell). The calculations were performed for the Σ_4 LA branch along the [100] axis, and the eigenvectors $e_\alpha(\kappa|\bar{\boldsymbol{q}}j)$ were evaluated for this branch from an empirical force-constant model fitted to the measured phonon frequencies. The quantity

$$W_{\kappa'}(\boldsymbol{q}) = \sum_{\beta\kappa} \left[I_\mu^{\beta,\kappa\kappa'}(\boldsymbol{q}) e_\beta(\kappa|\boldsymbol{q}j) M_\kappa^{-1/2} \right] \tag{5.4.10}$$

represents the coupling of charge fluctuations of type μ on site κ' to the ionic displacements in mode $(\boldsymbol{q}j)$. Fig. 20 shows this function along the [100] direction for the Σ_4 mode, and Nb atom 1 with an $(xy)\to(xy)$ fluctuation. It may be seen that it tends to have a peak at $\boldsymbol{q}=0.35\boldsymbol{a}^*$, showing that the lattice has a natural "geometrical" enhancement of the electron–phonon interaction at such $\boldsymbol{q}$-values. Das et al. show that thus, even assuming no structure is $\chi^0(\boldsymbol{q})$, the central peak intensity calculated from the imaginary part of the phonon Green's function using eq. (5.4.8) has a maximum at $\boldsymbol{q}=0.35\boldsymbol{a}^*$ (close to the position of the observed satellites in the neutron scattering experiments), owing to the $\boldsymbol{q}$ dependence of this electron–phonon coupling, provided χ (i.e. $n_{\text{d}}(E_{\text{F}})$) is numerically large. Of course using the actual $\chi^0(\boldsymbol{q})$ calculated from the energy bands which itself has a peak at $\boldsymbol{q}=0.35$a* (Myron and Freeman 1975, Myron et al. to be published; Ricco 1977) due to Fermi surface "nesting" features greatly enhances this peak intensity and in fact the peak in $\chi^0(\boldsymbol{q})$ can "fine-tune" the position at which the "charge density wave" (CDW) or periodic lattice distortion actually occurs. This may be the reason why the CDW wavevector changes on changing the electron/atom ratio by alloying Ti into, say 1T-TaS_2 (Wilson et al. 1975). On the other hand, the above calculation emphasizes the role of the electron–phonon matrix elements and also the role of the large "background" on which the actual peak in $\chi^0(\boldsymbol{q})$ rides, which is basically due to the high density of states at the Fermi level. A similar calculation has recently been carried out by Das and Sinha (to be published) for the d-band metals NbO_2 and VO_2 in the high-temperature (rutile structure) phase. Band structure calculations for VO_2 (Gupta et al. 1977) show that there is a very high density of states at the Fermi level due to an essentially pure d-band with d–d overlap along the metal atom chains parallel to the c-axis. The tetragonal symmetry splits the t_{2g}-orbital degeneracy (Goodenough 1971) at the atomic sites, but the d–d overlap mixes these orbitals again so that the d-band orbitals at the Fermi

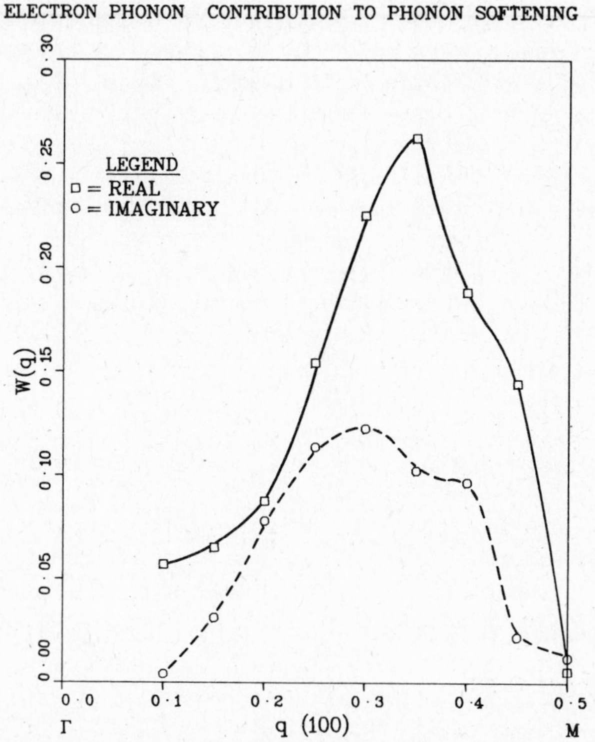

Fig. 20. $\boldsymbol{q}$-dependence of real and imaginary parts of quantity $W_I(\boldsymbol{q})$ (as discussed in the text) along ΓM for 2H–$NbSe_2$. The mode considered is the Σ_4 mode and the charge fluctuation is a d(xy)→d(xy) type monopole fluctuation on the Nb sites. From Das et al. (1978).

level were taken to be mixtures of all three t_{2g}-type orbitals. Force constants obtained from an analysis of phonon dispersion curves of TiO_2 (Traylor et al. 1971) were used to calculate the eigenvectors $e_\alpha(\kappa|\boldsymbol{q}j)$. The μ were taken to stand for transitions both between all t_{2g}-like orbitals on each Nb site in the unit cell, as well as transitions from xz (yz) orbitals to xz (yz) orbitals on nearest neighbors along the c-axis; and a parametrized form of the "overlap" contribution to the electron–phonon coupling $I^{\alpha,\kappa\kappa}_{\xi_1,\xi_2 l}(\boldsymbol{q})$ was taken, in addition to the part represented by eq. (5.1.2). The calculations showed that for NbO_2 the maximum instability occurred at the point P $(\frac{1}{4},\frac{1}{4},\frac{1}{2})$ while for VO_2 (owing to the different mass of the V atom and the consequent rearrangement of modes) it occurred around the point R $(\frac{1}{2},\frac{1}{2},0)$. Both of these facts are in accordance with experimental observations (Westman 1961, Brews 1970, Shapiro et al. 1974, Pynn et al.

1978). It is interesting to note that no soft mode was observed at the point P $(\frac{1}{4}, \frac{1}{4}, \frac{1}{2})$ in NbO_2 (Pynn et al. 1978). This implies a central peak driven by a relaxation effect of the type we have discussed above.

Eq. (5.4.4) shows the same behavior as in the models of Bhatt and McMillan (1975) or Yamada et al. (1974), i.e. as $\tilde{\omega}_{qj}\tau'$ goes from being $\ll 1$ to being $\gg 1$, we have a transition from a regime where the structure consists mainly of narrow peaks centered at the *softened* phonon frequencies ω_{qj} to a regime where the phonon peaks stop softening, broaden and a central peak appears. This theory involves always a "dynamical" central peak, i.e. there is in principle always some width in energy to the central peak, even though it may become very small by eq. (5.4.7). However, careful high-resolution studies of the central peak in many such systems, e.g. at the ω-phase transition in Nb–Zr alloys (Axe et al. 1977) show no detectable energy broadening down to the microvolt level, implying a purely static central peak. Such a static peak can arise due to Huang scattering from atomic displacements around impurities which would "condense" regions of the new structure around them as the instability temperature was approached.

The Huang scattering intensity for an isolated point defect is given for wavevector $\boldsymbol{Q}$, by

$$I(\boldsymbol{Q}) \propto \sum_{\alpha\beta} \left[Q_\alpha \chi^{\mathrm{L}}_{\alpha\beta}(\boldsymbol{q}) F_\beta(\boldsymbol{q}) \right]^2 \qquad (\boldsymbol{Q} = \boldsymbol{q} + \boldsymbol{G}), \tag{5.4.11}$$

where $\chi^{\mathrm{L}}_{\alpha\beta}(\boldsymbol{q})$ is the *real* part of the *static* susceptibility of the lattice,

$$\chi^{\mathrm{L}}_{\alpha\beta}(\boldsymbol{q}) \propto \sum_j e^*_\alpha(\boldsymbol{q}j) e_\beta(\boldsymbol{q}j)/\omega^2_{\boldsymbol{q}j} \tag{5.4.12}$$

and $F_\beta(\boldsymbol{q})$ is the $\boldsymbol{q}$-dependent force of the point defect on its neighbors. Thus,

$$I(\boldsymbol{Q}) \propto \sum_j (1/\omega_{\boldsymbol{q}j})^4 (\boldsymbol{Q}\cdot\boldsymbol{e}(\boldsymbol{q}j))^2 (\boldsymbol{F}(\boldsymbol{q})\cdot\boldsymbol{e}(\boldsymbol{q}j))^2 \tag{5.4.13}$$

and one may see that this diverges at the $\boldsymbol{q}$ for which the "virtual soft mode frequency" $\omega_{\boldsymbol{q}j} \to 0$. Note that this may happen, even if the actual neutron-observed phonon frequency has not softened to zero due to dissipative effects.

A central peak due to the "dirt effect" seems to have been observed in Nb (Shapiro and Pynn 1978) and dilute alloys of Zr in Nb (Wakabayashi 1978), under the pronounced dip in the LA [111] branch (see fig. 1). These authors find that (a) the intensity of the central peak appears to be very sensitive to impurity concentration and that (b) the $\boldsymbol{Q}$-dependence of the intensity is given quite well by the formula in eq. (5.4.13) where $\omega_{\boldsymbol{q}j}$ are the observed phonon frequencies.

6. *Conclusion*

While no claims are made here with regard to completeness of coverage, it is hoped that we have conveyed in this chapter the essence of the rich variety of phenomena observed in a study of phonons in d-band metals. Encompassing as it does the study of phonon spectra, phonon anomalies, lattice instabilities, charge density waves, metal–insulator transitions and superconductivity in these materials, this field has provided very fruitful ground for intensive experimental and theoretical investigations in the last few years. We have attempted to show the empirical correlations between the existence of phonon anomalies or lattice instabilities, high superconducting transition temperatures and the occurrence of d-bands around the Fermi level. We have also attempted to formulate the difficulties which must be overcome in a first-principles microscopic calculation of such phenomena, and to review, classify and evaluate the large number of theoretical approaches that have appeared in the last few years. It may fairly be said that the field is still in ferment and awaits further theoretical and experimental work in the next few years before some of the present controversies are resolved. There is little doubt that such work will not be long in coming.

Acknowledgements

The author wishes to express his sincere gratitude for the benefits he has received by collaboration in much of this work with Dr. R. P. Gupta, Dr. B. N. Harmon and Dr. S. G. Das. He also wishes to express his appreciation to Prof. C. Moser for his hospitality at the C.E.C.A.M. Workshop on Electron–Phonon Interactions and also to Prof. H. Bilz and the Max-Planck-Institut, Stuttgart for their hospitality during his visit there. He has also benefited greatly from discussions with Dr. W. Hanke, Dr. J. Hafner, Prof. F. M. Mueller, Dr. D. Pettifor, Prof. R. M. Pick, Dr. H. W. Myron, Dr. M. Gupta, Dr. J. Ashkenazi, Prof. A. J. Freeman, Prof. L. J. Sham, Dr. N. Wakabayshi, Prof. P. B. Allen, Dr. J. Cooke, Dr. G. Knapp, Dr. S. D. Bader, and Dr. D. Koelling. He also wishes to thank Prof. K. Scharnberg for useful discussions regarding the tight-binding method.

References

Allen, P. B. (1977), Phys. Rev. **16**, 5139.

Andersen, O. K. (1975), Phys. Rev. B **12**, 3060.

Ashkenazi, J., Dacorogna, M., and Peter, M., (1979) Solid State Comm. 29, 181.

Axe, J. D., A. Heidemann, W. S. Howells, S. C. Moss and R. Pynn (1977), Solid State Commun. **24**, 743.

Balkanski, M. (1978), editor, *Lattice dynamics*, *Proceedings of international conference on lattice dynamics*, Paris, 1977 (Flammarion, Paris).

Barisic, S., J. Labbé and J. Friedel (1970), Phys. Rev. Lett. **25**, 919.

Bhatt, R. N. and W. L. McMillan (1975), Phys. Rev. B **12**, 2042.

Bilz, H., B. Gliss and W. Hanke (1974), in *Dynamical properties of solids*, Ed. by G. K. Horton and A. A. Maradudin (North-Holland, Amsterdam), Vol. 1, p. 343.

Birnboim, A. and H. Gutfreund (1974), Phys. Rev. B **9**, 139.

Birnboim, A. and H. Gutfreund (1975), Phys. Rev. B **12**, 2682.

Born, M. and K. Huang (1954), *Dynamical theory of crystal lattices* (Oxford University Press, Oxford).

Brews, J. R. (1970), Phys. Rev. B **1**, 2557.

Brovman, E. G. and Yu. M. Kagan (1974), in *Dynamical properties of solids*, Ed. by G. K. Horton and A. A. Maradudin, (North-Holland, Amsterdam), Vol. 1, p. 191.

Butler, W. H. (1977), Phys. Rev. B **15**, 5267.

Butler, W. H., J. J. Olson, J. S. Faulkner and B. L. Gyorffy (1976), Phys. Rev. B **14**, 3823.

Butler, W. H., Pinski, F. J., and Allen P. B., (1979), Phys. Rev. B **19**, 3708.

Cooke, J. F. (1978), Bull. Am. Phys. Soc. **23**, 276.

Das, S. G. and S. K. Sinha, (19XX)

Das, S. G., S. K. Sinha and N. Wakabayashi (1978), in *Lattice dynamics*, *Proceedings of international conference*, Paris, 1977, Ed. by M. Balkanski (Flammarion, Paris), p. 596.

de Fontaine, D., N. E. Paton and J. C. Williams (1971), Acta Metall. **19**, 1153.

de Fontaine, D. and O. Buck (1973), Phil. Mag. **27**, 967.

Di Salvo, F. J., D. E. Moncton and J. V. Waszczak (1976), Phys. Rev. B **14**, 4321.

Dolling, G., T. M. Holden, E. C. Svensson, W. J. L. Buyers and G. H. Lander (1978), in *Lattice dynamics*, *Proceedings of international conference*, Paris, 1977, Ed. by M. Balkanski (Flammarion, Paris), p. 81.

Dye, D. H., D. P. Karim, J. B. Ketterson and G. W. Crabtree (1978), *Proceedings of Toronto Conference on physics of transition metals* (Inst. Phys. **39**).

Elyashar, N. and D. D. Koelling (1977), Phys. Rev. B **15**, 3620.

Endoh, Y., Y. Noda and Y. Ishikawa (1977), Solid State Commun. **23**, 951.

Evans, R., G. D. Gaspari and B. L. Gyorffy (1973), J. Phys. F (Metal Phys.) **3**, 39.

Evenson, W. E. and S. H. Liu (1968), Phys. Rev. Lett. **21**, 432.

Fawcett, E. (1978), editor, Proceedings of Toronto Conference on Physics of Transition Metals (Inst. Phys. 39).

Frölich, H. (1968), J. Phys. C: Proc. Phys. Soc. (London) **1**, 544.

Fujii, Y., G. Shirane and Y. Yamada (1975), Phys. Rev. B **11**, 2036.

Gale, S. J. and D. G. Pettifor (1977), Solid State Commun. **24**, 175.

Ganguly, B. N. and R. F. Wood (1972), Phys. Rev. Lett. **28**, 681.

Gaspari, G. D. and B. L. Gyorffy (1972), Phys. Rev. Lett. **28**, 801.

Gliss, B. and H. Bilz (1968), Phys. Rev. Lett. **21**, 884.

Golibersuch, D. C. (1967), Phys. Rev. **157**, 532.

Goodenough, J. B. (1971), J. Solid State Chem. **3**, 490.

Gupta, M. and A. J. Freeman (1976a), Phys. Rev. Lett. **37**, 364.

Gupta, M. and A. J. Freeman (1976b), Phys. Rev. B **14**, 5205.

Gupta, M. and A. J. Freeman (1976c), in *Superconductivity in d- and f-band metals, Proceedings of 2nd Rochester conference*, Ed. by D. H. Douglass (Plenum Press, New York), p. 313.

Gupta, M., D. E. Ellis, and A. J. Freeman (1977), Phys. Rev. B **16**, 3338

Hafstrom, J., G. S. Knapp and A. T. Aldred (1978), Phys. Rev. B **17**, 2892.

Hanke, W. (1971), in *Phonons*, Ed. by M. Nusimovici (Flammarion, Paris), p. 294.

Hanke, W. (1973), Phys. Rev. B **8**, 4591.

Hanke, W. and H. Bilz (1972), in *Inelastic scattering of neutrons* (International Atomic Energy Agency, Vienna).

Hanke, W. and L. J. Sham (1975), Phys. Rev. B **12**, 4501.

Hanke, W., J. Hafner and H. Bilz (1976), Phys. Rev. Lett. **37**, 1560.

Harley, R. T., R. D. Lowde, G. A. Saunders, R. Scherm and C. Underhill (1978), in *Lattice Dynamics, Proceedings of international conference*, Paris, 1977, Ed. by M. Balkanski (Flammarion, Paris), p. 726.

Harmon, B. N. and S. K. Sinha (1977), Phys. Rev. B **16**, 3919.

Harrison, W. A. (1966), *Pseudopotentials in the theory of metals* (Benjamin, New York).

Hausch, G. (1974), J. Phys. Soc. Japan **37**, 819.

Hayashi, E. and M. Shimizu (1969), J. Phys. Soc. Japan **26**, 1396.

Haydock, R., V. Heine and M. J. Kelly (1975), J. Phys. C (Solid State Phys.) **8**, 2591.

Hedin, L. and B. I. Lundqvist (1971), J. Phys. C **4**, 2064.

Hohenberg, P. C. and W. Kohn (1964), Phys. Rev. **136B**, 364.

Honig, J. M. and L. L. Van Zandt (1975), *Annual reviews of materials science*, Ed. by R. A. Huggins, Vol. 5, p. 225.

Iizumi, M. and G. Shirane (1975), Solid State Commun. **17**, 433.

John, W. (1973), J. Phys. F (Metal Phys.) **3**, L231.

Karim, D. P., J. B. Ketterson and G. W. Crabtree (1978), J. Low Temp. Phys. **30**, 389.

Keeton, S. C. and T. L. Loucks (1968), Phys. Rev. **168**, 672.

Keller, K. R. and J. J. Hanak (1967), Phys. Rev. **154**, 628.

Kellerman, E. W. (1940), Phil. Trans. Roy. Soc. A**238**, 513.

Klein, B. M. and D. A. Papaconstantopoulos (1974), Phys. Rev. Lett. **32**, 1193.

Klein, B. M. and D. Papaconstantopoulos (1976), in *Superconductivity in d- and f-band metals, Proceedings of 2nd Rochester conference*, Ed. by D. H. Douglass (Plenum Press, New York), p. 339.

Knapp, G. S., S. D. Bader, H. V. Culbert, F. Y. Fradin and T. E. Klippert (1975), Phys. Rev. B **11**, 4311.

Knapp, G. S., S. D. Bader and Z. Fisk (1976), Phys. Rev. B **13**, 3783.

Kohn, W. and L. J. Sham (1965), Phys. Rev. **140**, 1163.

Kress, W., P. Roedhammer, H. Bilz, W. D. Teuchert and A. N. Christensen (1978), Phys. Rev. B **17**, 111.

Liu, S. H., R. P. Gupta and S. K. Sinha (1971), Phys. Rev. B **4**, 1100.

Lottner, V., A. Kollma, T. Springer, W. Kress, H. Bilz and W. D. Teuchert (1978), in *Lattice dynamics, Proceedings of international conference*, Paris, 1977, Ed. by M. Balkanski (Flammarion, Paris), p. 247.

Mailfert, R., B. W. Batterman and J. J. Hanak (1967), Phys. Lett. **24A**, 315.

Maradudin, A. A. (1974), in *Dynamical properties of solids*, Ed. by G. K. Horton and A. A. Maradudin (North-Holland, Amsterdam), Vol. 1, p. 1.

Mattheiss, L. F. (1973), Phys. Rev. B **8**, 3719.

McMillan, W. L. (1968), Phys. Rev. **167**, 331.

Moncton, D. E., J. D. Axe and F. J. Di Salvo (1975), Phys. Rev. Lett. **34**, 734.

Moncton, D. E., J. D. Axe and F. J. Di Salvo (1977), Phys. Rev. Lett. **34**, 801.

Moncton, D. E., F. J. Di Salvo and J. D. Axe (1978), In *Lattice dynamics, Proceedings of International Conference*, Paris, 1977. Ed. by M. Balkanski (Flammarion, Paris), p. 561.

Moruzzi, V. L., A. R. Williams and J. F. Janak (1977), Phys. Rev. B **15**, 2854.
Moss, S. C., D. Y. Keating and J. D. Axe (1973), in *Phase transitions*, Ed. by L. E. Cross (Pergamon, New York), p. 179.
Moss, S. C., D. T. Keating and J. D. Axe (1975), Phys. Rev. Lett. **35**, 530.
Mueller, F. M. and H. W. Myron (1977),
Myron, H. W. and A. J. Freeman (1975) Phys. Rev. B **11**, 2735.
Myron, H. W., A. J. Freeman and S. C. Moss (1975), Solid State Commun. **17**, 1467.
Myron, H. W., J. Rath and A. J. Freeman (1977), Phys. Rev. B **15**, 885.
Nakagawa, Y. and A. D. B. Woods (1965) in *Lattice dynamics*, Ed. by R. F. Wallis (Pergamon Press, New York), p. 39.
Papaconstantopoulos, D. and B. M. Klein (1975), Phys. Rev. Lett. **35**, 110.
Papaconstantopoulos, D. A., L. L. Boyer, B. M. Klein, A. R. Williams, V. L. Moruzzi and J. F. Janak (1977), Phys. Rev. B **15**, 4221.
Peter, M., J. Ashkenazi and M. Dacorogna (1977), Helv. Phys. Acta **50**, 267.
Pettifor, D. G. (1972), J. Phys. C (Solid State Phys.) **5**, 97.
Pettifor, D. G. (1977), J. Phys. F (Metal Phys.) **7**, 1009.
Pick, R. M. (1971), in *Phonons*, Ed. by M. Musimovici (Flammarion, Paris), p. 20.
Pick, R. M., M. H. Cohen and R. M. Martin (1970), Phys. Rev. B **1**, 910.
Pickett, W. W. and B. L. Gyorffy (1976), in *Superconductivity in d- and f-band metals, Proceedings of 2nd Rochester conference*, Ed. by D. H. Douglass (Plenum Press, New York), p. 251.
Powell, B. M., P. Martel and A. D. B. Woods (1968), Phys. Rev. **171**, 727.
Prakash, S. (1978), in *Lattice dynamics*, *Proceedings of international conference*, Paris, 1977, Ed. by M. Balkanski (Flammarion, Paris), p. 30.
Price, D. L., R. P. Gupta and S. K. Sinha (1974), Phys. Rev. B **9**, 2573.
Pynn, R. and J. D. Axe, 1978
Pynn, R., J. D. Axe and P. M. Raccah (1978), Phys. Rev. B **17**, 2196.
Rajagopal, A. K. and M. H. Cohen (1972), Collect. Phenomena **1**, 9.
Ricco, B. (1977), Solid State Commun. **22**, 331.
Roedhammer, P., W. Reichardt and F. Holtzberg (1978), Phys, Rev. Lett. **40**, 465.
Schweiss, B. P., B. Renker, E. Schneider and W. Reichardt (1976), in *Superconductivity in d- and f-band metals, Proceedings of 2nd Rochester conference*, Ed. by D. H. Douglass, (Plenum Press, New York), p. 189.
Sham, L. J. (1974), in *Dynamical properties of solids*, Ed. by G. K. Horton and A. A. Maradudin (North-Holland, Amsterdam), Vol. 1, p. 301.
Sham, L. J. (1978), in *Lattice Dynamics*, *Proceedings of international conference*, Paris, 1977, Ed. by M. Balkanski (Flammarion, Paris), p. 557.
Shapiro, S. M. and R. Pynn (1978), Bull. Am. Phys. Soc. **23**, 313.
Shapiro, S., J. D. Axe, G. Shirane and P. M. Raccah (1974), Solid State Commun. **15**, 377.
Sharma, R. P. and J. C. Upadhyaya (1977), J. Phys. Chem. Solids **38**, 601.
Shirane, G. and J. D. Axe (1971a), Phys. Rev. B **4**, 2957.
Shirane, G. and J. D. Axe (1971b), Phys. Rev. Lett. **27**, 1803.
Shirane, G., J. D. Axe and R. J. Birgeneau (1971), Solid State Commun. **9**, 397.
Sinha, S. K. (1968), Phys. Rev. **169**, 477.
Sinha, S. K. (1969), Phys. Rev. **177**, 1256.
Sinha, S. K. (1973) in *CRC critical reviews of solid state sciences*, **3**, 273.
Sinha, S. K. and B. N. Harmon (1975), Phys. Rev. Lett. **35**, 1515.
Sinha, S. K. and B. N. Harmon (1976), In *Superconductivity in d- and f-band metals, Proceedings of 2nd Rochester conference*, Ed. by D. H. Douglass, (Plenum Press, New York), p. 269.

Sinha, S. K., R. P. Gupta and Harmon, 19XX
Sinha, S. K., Hafner and Bilz, 19XX
Sinha, S. K., R. P. Gupta and D. L. Price (1974), Phys. Rev. B **9**, 2564.
Smith, H. G. (1972), in *Superconductivity in d- and f-band metals*, Ed. by D. H. Douglass (AIP, New York).
Smith, H. G. and W. Glaser (1970), Phys. Rev. Lett. **25**, 1611.
Smith, H. G. and W. Glaser (1971), in *Proceedings of international conference on phonons*, Rennes, Ed. by M. A. Nusimovici (Flammarion, Paris), p. 145.
Smith, H. G., N. Wakabayashi and M. Mostoller (1976) in *Superconductivity in d- and f-band metals, proceedings of 2nd Rochester conference*, Ed. by D. H. Douglass (Plenum Press, New York), p. 223.
Stassis, C., Zaretsky, J., and Wakabayaski, N. (1978a), Phys. Rev. Letters **41**, 1726.
Stassis, C., Zaretsky, J., Arch, D., McMasters, O. D. and Harmon, B. N. (1978b), Phys. Rev. B **18**, 2632.
Stassis, C., Zaretsky, J., Arch, D., Harmon, B. H., and Wakabayashi, N. (1979), Phys. Rev. B **19**, 181.
Swittendick, A. C. (1975), *Hydrogen Energy, Part B,* ed., T. N. Vezeroglu (Plenum, New York), p. 201.
Tajima, K., Y. Endoh, Y. Ishikawa and W. G. Stirling (1976), Phys. Rev. Lett. **37**, 519.
Takahashi, H. (1968), Phys. Rev. 192, 474.
Terakura, K., (1978), J. Phys. C. 11, 469.
Testardi, L. R. and T. B. Bateman (1967), Phys. Rev. **154**, 402.
Traylor, S. G. and N. Wakabayashi, private communication
Traylor, J. G., H. G. Smith, R. M. Nicklow and M. K. Wilkinson (1971), Phys. Rev. B **3**, 3457.
Varma, C. M. and R. C. Dynes (1976) in *Superconductivity in d- and f-band metals, Proceedings of 2nd Rochester conference*, Ed. by D. H. Douglass (Plenum Press, New York), p. 507.
Varma, C. M. and W. Weber (1977), Phys. Rev. Lett. **39**, 1094.
Varma, C. M. and Weber, W., (1979), Phys. Rev. B **19**, 6142.
Varma, C. M., Blount, E. I., Vashishta, P. D. and Weber, W., (1979), Phys. Rev. B **19** 6130.
Venkataraman, G., L. W. Feldkamp and V. C. Sahni (1975), *Dynamics of perfect crystals* (MIT Press, Cambridge, Mass.).
von Barth, U. and L. Hedin (1972), J. Phys. C **5**, 1629.
Wakabayashi, N. (1977), Solid State Commun. **23**, 737.
Wakabayashi, N. (1978), Phys. Rev., to be published.
Wakabayashi, N., H. G. Smith and H. R. Shanks (1974), Phys. Lett. **50A**, 367.
Wakabayashi, N., H. G. Smith and R. M. Nicklow (1975), Phys. Rev. B **12**, 659.
Weber, W. (1973), Phys. Rev. B **8**, 5082.
Weber, W., H. Bilz and U. Schröder (1972), Phys. Rev. Lett. **28**, 600.
Westman, S. (1961), Acta Chem. Scand. **15**, 217.
Wilson, J. A., F. J. Disalvo and S. Mahajan (1974), Phys. Rev. Lett. **32**, 882.
Wilson, J. A., F. J. Disalvo and S. Mahajan (1975), Advan. Phys. **24**, 117.
Wilson, J. A. and S. Mahajan (1972), Solid State Commun. **22**, 551.
Wipf, H., Klein, M. V., Chandraskhar, B. S., Geballe, T. H. and Werwick, J. H. (1978) Phys. Rev. Letters 41, 1752.
Woods, A. D. B. and S. H. Chen (1964), Solid State Commun. **2**, 233.
Yamada, Y. (1975), AIP Conf. Proc. **24**, 79.
Yamada, H. and M. Shimizu (1967), J. Phys. Soc. Japan **22**, 1404.
Yamada, Y., T. Takatera and D. L. Huber (1974), J. Phys. Soc. Japan **36**, 641.
Zubarev, D. N. (1960), Sov. Phys.–Usp. **3**, 320.

CHAPTER 2

Phonons and the Superconducting Transition Temperature

P. B. ALLEN

Department of Physics
State University of New York
Stony Brook, New York 11794, USA

Dynamical Properties of Solids, edited by
G. K. Horton and A. A. Maradudin

Contents

1. *Introduction and experimental survey*

1.1. Introduction and plan

Since 1970 there has been a dramatic increase in experimental knowledge of the fascinating lattice dynamics of high transition temperature (T_c) superconductors. Coupled with this has been an explosion of theoretical papers on the difficult problems of electronic band structure and electron–phonon interactions in d-band metals, with the aim of understanding phonons and T_c. Simultaneously many new, exotic and idiosyncratic superconducting materials have been found. It seems fair to say that two separately difficult problems, first, lattice dynamics of d-band materials and second, understanding and raising T_c, have by their conjunction given new vitality to both fields. To be fair, it must also be said that the ultimate goal, raising T_c, has been achieved in only modest degree during this period. Superconductivity above 20 K was achieved in 1967 in $Nb_3Al_{0.75}Ge_{0.25}$ (Matthias et al. 1967); 23 K was achieved in Nb_3Ge in 1973 (Gavaler 1973, Testardi et al. 1974), and remains the highest known value at this writing. Further data are shown in table 1 and fig. 1. Values of T_c for elements can be found in Kittel (1976). The most extensive listing of data is by Roberts (1976).

The literature related to T_c and phonons is vast and rapidly changing, making a review risky. The conventional excuse, that no one person could adequately review the whole field, is tempting. Without further apology, I admit that the selection and treatment of topics given here reflects my idiosyncracies. I attempt both a snapshot of a field in rapid motion and an emphasis on firm results likely to survive. Lines of thought which seem promising are emphasized while those that appear (to me) to have outlived their usefulness are given only brief mention. From the wealth of good experiments and calculations, illustrative examples are chosen somewhat randomly. I ask indulgence of those who have been unfairly ignored; the offense was unintentional. A list of earlier reviews is given at the end, with the references. Material already discussed in these has been largely passed over. For example, relatively brief mention is given to much of the older work on A15-structure materials. However, the Landau theories of the martensitic phase transformation, which date from the work of Anderson

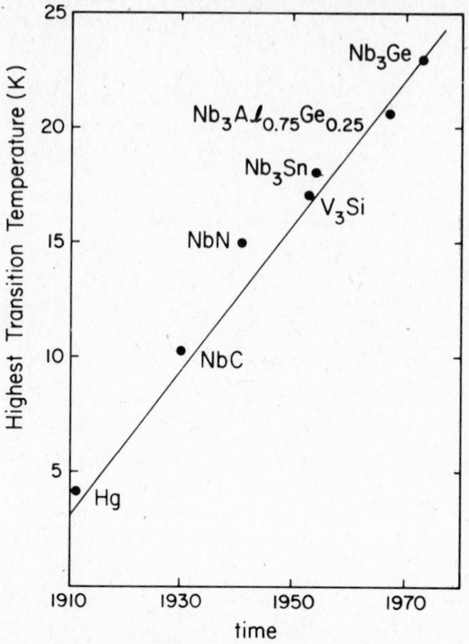

Fig. 1. Maximum known T_c as a function of time. See table 1 for references. The straight line was drawn by Friday (1973) before the discovery of superconductivity above 22 K in Nb_3Ge by Gavaler (1973). Given a steady increase of 3 degrees per decade, room-temperature superconductivity is predicted in the year 2888.

and Blount (1965), but have been poorly developed until recently, are given a detailed analysis in §5.3.

This review deals mainly with pure elements and ordered intermetallic compounds. There is a wealth of experimental information on alloys and amorphous metals. In fact the most detailed information presently available on phonons in alloys and amorphous metals comes from superconducting tunneling experiments. However, little enlightenment about the meaning of these data has come from theory. The task is formidable because it requires a theory for both electron and lattice dynamics. These subjects are

Table 1

Record high transition temperatures. Complete references are in the list of references, and in Review Article R8.

Year	T_c (K)	Compound	Author
1911	4.2	Hg	H. Kammerlingh Onnes
1930	10.3	NbC	W. Meissner and H. Franz
1941	15	NbN	Aschermann et al.
1953	17.1	V_3Si	G. F. Hardy and J. K. Hulm
1954	18.	Nb_3Sn	B. T. Matthias et al.
1967	20.1	$Nb_3Al_{0.75}Ge_{0.25}$	B. T. Matthias et al.
1973	23.	Nb_3Ge	J. R. Gavaler

still in the hands of their respective specialists and not even a tentative synthesis has been attempted.

Three unresolved questions are particularly emphasized in this article: (i) whether high T_c depends on anomalously low phonon frequencies (§ §1.2, 1.3, 3.2); (ii) whether phonon anomalies in d-band metals are better described as manifestations of band structure near E_F or chemical valence bond fluctuations (§4.2); and (iii) whether lattice instabilities go away at high T because electronic fluctuations or lattice fluctuations stabilize the undistorted phase (§5). I have done my best to present a fair account of at least two sides of each of these questions. It is inevitable that I have omitted, misunderstood or misrepresented some important contributions, and I apologize in advance for these errors.

1.2. The central question

The first experiment to dramatize the interesting lattice dynamics of high-T_c superconductors was the discovery by Batterman and Barrett (1964) of a structural transition in V_3Si at 21 K, not far above the superconducting transition at 18 K. Testardi et al. (1965) then showed that this transition is accompanied by extreme elastic softening of the crystal. A very similar transition was found soon after in Nb_3Sn (Mailfert et al. 1967, 1969). It is often speculated that such transitions may occur in other A15-structure superconductors. Progress in understanding these effects has been slowed by the difficulty of obtaining large single crystals. In fact, all high-T_c superconductors have the difficulty of being unstable and hard to prepare. Thus many of the standard spectroscopic techniques (Fermiology, inelastic neutron scattering, etc.) are forbiddingly difficult. Great experimental ingenuity has been required to find ways around these problems. Before 1970, the "Batterman–Barrett instability" was often regarded as a unique pathology of A15 metals. Since that time it has become reasonable to suppose that all high-T_c materials have unusual lattice dynamical properties; the Batterman–Barrett instability is a manifestation of a more general regularity. This generalized formulation received a major impetus from the neutron scattering measurement by Smith and Gläser (1970) of phonon dispersion in TaC and HfC, shown in fig. 2. Subsequently, Smith (1972a), in a review paper at the first Rochester meeting, assembled many additional examples of the same phenomenon: high-T_c materials (such as TaC with $T_c \simeq 11$ K) have both reduced frequencies and "dips" (or "wiggles" or "anomalies") in ω_Q, while closely related materials with low T_c (e.g. HfC, $T_c < 0.1$ K) have stiffer and smoother frequencies. I believe there is as yet no known counter-example to this regularity. *The obvious and central problem is to discover the microscopic physical mechanism for the anomalies and clarify the relation between the anomalies and* T_c. In particular, do the anomalously low-frequency phonons cause a significant increase in T_c, as argued by Testardi (1972)? Or, are anomalies and high T_c's

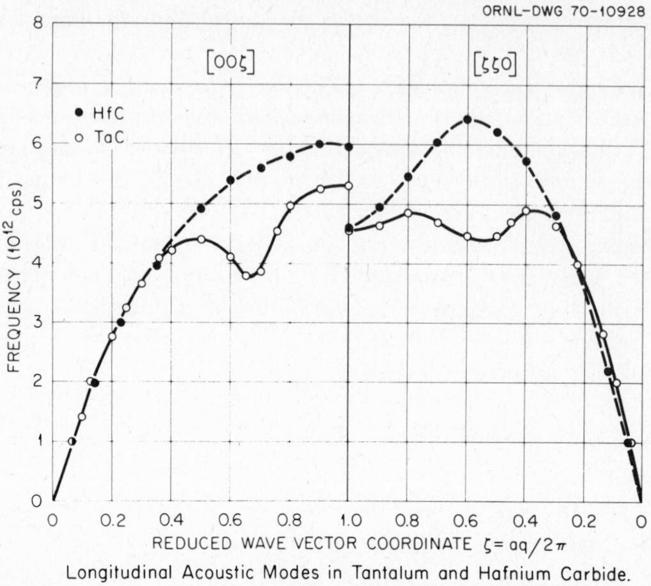

Fig. 2. Phonon dispersion curves measured by Smith and Gläser (1970) for a high-T_c superconductor (TaC) and a closely related poor superconductor (HfC). Only LA branches are shown. Similar anomalous behavior occurs in certain TA branches, and, to a smaller extent, in the LO branch.

simply both manifestations of the same strong electron–phonon coupling, with only weak or accidental bearing on each other? These questions lie in the realm where microscopic theory is just beginning to become competent. The theory of interatomic forces, although "solved" in the abstract, is still poorly developed except for especially simple materials (alkali metals and rare gas solids). For closely related reasons, our ability to understand the microscopic origin of high T_c is limited. Thus the central question is still largely open, in spite of much recent progress. It is my impression that although Testardi's arguments were at first greeted with some enthusiasm, the majority opinion has now turned toward the alternative side.

It is worth digressing here to attempt a clarification of the terms "soft phonon", and "phonon softening". There is no strict definition known to me for either term. It might be desirable to restrict usage of the term "soft phonon" to cases where a phonon $Q=(\boldsymbol{Q}\nu)$ has its frequency ω_Q driven to zero at a transition temperature T_0 as in the mean field behavior $(T-T_0)^{1/2}$ or with a modified exponent. Nearby phonons Q' will have frequencies $\omega_{Q'} \approx \omega_0((T-T_0)/T_0+\xi^2(\boldsymbol{Q}'-\boldsymbol{Q})^2)^{1/2}$. The question arises, should these phonons be called soft, and if so, how far away from $Q'=Q$ is it legitimate to call "soft"? Furthermore, the idealized behavior is seldom if ever observed. A first-order phase change usually occurs before ω_Q vanishes. Are these phonons "soft"? Suppose no transition takes place but we believe one ought to have occurred except that impurities or fluctuations have stabilized the undistorted phase? I favor a fairly generous allowance in all these cases. In this article the term "soft phonon" means a phonon with ω_Q depending strongly on T, associated with the proximity (in Q, T, purity) to a phase change. The term "phonon softening" is somewhat less charged with emotion and will be used more broadly to indicate an anomalous dependence of ω_Q on T (or some other parameter), not necessarily associated with a phase change.

To exemplify the current state of microscopic theory of phonons, consider the case of TaC shown in fig. 2. A phenomenological model was given by Weber (1973a, b) which involved two shells around each Ta atom, one of which had a negative spring constant. Variations of the parameters of Weber's model can cause the position in Q space of the dips to vary somewhat, and the depth of the minimum can be varied at will. By increasing the strength of the negative spring constant, $M\omega_Q^2$ at the minimum can be made to pass through zero and become negative. Instead of undergoing stable oscillations $u_Q \simeq \exp(\mathrm{i}\omega_Q t)$, the atomic displacements would grow like $\exp(|\omega_Q|t)$. Thus it is reasonable to suppose that the dip of ω_Q in TaC is a sign of an "incipient" lattice instability (i.e. one which is not quite present.) Weber's model (see also Weber et al. 1972) predicted anomalous behavior of the TA branch in the [110] direction with $[1\bar{1}0]$ polarization and was verified experimentally by Smith (1972b). Because of this impressive success, there have been efforts to translate Weber's model into microscopic mechanisms. The work of Hanke et al. (1976) accomplishes this in a qualitative way, but until microscopic theory becomes quantitative, the interpretation must be regarded as speculative. These authors provide qualitative support for Testardi's point of view that soft phonons have an enhanced contribution to T_c. There is not space in this article to give a detailed presentation of the current debates on this subject, but a further discussion is given in §4.2.

In fig. 3 a flow chart illustrates the various approximate casual relationships encountered in trying to untangle the relationship between phonons and T_c. At the level of primary concepts we have electron–phonon coupling and electronic response functions. The decision to label these as primary is somewhat arbitrary. At a deeper level we can ask what is the origin of these parameters, and our understanding is far from complete. We are not even sure of the degree to which the electronic screening alters the electron–phonon coupling (so there is a question mark on the highest

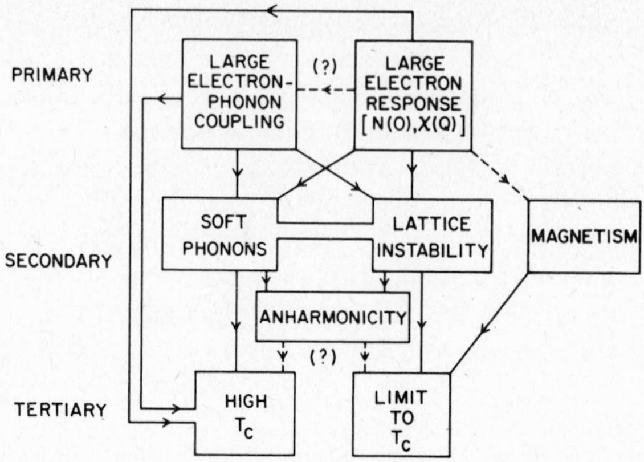

Fig. 3. Causative relations are indicated by arrows. High T_c is partly a direct consequence of the primary factors, and partly indirect through the phonon dispersion. Dashed lines with arrows represent less certain relationships.

dashed arrow.) Soft phonons and lattice instabilities are often different extremes of the same effect, so their boxes are joined. Anharmonicity is an inevitable side effect of either one, and thus quite pronounced in high-T_c materials, yet we do not fully understand its effect on T_c. Clearly the question of the origins and limitations to high T_c's is very complicated.

1.3. Phonon spectroscopy in A15 metals

Let us follow in roughly historical order the development of experimental knowledge about a particular metal of the A15 structure, namely Nb_3Sn. This is meant to serve as an introduction both to the various spectroscopic techniques and to the issues that are raised by the experiments. Fuller details can be found in several review articles listed at the end. The A15 crystal structure is cubic with 8 atoms per cell. Below 43 K, samples which are especially good transform to a slightly distorted tetragonal phase with no change in unit cell volume. This was shown by X-ray diffraction by Mailfert et al. (1967) whose data are shown in fig. 4. The transition is nearly second order but has a small discontinuity. The elastic energy for tetragonal strains ε of cubic crystals is $\frac{1}{6}(c_{11}-c_{12})\varepsilon^2$. The same coefficient, $c_{11}-c_{12}$, determines the velocity of TA phonons in the [110] direction with [1$\bar{1}$0] polarization (T_1 branch). It is reasonable to suppose that near the transition, the restoring force for tetragonal distortions and thus the TA phonon velocity may get very small. This was in fact found. Fig. 5 shows the data of Keller and Hanak (1967). Thus the Batterman–Barrett instability is an almost perfect example of a "soft mode" phase

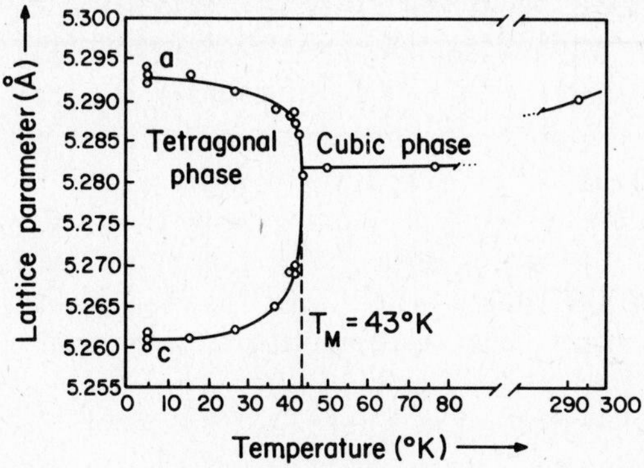

Fig. 4. Lattice parameters *a* and *c* versus temperature in the cubic and tetragonal phases of Nb_3Sn (Mailfert et al. 1969).

transition. However, it was pointed out by Anderson and Blount (1965) that symmetry forbids tetragonal strain to appear as the primary order parameter in a second-order phase transition in A15 structure. They suggested that a hidden order parameter, probably a sublattice distortion, might occur. Shirane and Axe (1971b) found that the only sublattice distortion measurable by neutrons has the same symmetry as the tetragonal distortion. This distortion consists of a dimerization of the two chains

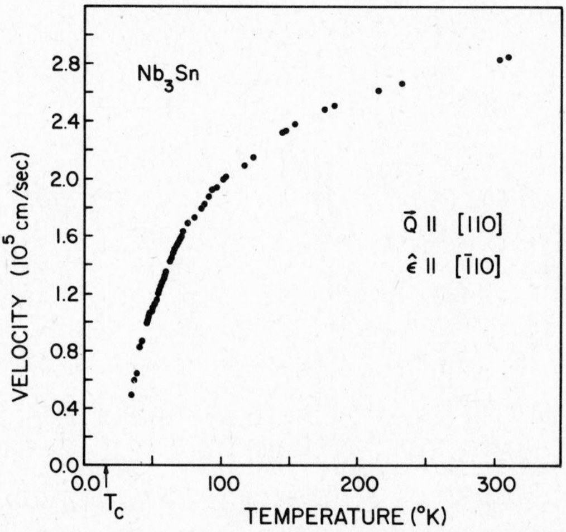

Fig. 5. Sound velocity in Nb_3Sn versus temperature for TA phonons polarized [$1\bar{1}0$] and propagating in the [110] direction (from Keller and Hanak 1967; redrawn). The sample used here has a slightly lower instability temperature than the one used in fig. 4.

orthogonal to the tetragonal axis, with a distortion of 0.003 lattice constants at 4.5 K. The corresponding optical mode should have an interesting temperature dependence, but has unfortunately not been measured. A more detailed discussion of this phase transition is given in §5.3.

The soft mode behavior of long-wavelength acoustic phonons in Nb_3Sn has many possible implications for superconducting T_c's. The onset of superconductivity arrests the structural transition and the T-dependence of sound velocities. Testardi (1971) has used this to predict strain dependence of T_c. An important but difficult question is how much of the observed high T_c ($\simeq 18$ K) arises specifically because of the coupling of electrons to the soft phonons connected with the structural instability? It is necessary to have information on phonon dispersion at short as well as long wavelengths, because (as shown in §§2 and 3) T_c depends on a Brillouin zone average of contributions from all phonons, which gives vanishingly small weight to long wavelengths. Unfortunately, the largest known single crystal, grown by Hanak, is too small for extensive neutron scattering. Of the 24 total branches (8 atoms per cell), only a few transverse acoustic branches have been seen by neutrons (Shirane and Axe 1971a, Axe and Shirane, 1973a.) These are shown in fig. 6. At small values of Q, the neutron results agree with ultrasonic measurements. Somewhat surprisingly, the softening which occurs in the sound velocity as T is lowered is

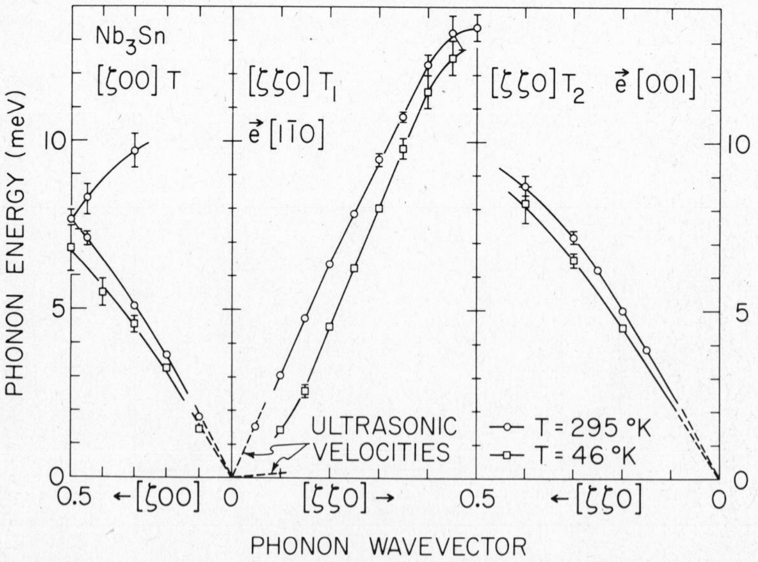

Fig. 6. Dispersion relations for TA phonons in Nb_3Sn at T = 295 and 46 K (Axe and Shirane 1973a).

reflected in the behavior of the phonons all the way to the Brillouin zone boundary. The effect is quite dramatic out to 20% of the zone boundary in the [110] direction. However, it is perhaps equally interesting that all the TA phonons, not just the long-wavelength [110] phonons implicated in the structural transition, have anomalous thermal behavior.

To get around the difficulty of not having large single crystal specimens, there are several methods for directly measuring the phonon density of states using polycrystalline samples. A method which shows promise for the future is Raman scattering (Spengler et al. 1978). Incoherent neutron scattering is a better known alternative. For a nucleus like ^{51}V, which scatters incoherently, the quantity which can be measured is

$$G(\omega) = \sum_i (\sigma_i / M_i) F_i(\omega) / \sum_j (\sigma_j / M_j), \tag{1.1}$$

where σ_i and M_i are the cross section and mass of the ith nucleus, and $F_i(\omega)$ is the amplitude-weighted density of states

$$F_i(\omega) = (1/N) \sum_Q |\varepsilon_Q^i|^2 \delta(\omega - \omega_Q), \tag{1.2}$$

where Q runs over all phonon eigenstates in the Brillouin zone. Knowledge of $F_i(\omega)$ would be very pertinent to understanding T_c. Unfortunately, most nuclei have finite cross sections for coherent scattering, which gives rise to interference terms in the inelastic cross section for polycrystalline materials, in addition to the desired term (1.1). Gompf et al. (1972) have tested approximate methods for experimentally cancelling the interference term by averaging over many scattered angles. These methods have been used by Schweiss et al. (1976) to study various high-T_c superconductors. Their data for Nb_3Sn are shown in fig. 7. The most interesting feature is the lowest peak which occurs at about 9 meV at $T = 5.6$ K and shifts to 20% higher energy by $T = 297$ K. This is a very dramatic change, especially considering that most materials exhibit a decrease (or softening) of phonon energies as T is increased. The value of the peak frequency, 9 meV, is about right to correspond to zone boundary TA phonons as seen in fig. 6, and the magnitude and direction of the thermal shift also corresponds to that seen in fig. 6. However, fig. 7 reveals more than does fig. 6. Numerical integration of the $G(\omega)$ curves of fig. 7 shows about a 7% decrease in r.m.s. phonon frequency in going from 297 to 5.6 K. Clearly more than just a single branch (i.e. one out of 24 branches) is affected by the softening. This type of behavior is new to solid state physics, and there is little theoretical understanding of why the whole phonon spectrum should be especially sensitive to temperature. It is not an isolated pathology of Nb_3Sn, but has

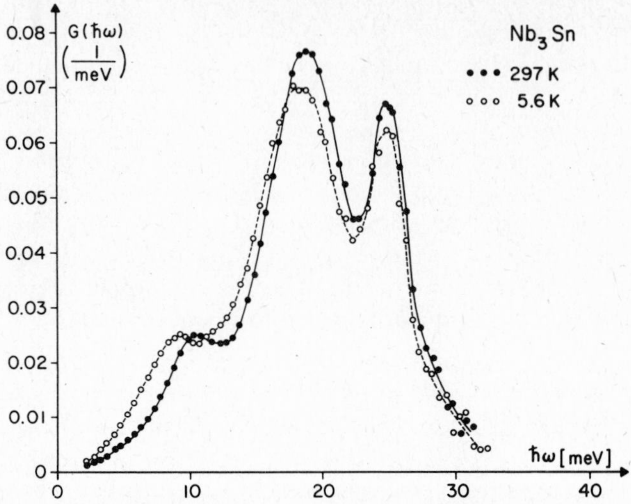

Fig. 7. Weighted density of states $G(\omega)$ for phonons in Nb_3Sn at 297 and 5.6 K. The measurements, on polycrystalline samples, are by Schweiss et al. (1976).

also been observed (Schweiss et al. 1976) in V_3Ge, V_3Ga, V_3Si and Nb_3Al. Of these, only V_3Si is known to undergo a structural transition. Other evidence for anomalous T-dependence of phonon spectra in A15 metals has been derived from Mössbauer spectroscopy (Kimball et al. 1976a, b) and specific heat (Knapp et al. 1975, 1976). In §5 it is argued that there are two distinct possible mechanisms: (a) intrinsic T-dependence of force constants (which can be expected to arise from Fermi smearing when electronic density of states $N(\varepsilon)$ is large and rapidly varying), or (b) anharmonic renormalization (which comes from Bose factors rather than Fermi factors in perturbation theory.) Several arguments supporting choice (b) were given by Testardi (1972). Choice (a) has been more commonly invoked, and derives support from the fact that magnetic susceptibilities (χ) and Knight shifts (K_s) of these materials have anomalous T-dependence. However, Nb_3Al has the largest thermal shift of phonon spectrum yet measured, and no anomalies in χ or K_s. The relation of large phonon thermal shifts to T_c is also still obscure, although it is surely related to large electron–phonon coupling. Similar dramatic thermal shifts have been seen in $SnMo_6Se_8$, a high critical field Chevrel phase superconductor, by neutron scattering (Bader et al. 1976, Schweiss et al. 1976), following earlier Mössbauer work (Kimball et al. 1976b). The former authors argue that here the effect is *not* related to strong electron–photon coupling, but is an anharmonic effect on the Sn atom. The Sn atom may be mostly irrelevant to T_c, which is believed to be associated with Mo d-electrons. However, the regularity is too striking to dismiss. As a final example, elemental Nb also has an anomalous thermal shift, less prominent than in Nb_3Sn, but of the same sign, as has been measured up to 1030 K by Powell et al. (1972). It begins to appear that anharmonic effects are large in nearly all high-T_c superconductors. It is not easy to cite a counter-example, and this regularity seems almost as pervasive as the correlation between high T_c and anomalous dispersion, cited in §1.2.

In order to make a firm connection between phonons and T_c, it is necessary to know more than phonon spectra; the electron–phonon coupling strength is also crucial. The most direct and elegant experimental measure of this is via tunneling spectroscopy and the function $\alpha^2F(\omega)$. This requires a more technical introduction which is postponed until §2. Only limited data are available for high-T_c compounds. In fig. 15 the measured $\alpha^2F(\omega)$ for Nb_3Sn will be shown. Subject to the qualifications listed in §2.3, the data of fig. 15 can be compared with fig. 7, and the relevant coupling strengths can be deduced. Another very elegant method for experimentally determining the electron–phonon coupling was found by Axe and Shirane (1973b). They made careful measurements of the inelastic neutron line shapes $S(Q,\omega)$ of [110] TA phonons polarized [1$\bar{1}$0] in Nb_3Sn. This is the phonon branch whose $Q \to 0$ slope is governed by $c_{11} - c_{12}$ (fig. 5). The line shapes for the phonon $Q = (2\pi/a)(0.18, 0.18, 0)$ are shown in fig. 8 at $T = 6$ and 26 K. The high-temperature line is broad, but narrows suddenly as T is decreased below $T_c \simeq 18$ K. The only reasonable mechanism for the extra broadening at $T > T_c$ is electron–phonon decay processes. These processes

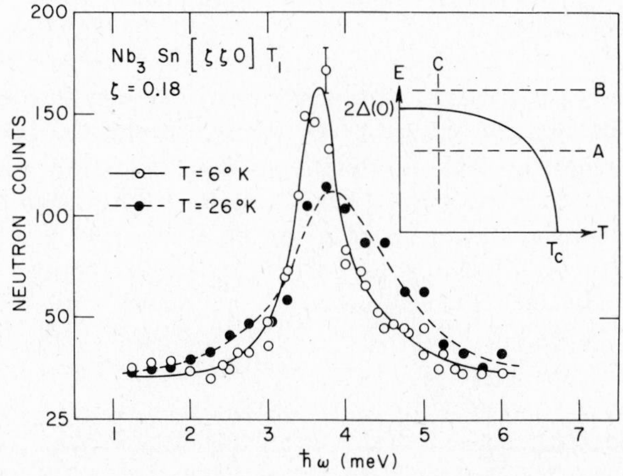

Fig. 8. Lineshape of [110] TA phonon (polarized [1$\bar{1}$0]) in Nb_3Sn, showing a broad line at 26 K and a narrow line at 6 K (Axe and Shirane 1973b). The higher temperature linewidth is a direct measure of electron–phonon coupling strength.

are "frozen out" in the superconducting state because the phonon energy ($\simeq$4 meV) is insufficient to break Cooper pairs ($2\Delta\simeq 6.9$ meV). The extra linewidth above T_c provides a direct measure (Allen 1972) of the coupling strength of this phonon to Fermi surface electrons (see §3). From this experiment, there can be no doubt that the phonon branch which goes unstable at the structural transition is quite strongly coupled to Fermi surface electrons.

The question as to how much T_c is enhanced by coupling to anomalously low-frequency phonons remains open, and probably does not have a single answer for all systems. Chu et al. (1977) have studied the effect of pressure simultaneously on superconducting (T_c) and structural (T_0) transition temperatures, for several different systems. If T_c were significantly enhanced by soft phonons, then pressure should always change T_c and T_0 in opposite directions. Decreasing T_0 would soften the phonons at low T and thus enhance T_c, and vice versa. Such behavior is observed in a number of cases, but counter-examples occur as well.

Much remains to be learned experimentally about phonons in Nb_3Sn, but it is impressive to see how much has been learned by innovative experiments on less than perfect specimens in the last few years. Theory is currently far behind experiment in this area, but motivated by the wealth of fascinating data it is making rapid progress. The subsequent sections explore the theory and its application to simpler materials such as elemental Nb. In §5 we shall return to the analysis of the complexities displayed by materials like Nb_3Sn.

2. *Theory of* T_c

It should be said at the outset that although we believe we have an excellent microscopic theory for superconductivity, including T_c, nevertheless, we are *not* able to calculate T_c reliably for an arbitrary material except in favorably simple cases. These simple cases, primarily elements, are discussed in §3. This section presents the underlying microscopic theory.

The characteristic magnitude of the temperature T_c for the transition to the superconducting state is $\lesssim 10$ K. The current maximum known value is 23 K. This contrasts with most other phase transitions in matter such as magnetic and structural transitions which are characterized by temperatures $T_c \lesssim 10^2$–10^3. This discrepancy in temperature scales suggests that there is something especially delicate about the superconducting state. In the early 1950's, both Bardeen and Fröhlich correctly guessed that electron–phonon interactions were crucial to the existence of superconductivity. There are two distinct energy scales which can enter, the Debye temperature Θ_D for phonons, and the Fermi energy E_F for electrons. There is also the coupling matrix element between electrons and phonons, $M_{kk'}$, which has order of magnitude $(\Theta_D E_F)^{1/2}$ as is shown in §3. The early theories of Fröhlich and Bardeen suggested that T_c should be of order Θ_D, one order of magnitude larger than observed. A correct theory emerged in 1957 (Bardeen, Cooper and Schrieffer, or BCS, 1957). This theory helps explain the low values of T_c by showing that there is an exponential reduction of T_c below the Fröhlich–Bardeen value,

$$T_c \cong \Theta_D \exp\left(- \frac{1}{N(0)\,V} \right), \tag{2.1}$$

where $N(0)$ is the (single spin) density of electron state at E_F (which is taken to be the zero of energy for electrons) and V is the average effective attractive interaction between electrons at the Fermi surface. The product $N(0)\,V$ is so important that it has a name of its own, λ:

$$\lambda = N(0)\,V = N(0)\left\langle \frac{|M_{kk'}|^2}{\omega_{k-k'}} \right\rangle_{FS}. \tag{2.2}$$

Here we are using the result (to be proved in §2.1) that V is given by the square of $M_{kk'}$ divided by the phonon energy $\omega_{k-k'}$. The brackets $\langle\ \rangle_{FS}$ denote a Fermi surface average. The order of magnitude of λ is found to be 1 by noting that $N(0) \simeq E_F^{-1}$, $\omega_{k-k'} \simeq \Theta_D$ and $|M_{kk'}|^2 \simeq \Theta_D E_F$. In fact, modern experiments show that λ often exceeds 1. Eq. (2.1) then gives $T_c \simeq \Theta_D \exp(-1)$ which fails to explain the low observed values of T_c. The main task of this section is to discuss the modern improvements on eq. (2.1), which take into account the so-called "strong-coupling effects" – i.e. deviations from BCS theory which appear for materials with $\lambda \gtrsim 0.7$. The modern theory of T_c differs from (2.1) mainly in having a more complicated function of λ as an argument of exponential. The theory usually gives lower T_c than (2.1), but has been recently shown to permit T_c to be as high (or higher, for large enough λ!) than Θ_D. Thus the difficulty in finding higher T_c's is not completely an intrinsic feature of the theory, but must partly reflect external constraints on the magnitude of λ. The source of this constraint is not completely understood, but is believed to be intimately related with crystal instabilities, such as discussed in §1 for A15 metals, and in §5 in more generality.

The plan of this section is to begin in §2.1 with BCS theory leading to eq. (2.1). Then the rudiments of the Eliashberg (1960) theory are presented in §2.2, followed by a discussion in §2.3 of the tunneling experiments which motivate and confirm this theory. In §2.4 the modern theory of T_c is reviewed, and in §2.5 the question of the optimum phonon spectrum for T_c

is dealt with. In order to establish the notation, we start by writing out the Hamiltonian for interacting electrons and phonons.

The system consists of Bloch electrons described by $\mathcal{H}_{\mathrm{e}}$, phonons described by $\mathcal{H}_{\mathrm{p}}$ and an interaction $\mathcal{H}_{\mathrm{ep}}$. The Bloch electrons have energy ε_k, wavefunction ψ_k and quantum numbers $k=(\boldsymbol{k}n)$, where k is used as a shorthand for wavenumber $\boldsymbol{k}$ (lying in the first Brillouin zone) and band index n:

$$\mathcal{H}_{\mathrm{e}}=\sum_k \varepsilon_k c_k^+ c_k. \tag{2.3}$$

The phonons have energy ω_Q, polarization vector $\hat{\varepsilon}_Q^a$ and quantum numbers $Q=(\boldsymbol{Q}\nu)$, where Q is shorthand for wavenumber $\boldsymbol{Q}$ and branch index ν:

$$\mathcal{H}_{\mathrm{p}}=\sum_Q \omega_Q\left(a_Q^+ a_Q+\frac{1}{2}\right). \tag{2.4}$$

The operators c_k (c_k^+) and a_Q (a_Q^+) are destruction (creation) operators for electrons and phonons, respectively. The interaction $\mathcal{H}_{\mathrm{ep}}$ is written as in terms of the coordinate $\boldsymbol{R}$ and displacement $\boldsymbol{u}$ of the atom located near site la:

$$\boldsymbol{R}(la)=\boldsymbol{l}^a+\boldsymbol{u}(la), \tag{2.5}$$

$$\mathcal{H}_{\mathrm{ep}}=\sum_{ila} u_\alpha(la)\,\partial V_a(\boldsymbol{r}_i-\boldsymbol{R}(la))/\partial R_\alpha(la). \tag{2.6}$$

The atoms are numbered by two indices, l which numbers the cells and a which numbers atoms within the cell. These atoms have displacements $\boldsymbol{u}(la)$ about their average positions $\boldsymbol{l}^a$. The meaning of $\partial V_a/\partial R_\alpha(la)$ is the *change*, per unit displacement of the la atom in the α direction, of the *total self-consistent* crystal potential felt by an electron at $\boldsymbol{r}_i$. In the absence of Coulomb interactions between conduction electrons this would be the gradient of the bare electron–atom core potential. It is explicitly assumed that displacements $\boldsymbol{u}(la)$ are small enough that a linear approximation is justified. Consistent with this, the total self-consistent potential can be constructed by screening the gradient of the bare potential by an appropriate (non-local) dielectric function $\varepsilon^{-1}(\boldsymbol{r},\boldsymbol{r}')$. The Coulomb interaction thus appears nowhere explicitly in the Hamiltonian (2.3–2.6), but is everywhere implicitly included. This procedure is sometimes known as the Fröhlich Hamiltonian. Finally the electron–phonon part (2.6) can be written in second-quantized form:

$$\boldsymbol{u}(la)=\sum_{Q}(2M_a\omega_Q)^{-1/2}\hat{\varepsilon}_Q^a(a_Q+a_{-Q}^+)e^{\mathrm{i}\boldsymbol{Q}\cdot\boldsymbol{l}}, \tag{2.7}$$

$$M_{kk'}^{\nu}=\sum_{a}(2M_a\omega_{\boldsymbol{k}-\boldsymbol{k}'\omega})^{-1/2}\langle k'|\hat{\varepsilon}_{\boldsymbol{k}-\boldsymbol{k}'\nu}^a\cdot\partial V_a(\boldsymbol{r}-\boldsymbol{\tau}_a)/\partial\boldsymbol{\tau}_a|k\rangle, \tag{2.8}$$

$$\mathcal{H}_{\mathrm{ep}}=\sum_{kk'}M_{kk'}c_{k'}^+c_k(a_{k'-k}^+a_{k-k'}^+). \tag{2.9}$$

Here $\boldsymbol{l}^a$ has been written as $\boldsymbol{l}+\boldsymbol{\tau}_a$ where $\boldsymbol{l}$ locates the cell and $\boldsymbol{\tau}_a$ the atom in the cell. The electron–phonon matrix element $M_{kk'}$ depends on phonon branch ν as well as electron band numbers n and n'. This is exhibited explicitly in (2.8) but is suppressed in (2.9) where a sum over ν is meant to be implied. In these equations M_a is the mass of the ath atom and $\hbar$ has been set to 1.

2.1. BCS theory

The fundamental new feature of the superconducting state is a gap, or minimum energy, 2Δ, in the electronic excitation spectrum. This is illustrated in fig. 9 by a graph of the density of states. A normal metal has a dispersion relation ε_k which gives a density of states $N_{\mathrm{n}}(\varepsilon)$ which we can assume to be slowly varying (i.e. constant) near $\varepsilon=E_{\mathrm{F}}=0$ when ε varies on the scale of Θ_{D}. The dispersion relation for the superconductor is modified to $(\varepsilon_k^2+\Delta^2)^{1/2}=E_k$, which gives a drastically altered density of states $N_{\mathrm{s}}(\varepsilon)$

$$N_{\mathrm{n}}(\varepsilon)=\sum_{k}\delta(\varepsilon_k-\varepsilon)\simeq N(0),$$

$$N_{\mathrm{s}}(\varepsilon)=\sum_{k}\delta(E_k-\varepsilon)\simeq N(0)\left[\varepsilon/(\varepsilon^2-\Delta^2)^{1/2}\right]\Theta(\varepsilon^2-\Delta^2), \tag{2.10}$$

where $\Theta(x)$ is the unit step function. In order to sustain this gap in the excitation spectrum, the metal must develop a rigid organization (or "condensation") of electrons in momentum space, with each electron $k=(\boldsymbol{k}n)$ pairing off with another electron $-k=(-\boldsymbol{k}n)$ of opposite spin. These are called Cooper pairs, and the gap Δ is roughly speaking the binding energy of a Cooper pair. As the temperature is raised from 0, an increasing thermal population of excitations is created until there is no more free energy to be gained, and the system goes "normal" at a transition temperature $k_{\mathrm{B}}T_{\mathrm{c}}=0.57\Delta$. The gap parameter Δ is closely analogous to the Weiss molecular field in the elementary theory of magnetism

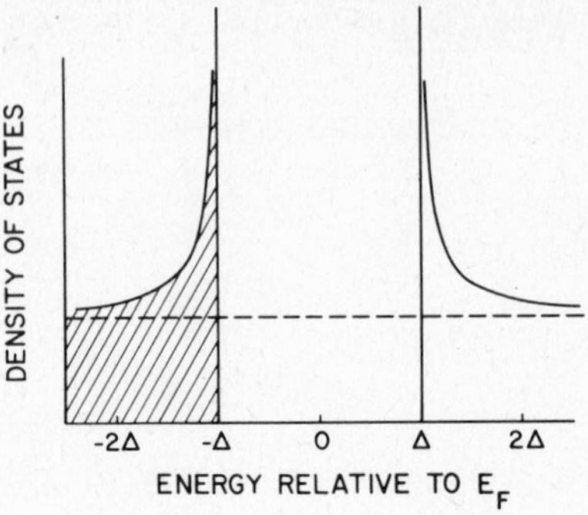

Fig. 9. Density of states versus energy in a BCS superconductor (solid line) and the corresponding normal metal (dashed line). The shaded region represents states occupied in the superconductor at $T=0$.

(Anderson 1958, Kittel 1963). As increasing numbers of excitations are created, the binding energy Δ of the remaining paired electrons diminishes, and Δ goes to zero like $(1-T/T_c)^{1/2}$ as T approaches T_c.

BCS theory is not equipped to handle the actual Hamiltonian (2.3), (2.4), (2.9), but instead requires a pairwise interaction which scatters Cooper pairs from k to k':

$$\mathcal{H}_{\text{eff}} = -\sum_{kk'} V(kk') c^{+}_{k'\uparrow} c^{+}_{-k'\downarrow} c_{k\uparrow} c_{-k\downarrow}. \tag{2.11}$$

Thus the first step is to replace the true Hamiltonian by an effective Hamiltonian which replaces the phonon parts (2.4) and (2.9) by an effective electron–electron interaction. The result, known as the Bardeen–Pines (1955) interaction, was first derived by Fröhlich (1952); a derivation is given by Kittel (1963). It is valid only in lower-order perturbation theory and fails to give quantitatively correct answers for materials with strong electron–phonon coupling:

$$V_{\text{BP}}(kk') = \sum_{\nu} |M^{\nu}_{kk'}|^2 2\omega_{k-k'\nu} \Big/ \left[\omega^2_{k-k'\nu} - (\varepsilon_k - \varepsilon_{k'})^2\right]. \tag{2.12}$$

A minus sign is inserted in (2.11) and (2.12) to conform to the BCS convention that positive V means attractive interaction. This interaction is attractive when the electron energy difference $|\varepsilon_k - \varepsilon_{k'}|$ is less than the phonon energy.

To solve this problem it is necessary to search for a ground state that breaks some symmetry of the original Hamiltonian. In a ferromagnet one examines states with non-vanishing values of magnetization $\langle \boldsymbol{M} \rangle$, even though $\mathcal{H}$ is independent of the direction of spin quantization. The BCS idea is exactly analogous, although more subtle: one examines states with non-vanishing values of $\langle c_{k\uparrow}c_{-k\downarrow}\rangle$. The Hamiltonian has gauge symmetry: when $c_{k\uparrow}$ is replaced by $c_{k\uparrow}e^{-i\phi}$ (and $c_{k\uparrow}^{+}$ by $c_{k\uparrow}^{+}e^{+i\phi}$), with ϕ independent of k, $\mathcal{H}$ is unchanged. Just as the ferromagnetic state is characterized by a fixed and measurable direction of $\langle \boldsymbol{M}\rangle$, so the superconductor has a fixed and measurable value of $\phi=-\frac{1}{2}\mathrm{Im}\langle\log c_{k\uparrow}c_{-k\downarrow}\rangle$. Actually it is only differences of ϕ between two different metals that can be measured, using the Josephson effect.

In BCS theory, the value of $\langle c_k c_{-k}\rangle$ is self-consistently adjusted in order to minimize the total free energy. The minimization results in a non-linear integral equation

$$\Delta_k \equiv \sum_{k'} V_{\mathrm{BP}}(k,k')\langle c_{k'\uparrow}c_{-k'\downarrow}\rangle = \sum_{k'} V_{\mathrm{BP}}(k,k')\frac{\Delta_{k'}}{2E_{k'}}\tanh\frac{E_{k'}}{2k_{\mathrm{B}}T}. \tag{2.13}$$

This equation has solutions only below a critical temperature T_{c}. As this temperature is approached from below, Δ_k goes to zero. To locate the value of T_{c}, (2.13) can be linearized by setting E_k equal to $|\varepsilon_k|$. The linear equation has no solutions except when T equals T_{c}. To simplify the algebra sufficiently to solve the problem, BCS replaced (2.9) by a simpler factorizable "square-well" model interaction

$$V_{\mathrm{BCS}}(kk') = V_{\mathrm{BP}}\Theta(\omega_{\mathrm{D}}-|\varepsilon_k|)\Theta(\omega_{\mathrm{D}}-|\varepsilon_{k'}|), \tag{2.14a}$$

where ω_{D} is the Debye frequency and V_{BP} is the Fermi surface average of $V_{\mathrm{BP}}(kk')$,

$$\begin{aligned} V_{\mathrm{BP}} &= \langle V_{\mathrm{BP}}(kk')\rangle_{\mathrm{FS}} \\ &\equiv \sum_{kk'} V_{\mathrm{BP}}(kk')\delta(\varepsilon_k)\delta(\varepsilon_{k'}) \Big/ \sum_{pp'}\delta(\varepsilon_p)\delta(\varepsilon_{p'}). \end{aligned} \tag{2.15}$$

This equation defines the Fermi surface average used in eq. (2.2). The final result is a famous formula

$$(T_{\mathrm{c}})_{\mathrm{BCS}} = \omega_{\mathrm{D}}\exp(-1/\lambda), \tag{2.1a}$$

where λ, defined in (2.3), corresponds identically to $N(0)V_{\mathrm{BP}}$.

The model was carried one step further by Bogoliubov, Tolmachev and Shirkov (BTS) (1958) who considered the effect of the Coulomb repulsion, using a square-well model for the Coulomb effect as well, with a high-energy cutoff ω_{pl}, the electronic plasma energy:

$$V_{\mathrm{c}}(kk') = -V_{\mathrm{c}}\Theta(\omega_{\mathrm{pl}}-|\varepsilon|)\Theta(\omega_{\mathrm{pl}}-|\varepsilon'|). \tag{2.14b}$$

The minus sign allows V_c to be a positive number. The interaction parameter $N(0)V_c$ has been given the name μ. The solution of Bogoliubov's two-square-well model is

$$(T_c)_{BTS} = \omega_D \exp[-1/(\lambda-\mu^*)], \tag{2.16}$$

$$\mu^* = \mu/[1+\mu \ln(\omega_{pl}/\omega_D)], \tag{2.17}$$

$$\mu = N(0)V_c. \tag{2.18}$$

This gives the important result that the Coulomb repulsive parameter is reduced from μ (a number of order 1.0) to μ^* (a number of order 0.1). The physical origin of this helpful reduction is obscure in BCS theory, and will be clarified in the next section on Eliashberg theory.

Before proceeding to the revisions of BCS theory, it should be mentioned that one aspect is totally omitted from this review, namely the anisotropy of the k-dependent gap, Δ_k. It is common to ignore this and consider only an average gap Δ, as is done in this and subsequent sections. One justification is that impurities have the effect of washing out anisotropy of Δ_k if the impurity-induced electron scattering rate $1/\tau_k$ exceeds $\Delta/\hbar$, as is likely in most high-T_c materials. Another justification is that the effect of anisotropy on T_c is usually quite small. It is known that for small anisotropy, T_c is enhanced above eq. (2.1), but the effect is second order, i.e. proportional to the *square* of the r.m.s. relative anisotropy $\sqrt{\langle\Delta^2\rangle/\langle\Delta\rangle^2-1}$ (Markowitz and Kadanoff 1963). This result continues to hold in strong-coupling theory (Butler and Allen 1976). Increases of T_c in high-T_c materials by more than a few percent seem unlikely from this mechanism. A recent suggestion to the contrary (Farrell and Chandrasekhar 1977) should be noted, however, and also a counterargument by Gurvitch et al. (1977).

2.2. Eliashberg theory

The Bardeen–Pines interaction (2.12) does not constitute a truly correct starting point for a theory of superconductivity. The true interaction between electrons is strongly time dependent, while (2.12) attempts to mimic this by a rapid dependence on ε_k and $\varepsilon_{k'}$. A physical consequence of the time-dependent interaction is that the binding energy of the Cooper pair $\Delta(t)$ is time dependent. If a superconductor is probed with an a.c. external field at frequency ω, the response will determine a gap $\Delta(\omega)$ which depends on the probe frequency, being the Fourier transform of $\Delta(t)$. *Thus the fundamental quantity of Eliashberg* (1960) *theory is the function* $\Delta(\omega)$ *which is a complex and frequency-dependent generalization of the BCS gap*

Δ. This gap is determined by a complex, frequency-dependent interaction, the Eliashberg interaction $V_E(kk',\omega)$. It is interesting to derive this interaction first as a function of space and time: the effective potential $-V_E(\boldsymbol{x}t,\boldsymbol{x}'t')$ represents the change in energy of an electron at $(\boldsymbol{x}t)$ due to the presence of an electron at $(\boldsymbol{x}'t')$. The minus sign gives the BCS sign convention.

The calculation of V_E has three parts (Butler et al. 1979): (i) The presence of an electron at $(\boldsymbol{x}'t')$ causes an impulsive force F on the lattice, (ii) the impulse causes a lattice displacement u at subsequent times; (iii) at time t the lattice displacements $u(t)$ lower the energy of the electron at $\boldsymbol{x}$. First, the impulsive force $F_{\alpha' l'}$ on the atom at $\boldsymbol{l}'$ in the α' direction is

$$F_{\alpha'}(l',\tau) = -\nabla_{l'\alpha'} V(\boldsymbol{x}'-\boldsymbol{l}')\delta(\tau - t'). \tag{2.19}$$

The notation is the same as eqs. (2.3–2.4), except that for simplicity a single atom per cell is assumed. Second, the lattice displacement $\boldsymbol{u}(l,t)$ of the lth atom at a later time t is determined by $D_{\alpha\alpha'}(l,l',t-t')$, the phonon Green's function or displacement–displacement correlation function

$$u_\alpha(l,t) = \sum_{l'} D_{\alpha\alpha'}(l,l',t-t')F_{\alpha'}(l',t), \tag{2.20}$$

$$\mathbf{D}(l,l';t-t') = -\frac{i}{h}\theta(t-t')\langle[\boldsymbol{u}_l(t),\boldsymbol{u}_{l'}(t')]\rangle, \tag{2.21}$$

where $\langle\ \rangle$ means thermal ensemble average and $\boldsymbol{u}_l(\tau)$ is the Heisenberg operator $\exp(\mathrm{i}\mathcal{H}\tau)\boldsymbol{u}_l\exp(-\mathrm{i}\mathcal{H}\tau)$. Third, the energy shift of an electron $(\boldsymbol{x}t)$ is

$$V_E(\boldsymbol{x}t,\boldsymbol{x}'t') = -\sum_l \nabla_l V(\boldsymbol{x}-\boldsymbol{l})\langle\boldsymbol{u}_l(t)\rangle. \tag{2.22a}$$

Finally, putting the previous three equations together, the Eliashberg interaction is

$$V_E(\boldsymbol{x}t,\boldsymbol{x}'t') = -\sum_{ll'} \nabla_l V(\boldsymbol{x}-\boldsymbol{l})\cdot\mathbf{D}(ll';t-t')\cdot\nabla_{l'}V(x'-l'). \tag{2.22b}$$

The phonon Green's function for a perfect harmonic lattice is

$$\mathbf{D}(l,l';t-t') = -\theta(t-t')\sum_Q \frac{\hat{\varepsilon}_Q\hat{\varepsilon}_Q}{M\omega_Q}\sin(\omega_Q(t-t'))\exp[\mathrm{i}Q(\boldsymbol{l}-\boldsymbol{l}')]. \tag{2.23}$$

It is revealing to consider the space and time dependence of the resulting interaction (2.22–2.23). The Green's function (2.23) will usually not be large unless the atoms $\boldsymbol{l}$ and $\boldsymbol{l}'$ are close together. For small times $(t-t')$, the sine function is small. The optimum time $(t-t')$ corresponds to one-quarter of a typical phonon period. For longer times or large separations the various wavevectors in the sum begin to interfere destructively. For very short times, $\sin\omega\tau$ can be replaced by $\omega\tau$, the frequency cancels out of (2.23), the branch sum $\sum_\nu \hat{\varepsilon}_{Q\nu}\hat{\varepsilon}_{Q\nu}$ is the unit tensor, and the Q sum gives $\delta_{ll'}$, a local interaction, as required by causality, and vanishing as $t-t'$ goes to zero. The interaction is also local for an Einstein spectrum (corresponding to independent atoms with no signal propagation). The most favorable situation for binding Cooper pairs is apparently when the electrons at $\boldsymbol{x}t$ and $\boldsymbol{x}'t'$ interact via the same atom $\boldsymbol{l}$ with a time-retardation $\pi/(2\omega_{\mathrm{D}})$. Thus it is often said that the Eliashberg interaction is local in space but non-local in time. The non-locality in time has an extremely important benefit – it strongly suppresses the repulsive but nearly instantaneous Coulomb interaction between electrons, because the two members of the Cooper pair can be widely separated in space at any given moment of time. This suppression is given by the logarithm of the ratio of Coulomb to phonon time scales, as shown in eq. (2.17). The screened Coulomb interaction has a time scale fixed by plasma oscillations.

Let us now transform the interaction (2.22) to wavevector and frequency space. We take the Bloch-wave matrix elements corresponding to scattering a Cooper pair $\psi_{k\uparrow}(x)\psi_{-k\downarrow}(x')$ to $\psi_{k'\uparrow}(x)\psi_{k'\downarrow}(x')$, and we Fourier transform $t-t'$ into ω. This gives

$$V_{\mathrm{E}}(kk'\omega) = -\sum_\nu |\langle k'|\nabla V(x)|k\rangle\cdot\hat{\varepsilon}_{\boldsymbol{k}-\boldsymbol{k}'\nu}|^2 D_\nu(\boldsymbol{k}-\boldsymbol{k}',\omega), \tag{2.24}$$

$$D_\nu(Q,\omega) = \left[M\left(\omega^2-\omega_{Q\nu}^2\right)\right]^{-1}. \tag{2.25}$$

Finally, using (2.8), this can be written

$$V_{\mathrm{E}}(k,k';\omega) = \sum_\nu |M_{kk'}^\nu|^2 2\omega_{k-k'\nu}/\left(\omega_{k-k'\nu}^2-\omega^2\right). \tag{2.26}$$

This is identical to the Bardeen–Pines result if ω is replaced by $(\varepsilon_k-\varepsilon_{k'})$, a replacement which is not justifiable in the more accurate Eliashberg theory. Nevertheless, the close similarity of the theories is apparent. Just as in BCS theory, it is only the Fermi surface average of (2.26) which enters the usual isotropically averaged version of Eliashberg theory. Using the notation of (2.15), this is

$$\lambda(\omega) \equiv N(0)\langle V_{\mathrm{E}}(kk',\omega)\rangle_{\mathrm{FS}} = \int_0^\infty \mathrm{d}\Omega\, \alpha^2 F(\Omega) \frac{2\Omega}{\Omega^2-\omega^2}, \tag{2.27}$$

$$\alpha^2 F(\Omega) \equiv N(0) \frac{\sum_{kk'\nu} |M_{kk'}^{\nu}|^2 \delta(\Omega-\omega_{k-k'\nu})\delta(\varepsilon_k)\delta(\varepsilon_{k'})}{\left(\sum_p \delta(\varepsilon_p)\right)^2}. \tag{2.28}$$

Eq. (2.23) introduces the very important electron–phonon spectral function $\alpha^2F(\Omega)$. All of the information about ε_k, ω_Q and $M_{kk'}$ needed to calculate T_c and $\Delta(\omega)$ is contained in this function. The meaning of α^2F is that it counts, at fixed frequency Ω, how many phonons with $\omega_Q=\Omega$ there are, and weights each phonon by the strength and number of electron transitions from k to $k+Q$ across the Fermi surface which this phonon can participate in. The transition temperature, T_c, and all thermodynamic properties of superconductors are completely determined by $\lambda(\omega)$, the Hilbert transform of $\alpha^2F(\omega)$, and one additional number, μ^*, the effective Coulomb repulsion.

An important measure of the strength of the electron–phonon coupling is the value of $\lambda(\omega)$ at zero frequency. This turns out to be identical to λ or $N(0)V_{\mathrm{BP}}$ of eq. (2.2). Two important formulas for λ follow from (2.27) and (2.28):

$$\lambda = \lambda(0) = 2\int_0^\infty \mathrm{d}\Omega\, \alpha^2F(\Omega)/\Omega, \tag{2.29}$$

$$\lambda = N(0) \frac{2\sum_{kk'\nu} (|M_{kk'}|^2/\omega_{k-k'\nu})\delta(\varepsilon_k)\delta(\varepsilon_{k'})}{\left(\sum_p \delta(\varepsilon_p)\right)^2}. \tag{2.30}$$

Unfortunately it is beyond the scope of this article to present the mathematical details of Eliashberg theory. In §2.3 the verification of theory by tunneling experiments is described. In §§2.4 and 2.5 the results of theory relevant to T_c are discussed.

2.3. Tunneling experiments

No discussion of Eliashberg theory can omit a discussion of tunneling experiments, which have provided both motivation for and verification of the theory. The simplest tunneling geometry (used by Giaever (1960) in his

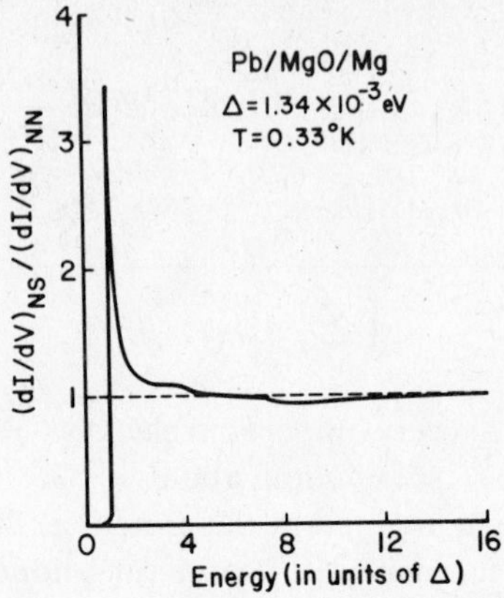

Fig. 10. Differential conductivity of a superconducting–normal tunnel junction, with lead as the superconducting electrode (Giaever et al. 1962). The overall shape confirms the BCS prediction (eq. (2.10)). The anomalous structure around 4 and 8 times Δ is the first experimental observation of the Eliashberg frequency-dependent gap (eq. (2.31)).

original experiments) has a normal metal as one electrode and a superconductor as the other. The differential conductivity, dI/dV, according to the golden rule, measures the increment of new states available to tunnel into per unit energy increment, $dE = e\,dV$. Assuming a featureless density of states in the normal electrode, dI/dV measures the density of states $N_s(E)$ at $E = eV$ in the superconducting electrode. The early experiments confirmed the behavior predicted by eq. (2.10) and provided an elegant method of measuring Δ. Later experiments (Giaever et al. 1962) began to show extra structure at higher voltages which correlated with phonon energies (see fig. 10). This has a natural explanation in Eliashberg theory, and motivated rapid further development of the subject. As is seen from eqs. (2.26), or (2.27) the Eliashberg interaction has resonances at frequencies ω which correspond to phonon frequencies (in particular peak frequencies in the phonon density of states distribution $F(\Omega)$, or more rigorously, the spectral function $\alpha^2F(\Omega)$. Eliashberg theory then shows that these resonances generate structure in the complex, frequency-dependent gap, which can be observed with an a.c. probe. Infrared absorption (Joyce and Richards 1970) sees this structure as is shown in fig. 11. Such experiments are technically very difficult, and have not yet achieved resolution competitive with tunneling experiments. Tunneling, although apparently a d.c. probe, is equivalent for this purpose to an a.c. probe at frequency $\hbar\omega = eV$, giving the result (Scalapino et al. 1966, McMillan and

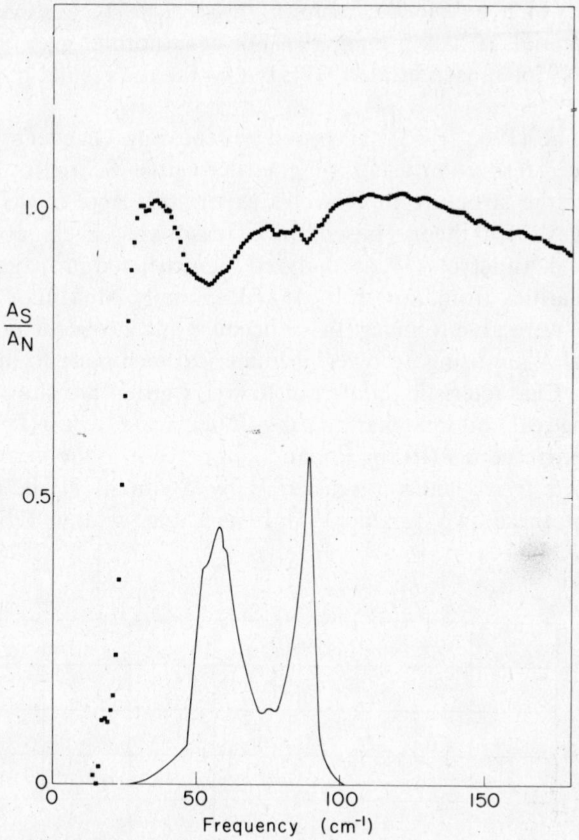

Fig. 11. Ratio of infrared absorbance of normal and superconducting Pb as a function of frequency. The structure correlates with peaks of $\alpha^2F(\omega)$ (Joyce and Richards 1970).

Rowell 1969)

$$\left.\frac{dI}{dV}\right|_{eV=\hbar\omega} \alpha N_s(\omega) = N(0)\,\mathrm{Re}\left\{|\omega|/(\omega^2-\Delta^2(\omega)^{1/2}\right\}. \tag{2.31}$$

Thus the phonons responsible for superconductivity leave their signature in $\Delta(\omega)$ which is measureable by tunneling. The measurement is difficult except in "strong-coupling" materials where Δ/ω_{ph} is comparatively large. This is because the structure in $\Delta(\omega)$ occurs at $\omega \approx \omega_{ph}$; according to (2.31), when Δ/ω_{ph} is small, this structure appears as a correction of order $(\Delta/\omega_{ph})^2$ to the normal state conductance. Various new techniques have been tested which enable the phonon structure to be resolved in weak-coupling materials (Chaikin et al. 1977) or even normal materials (Yanson and Shalov 1976, Jansen et al. 1977), but so far the results have not been definitive.

Rowell et al. (1962, 1963) succeeded in repeating Giaever's experiments with enhanced resolution using a higher derivative. Schrieffer et al. (1963) showed that the structure in Rowell's experiment agreed well with solutions of Eliashberg theory based on a reasonable guess about $\alpha^2F(\Omega)$. Scalapino and Anderson (1964) analyzed the predicted propagation of Van Hove singularities from $F(\omega)$ into $\mathrm{d}I/\mathrm{d}V$. Finally McMillan and Rowell (1965, 1969) were able to invert the procedure and *extract* $\alpha^2F(\Omega)$ from the experimental $N_s(\omega)$ using a powerful numerical technique to invert Eliashberg theory. Characteristic data from Rowell's group are shown in fig. 12. The resulting real and imaginary parts of $\Delta(\omega)$, are shown in fig. 13. Fig. 14 shows the extracted $\alpha^2F(\Omega)$ for Pb, and compares it to the density of states $F(\omega)$ deduced from neutron scattering by Stedman et al. (1967). The similarity of these two functions is remarkable, and will be discussed further in §3.

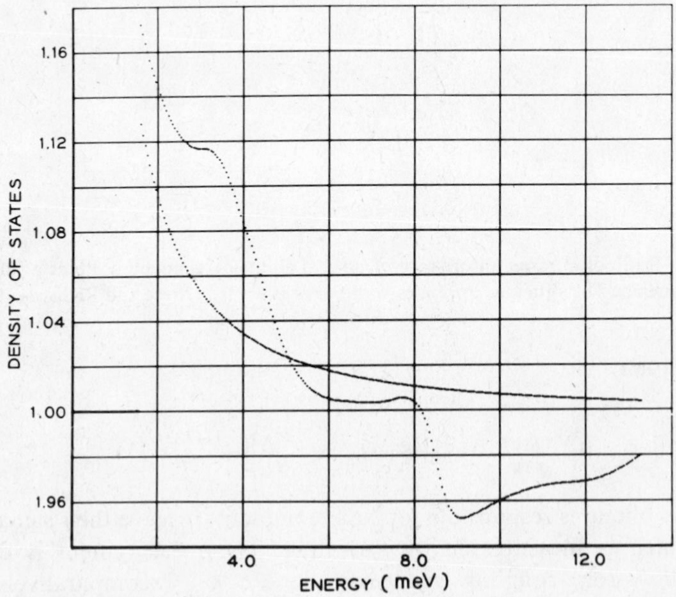

Fig. 12. Density of states $N(E)$ versus energy $E-\Delta_0$ for superconducting Pb (eq. (2.31)) compared with the BCS prediction (smooth curve; eq. 2.10). The data for Pb are extracted from $\mathrm{d}I/\mathrm{d}V$ measured for a Pb–I–Pb junction (McMillan and Rowell 1969). The anomalous structure in $\Delta(\omega)$ induced by phonons is more prominently displayed than in fig. 10.

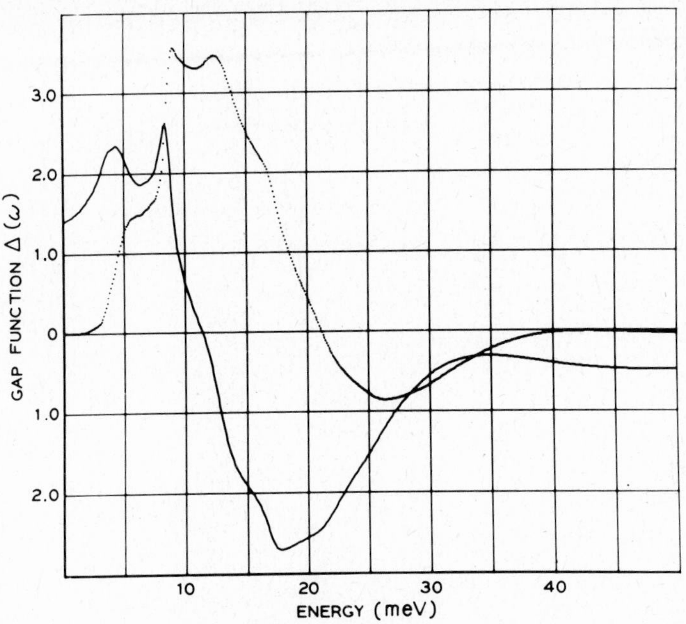

Fig. 13. Real (solid line) and imaginary (dotted line) parts of the energy gap $\Delta(\omega)$ for Pb plotted versus $\omega-\Delta_0$. The values of $\Delta(\omega)$ were extracted from the data of fig. 12 by numerical inversion of Eliashberg theory (McMillan and Rowell 1969).

The tunneling technique has become the most powerful spectroscopic tool for probing the relation between phonons and T_c. As discussed by McMillan and Rowell (1969), tunneling has provided detailed verification of the accuracy of Eliashberg theory. The experimental spectrum $N_s(\omega)$ for $\omega<\omega_D$ is sufficient to yield $\alpha^2F(\Omega)$ by Eliashberg theory. This alone does not "prove" Eliashberg theory, although the reasonableness of $\alpha^2F(\Omega)$ provides strong plausibility. More impressively, these results in turn allow $\Delta(\omega)$ to be correctly predicted for larger ω (where harmonic structure occurs.) The correctness of Eliashberg theory can now hardly be doubted. Migdal's (1958) theory suggests that Eliashberg theory is correct to leading order in the small parameter $(m/M)^{1/2}$ (electron to ion mass ratio). The tunneling experiments seem to confirm this expectation, making Eliashberg theory the most quantitatively successful many-body theory in contemporary physics.

Nearly all the reliable tunneling measurements of $\alpha^2F(\Omega)$ published so far involve sp-metals. Spectra for Ta (Shen 1970), Nb (Robinson et al. 1976, Bostock et al. 1976), La (Lou and Tomash 1972, Wuhl et al. 1973), Nb_3Sn (Shen 1972a) and PdH (Eichler et al. 1975, Silverman and Briscoe

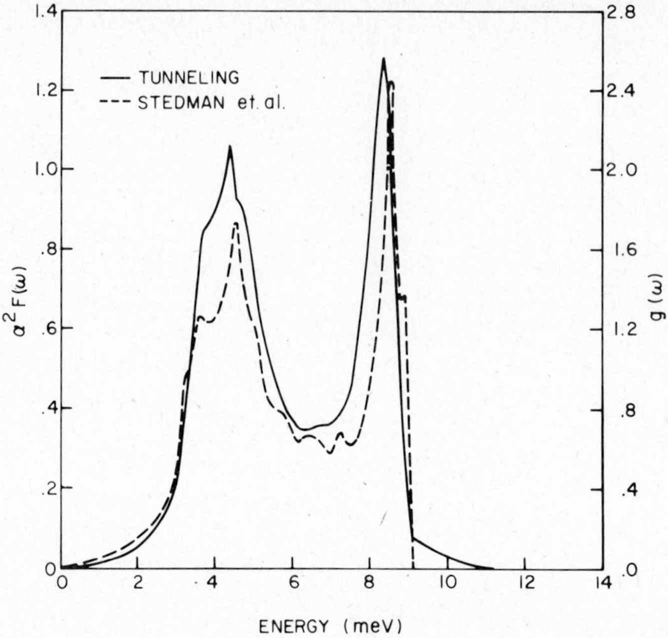

Fig. 14. Electron–phonon spectral function $\alpha^2F(\Omega)$ (solid line) and density of states $F(\Omega)$ (dashed line) *versus* energy for Pb. The values of $\alpha^2F(\Omega)$ were found from the data of fig. 12 by numerical inversion of Eliashberg theory. The values of $F(\Omega)$ were computed from neutron scattering measurements of ω_Q by Stedman et al. (1967). (Figure courtesy of J. M. Rowell.)

1975) have also been reported, but except for Ta (and maybe La), the resolution of these experiments is in some doubt. It is unfortunate that these experiments are so difficult because they provide extremely detailed information of great help in explaining the microscopic origin of T_c. For example, the tunneling spectrum for Ta by Shen (1970) largely laid to rest the previously persistent speculation that superconductivity in d-band elements might be caused by mechanisms other than electron–phonon, or governed by equations other than Eliashberg's. These speculations have been raised anew by Bostock et al. (1976) who find for Nb that dI/dV is best fitted by a rather ordinary $\alpha^2F(\Omega)$ spectrum but a negative value of μ^*. Similar data are obtained by Gärtner and Hahn (1976). Presumably $\mu^*<0$ would indicate some type of attractive interaction which propagates rapidly compared with the phonon time scale. However, Robinson et al. (1976) fit their data very well with a positive value of μ^*. The situation was discussed in some detail by Shen (1972b), who also found $\mu^*<0$ and attributed this to surface contamination and the difficulty of growing an insulating oxide barrier to Nb. Shen felt that the negative value of μ^* was highly unlikely to be a true intrinsic effect. The data of Robinson et al. seem to confirm this.

Recent theoretical work by Arnold et al. (1978) helps clarify some of these difficulties. They find that the presence of a conducting layer (either normal or superconducting with reduced T_c) between the superconducting electrode and the oxide barrier, causes a distortion of the dI/dV versus V spectrum. Using a new theory for proximity-effect tunneling by Arnold (1978), they have reanalyzed the data of Gärtner and Hahn (1976) and find μ^* positive and λ increased to a value near 1.0. This analysis involves a new parameter related to the thickness of the metallic Nb–suboxide layer believed to exist adjacent to the barrier in the junctions of Gärtner and Hahn. The reanalyzed values of λ and μ^* apparently depend on the somewhat arbitrary choice made for this parameter. Bostock and MacVicar (1978) claim that the negative value of μ^* in their junctions cannot be explained in this way, and the subject remains controversial.

In fig. 15 the tunneling data of Shen (1972a) for Nb_3Sn are shown. It is especially interesting to compare the measured $\alpha^2F(\Omega)$ with the density of states $F(\Omega)$ shown in fig. 7 as measured by neutron time-of-flight on polycrystalline samples. The general similarity is clear but there are also striking differences. The low-frequency (8 meV) peak is relatively larger in α^2F than in F, while the high-frequency (25 meV) peak is much lower. This could be attributed to a stronger frequency dependence of $\alpha^2(\omega)$ than is seen in most metals. However, at least part of the explanation is likely to lie in the probable surface contamination or damage in Nb_3Sn films just below the oxide barrier. The tunneling electron current probes the electrode to a depth given by a mean free path,

$$l(\omega)=v_F\tau(\omega) \tag{2.32}$$

(McMillan and Rowell 1969). Higher-energy electrons have shorter phonon-limited mean free paths, and this effect is severe in d-band metals where v_F is small. Thus in Nb the anticipated mean free path (Shen 1972b) is $\simeq 30$ Å at energies which probe the higher peak in $\alpha^2F(\Omega)$. This effect makes the higher-energy parts of $\alpha^2F(\Omega)$ more sensitive to surface contamination than the lower-energy parts. Probably this is the main factor accounting for the suppression of the high-energy peak in Nb_3Sn in fig. 15. For the same reason, Shen feels that the relatively low value, $\mu^*\simeq 0.06$, which gives the best fit to his data, is not to be relied upon. A somewhat arbitrary choice $\mu^*=0.11$, would require an increase in the magnitude of α^2F as shown on the right-hand scale in fig. 15.

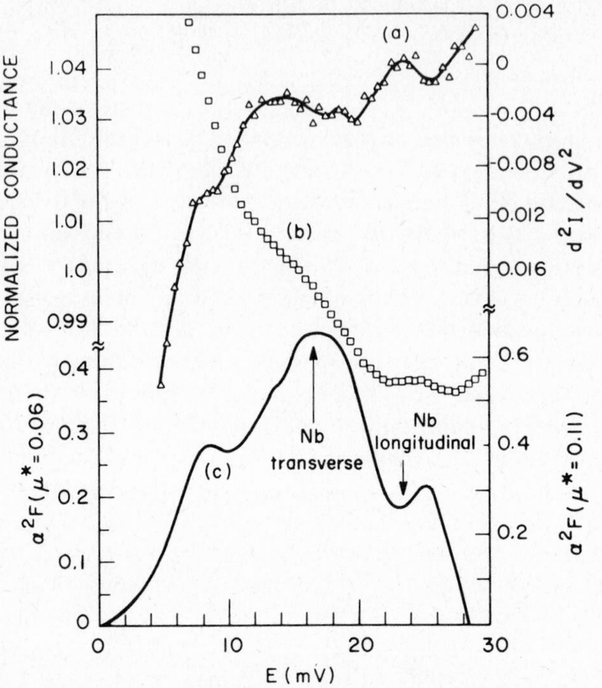

Fig. 15. Curve (a) is the second derivative d^2I/dV^2 and curve (b) is the differential conductance dI/dV plotted versus $\omega-\Delta_0$. The measurement (Shen 1972a) is on a Nb_3Sn–I–Pb tunnel junction. Curve (c) is $\alpha^2F(\Omega)$ for Nb_3Sn extracted from these data by numerical inversion of Eliashberg theory. The left-hand scale for α^2F is used if $\mu^*=0.06$ is accepted, while the right-hand scale corresponds to $\mu^*=0.11$.

2.4. McMillan-type equations

Eliashberg theory gives an integral equation whose solution yields T_c. Two functions, $\lambda(\omega)$ and $\mu(\omega)$, must be given as input. The first of these functions, $\lambda(\omega)$, defined in eq. (2.22), is the effective frequency-dependent phonon attraction, $N(0)V_E(\omega)$. The second is the analogous function, $N(0)V_c(\omega)$, for the Coulomb repulsion. The characteristic frequency scale of the former is ω_D and of the latter is ω_{pl}, the electronic plasma frequency. Because the latter frequency is very high, there is no harm done in replacing $\mu(\omega)$ by a "square-well" model, $\mu\theta(\omega_{pl}-|\omega|)$, where $\mu=\mu(0)$ is a number of order 1. A model has been widely used where $\lambda(\omega)$ is similarly replaced by $\lambda\theta(\omega_D-|\omega|)$, where $\lambda=\lambda(0)$. These models are much like the two-square-well model of BCS theory, eqs. (2.13) and (2.15), and yield similar results for T_c, namely

$$(T_c)_{NW} = \omega_D \exp\{-1/[\lambda/(1+\lambda) - \mu^*]\}, \tag{2.33}$$

a result apparently first given by Nakajima and Watabe (1963). The only alteration of the BCS formula (2.17) is that λ is "renormalized" to $\lambda/(1+\lambda)$. The same renormalization factor $(1+\lambda)$ occurs in the low-temperature specific heat coefficient $\gamma = (2\pi^2 k_B^2/3)N(0)(1+\lambda)$ and in the electronic mass as measured by cyclotron resonance or de Haas–van Alphen effect (Grimvall 1976). The occurrence of the factor $1+\lambda$ is called a strong-coupling effect, although in fact the fractional effect of the extra $1+\lambda$ on T_c is as large for weak-coupled superconductors (small λ) as for strong-coupled materials (large λ).

The first paper to present extensive numerical solutions of Eliashberg theory to determine T_c, taking the frequency dependence of $\lambda(\omega)$ into account (using realistic forms for $\alpha^2F(\omega)$), was the classic paper of McMillan (1968). He chose $\alpha^2F(\omega)$ to have the shape of $F(\omega)$ for Nb, and solved numerically for T_c for various values of λ between 0.1 and 1.5 and several values of μ^* between 0 and 0.25. His results fitted nicely to a two-square-well formula like (2.33) with small adjustment of parameters. His final answer was the famous "McMillan equation"

$$(T_c)_{McM} = (\omega_D/1.45)\exp[-1.04(1+\lambda)/(\lambda - \mu^* - 0.62\lambda\mu^*)]. \tag{2.34}$$

Numerous attempts have been made to test or improve on this equation. It is helpful to introduce the various frequency moments of $\alpha^2F(\omega)$. More correctly, the most natural moments are those of the normalized weight function g:

$$g(\omega) = (2/\lambda\omega)\alpha^2F(\omega),$$

$$\langle\omega^n\rangle = \int_0^\infty d\omega\, \omega^n g(\omega), \tag{2.35}$$

$$\bar{\omega}_n = \langle\omega^n\rangle^{1/n}. \tag{2.36}$$

Three of these moments have special significance, namely $\langle\omega^2\rangle$, $\langle\omega\rangle$ and $\bar{\omega}_0$ (which is also called ω_{log}, and can be defined as the limit as $n \to 0$ of $\bar{\omega}_n$,

$$\omega_{log} = \lim_{n\to 0} \bar{\omega}_n = \exp\left\{\int_0^\infty d\omega \log\omega\, g(\omega)\right\}. \tag{2.37}$$

The prefactor, ω_D, of (2.34), is unfortunately not a good enough measure of the effective phonon frequency which determines T_c. It is clear from theory that since only $\alpha^2F(\omega)$ and μ^* enter Eliashberg theory, the prefactor of (2.34) should be a functional of α^2F. Dynes (1972) showed that the use of $\langle\omega\rangle/1.2$ in place of $\omega_D/1.45$ improved the agreement with experiment. Kirzhnits et al. (1973) and other authors have suggested that $\omega_{\log}$ is fundamentally more correct. Allen and Dynes (1975a, b) showed that $\omega_{\log}/1.2$ in place of $\omega_D/1.45$ in (2.34) gives excellent agreement with both experiment and exact numerical solutions of the theory for all cases where $T_c/\omega_{\log}$ is less than 0.09 (corresponding to $\lambda < 1.4$). Thus the McMillan equation is now extremely well verified over most of the experimentally observed range of values of λ, provided a minor modification is made in the prefactor.

For very strongly-coupled materials, with λ larger than 1.4, an interesting deviation from McMillan's equation is observed. This was first seen clearly in tunneling measurements (Dynes and Rowell 1975) on alloys of Pb with Tl and Bi. The experimental values of T_c, $\langle\omega\rangle$ and λ, for this alloy series, are shown in fig. 16 where they are compared with the prediction of eq. (2.34). McMillan's equation systematically underestimates T_c when $\lambda > 1.5$, and this effect cannot be ascribed to variability in μ^* or any other simple cause. A reinvestigation of Eliashberg theory (Allen and Dynes 1975a, hereafter denoted AD) showed that the discrepancy for large λ represents the onset of cross-over to a new regime where (2.34) no longer holds. For large enough λ, it can be shown that T_c must behave as $\simeq 0.18\sqrt{\lambda\langle\omega^2\rangle}$. The meaning of "large enough λ" is $\lambda \gtrsim 10\omega_{\max}^2/\langle\omega^2\rangle$, not $\lambda \gtrsim 10$ as stated by AD (Leavens 1976, and private communication.) The suggestion by Leavens (1976) of a more fundamental error in AD is incorrect. Louie and Cohen (1977) have verified the principal conclusions of AD. A very careful and interesting study of the asymptotic regime has been made by Wu et al. (1977). Unfortunately, the asymptotic regime of very large λ is far beyond the range of known values of λ, which are so far never larger than 3.0. The failure of McMillan's equation (2.34) when $\lambda > 1.5$ can be traced to two sources. First, the two-square-well model, which yielded the prototype equation (2.33), gives an incorrect picture of the significance of coupling to phonons of energy less than $\pi k_B T_c$ (Karakozov et al. 1975). Second, McMillan's exact numerical solutions were confined to values of $\lambda < 1.5$. Correction factors to (2.34) have been given by AD which enable the McMillan equation to be used for a much wider range of parameters. Louie and Cohen (1977) give a more accurate but somewhat more complicated) procedure. Finally it should be said that the exact solution for T_c given $\alpha^2F(\Omega)$ and μ^* can now be readily accomplished using Eliashberg theory and a modest computer, making approximate equations considerably less necessary (Owen and Scalapino 1971, Bergmann and Rainer 1973). There is also some virtue in a very simple

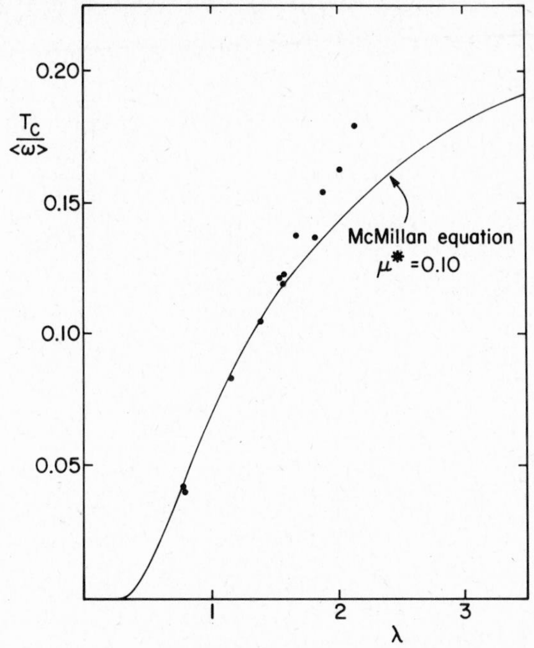

Fig. 16. T_c normalized to mean phonon energy $\langle\omega\rangle$ (eq. (2.35)) plotted versus λ. The values of $\langle\omega\rangle$ and λ are from tunneling measurements by Dynes and Rowell (1975) on alloys of Pb with Tl and Bi. The solid line is (2.34) with $\Theta_D/1.45$ replaced by $\langle\omega\rangle/1.2$ and with μ^* set equal to 0.1. The systematic discrepancy for $\lambda > 1.5$ represents a breakdown of eq. (2.34) as discussed in the text.

approximate equation consisting of a straight line fit to the data of fig. 16. This was first noticed by Leavens and Carbotte (1974), and has been restated more simply by Rowell (1976). Leavens (1975, 1976, 1977) has emphasized that this empirical linear relation can be written as $T_c \propto A = \lambda\langle\omega\rangle/2$, where A is the "area under $\alpha^2F(\Omega)$", as can be seen from eq. (2.35). This has the virtue of being a single parameter equation for T_c. However, there is also some virtue in thinking of A as the product of two parameters, λ and $\langle\omega\rangle$, as the latter have independent physical significance (and can be estimated from other experiments), whereas the former does not.

In summary, the original McMillan equation remains reputable and widely accepted. Provided λ does not exceed 1.5, the principal uncertainty is the choice of prefactor. If ω_{log} is known, $\omega_{log}/1.2$ should be used instead of $\Theta_D/1.45$. Generally the latter choice of prefactor is an overestimate and yields T_c too large (if λ is fixed) or λ too small (if T_c is fixed.) The situation

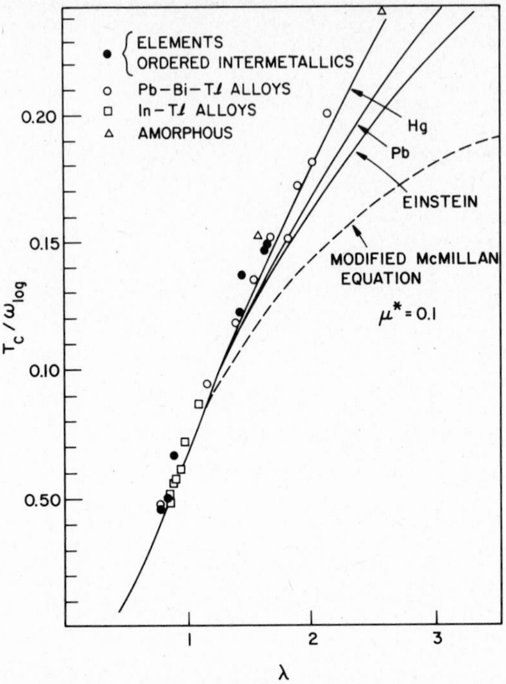

Fig. 17. T_c normalized to $\omega_{\log}$ (eq. (2.37)) plotted *versus* λ. The data represent most of the available tunneling experiments. The dashed line is eq. (2.34) with $\Theta_D/1.45$ replaced by $\omega_{\log}/1.2$ and $\mu^*=0.1$. The solid lines represent exact numerical solutions of Eliashberg theory (Allen and Dynes 1975a, b) for three different shapes of $\alpha^2F(\Omega)$, namely Einstein (delta function), Pb (see fig. 14) and Hg (see fig. 19). The clustering of the large λ data on the Hg curve is partly accidental and partly because large λ amorphous films have moment distributions of $\alpha^2F(\omega)$ similar to Hg.

is illustrated in fig. 17 where values of $T_c/\omega_{\log}$ are plotted *versus* λ for most of the metals where these parameters are known from tunneling. The agreement with McMillan's equation is excellent (in the regime where it applies), and the agreement with exact microscopic theory is even better.

In addition to anisotropy effects mentioned in §2.1, another effect is omitted in the theories given here: the possibility of structure in electronic properties like $N(\varepsilon)$ for energies ε within a phonon energy of the Fermi level. In the usual formulation of Eliashberg theory (as given here) it is assumed that any such structure is negligible relative to the sharp ε-variation from factors like $(\varepsilon\pm\omega\pm\omega_Q)^{-1}$. If additional sharp ε-variation occurs, Eliashberg theory becomes much harder to solve. However, a plausible model has been made by Horsch and Rietschel (1977). In this model, additional energy-variation appears only in factors $N(\varepsilon)$ and $\Delta(\varepsilon,\omega)$ turns out to be independent of ε. Horsch and Rietschel find a possible resonant effect which can enhance T_c if structure in $N(\varepsilon)$ is correctly tuned in the energy range $\varepsilon_F \pm \omega_D$. Improved solutions of this model have been made by Lie and Carbotte (1978). There is as yet no experimental evidence for this effect. It seems that significant enhancement could occur only by a happy accident, and would not explain, for example, why many A15 metals have a high T_c. It is also unclear whether such enhancement could survive in the presence of the relatively large disorder present, for example, in high-T_c Nb_3Ge films.

2.5. "Optimum" phonons–the analysis of Bergmann and Rainer

Bergmann and Rainer (1973 – hereafter denoted BR) have asked and answered a fundamental question about T_c and phonons, namely, what is the effect, ΔT_c on T_c, of a small change $\Delta\alpha^2 F$ in $\alpha^2 F(\Omega)$ at some frequency $\Omega = \omega$. The answer is expressed mathematically by the functional derivative $\delta T_c / \delta\alpha^2 F(\omega)$ defined by

$$\Delta T_c = \int_0^\infty d\Omega \left[\delta T_c / \delta\alpha^2 F(\Omega) \right] \Delta\alpha^2 F(\Omega). \tag{2.38}$$

The meaning of $\delta T_c / \delta\alpha^2 F(\omega)$ is the change in T_c caused by a change in $\alpha^2 F(\Omega)$ of unit weight located at $\Omega = \omega$ (i.e. $\Delta\alpha^2 F$ is chosen to be $\delta(\omega - \Omega)$). A prescription for calculating $\delta T_c / \delta\alpha^2 F$ is given by BR. They have found that this functional derivative is never negative, but goes to zero linearly with ω at zero frequency, with a slope roughly inversely proportional to T_c. This should be contrasted with the change in λ, namely $\delta\lambda / \delta\alpha^2 F(\omega)$, which is easily seen from (2.29) to be $2/\omega$. Thus very-low-frequency phonons, which have a diverging effect on λ, have a vanishing effect on T_c. This contradicts McMillan's equation, and requires that low-frequency phonons make the same contribution to T_c that they make to the quantity $\sqrt{\lambda\langle\omega^2\rangle}$. The linear behavior of $\delta T_c / \delta\alpha^2 F(\omega)$ persists up to approximately $\omega_M = 2\pi k_B T_c$, where a maximum occurs followed by a gentle fall-off. If the value of T_c is high enough that ω_M exceeds the maximum phonon frequency (which has never been observed in practice but is permitted in principle) then $\delta T_c / \delta\alpha^2 F$ is linear throughout the range of Ω and the proportionality between T_c and $\sqrt{\lambda\langle\omega^2\rangle}$ can be derived by integration. Therefore this result first published by AD is implicit in the paper of BR, as was recognized by Rainer and by Garland (private communication).

The vanishing of $\delta T_c / \delta\alpha^2 F$ for small ω can be understood as an example of Anderson's (1959, 1961) theorem, namely that static perturbations invariant under time reversal do not break Cooper pairs and thus have no first-order effect on T_c. The long-wavelength, low-frequency phonons (with

$\omega \ll \omega_M = 2\pi T_c$) are essentially static perturbations as far as superconducting pairing is concerned. This result was somewhat surprising at the time, because it had become widely believed that low-frequency phonons were harmful to T_c (Appel 1968, Allen 1973, Medvedev et al. 1973). Use of square-well constraints on $\Delta(\omega)$ gives unrealistic answers to questions of this type; in actuality, $\Delta(\omega)$ has considerable freedom to adjust to minimize the free energy and maximize T_c. Talbot and Leavens (1978) have found that thermally excited phonons have a significant adverse effect on T_c, but at each frequency ω this is more than cancelled by the favorable effect of spontaneous phonon emission and absorption. The conclusion of Bergmann and Rainer was correctly guessed by Barisić (1972, appendix B), but the analysis given there seems only accidentally correct.

Fig. 18 shows the behavior of $\delta T_c/\delta\alpha^2F(\omega)$ calculated using $\alpha^2F(\Omega)$ for Pb scaled by various constants to increase or decrease T_c and λ. The family of $\delta T_c/\delta\alpha^2F(\omega)$ curves would be approximately a single universal curve if normalized by dividing by T_c and plotted versus ω/T_c instead of ω (see fig. 1 of BR for the latter scaling.) In particular the shape of $\delta T_c/\delta\alpha^2F(\omega)$ is almost completely insensitive to the shape of $\alpha^2F(\Omega)$. The curve of fig. 18 labelled $T_c = 0.1$ is approximately correct for actual Pb with $\lambda = 1.5$. The maximum of $\delta T_c/\delta\alpha^2F$ (occurring at ω_M) tells what frequency is most favorable for a further increase of T_c caused by alteration of $\alpha^2F(\Omega)$. For Pb, the TA peak of α^2F occurs exactly at the most favorable location; this implies it would not be advantageous to attempt to soften the TA branch (as can often be done by introducing large amounts of disorder, or a large surface to volume ratio.) If the area under α^2F were kept fixed, T_c could only diminish by moving the TA peak. For materials with lower T_c, increases of T_c would occur, but decreases would occur for materials with higher T_c. These insights quantify and to some extent dampen the earlier hope that T_c could be raised to high values by using thin film or amorphous specimens, with low TA frequencies. However, the argument depends critically on several quantities, namely (1) the ratio T_c/ω_{TA}, and (2) whether the area under a $\alpha^2F(\Omega)$ remains fixed. The latter is not possible to know *a priori*.

As has been outlined in this section, Eliashberg theory appears to be quite thoroughly verified and understood thanks to tunneling spectroscopy. The purely mathematical aspects of the connection between T_c and material properties appear to be solved. This by no means implies that T_c and its relation to phonons is understood. The remaining sections will explore a few aspects of the difficult and largely unsolved problems of understanding the remarkable variation of material properties displayed by nature and uncovered by experimental ingenuity.

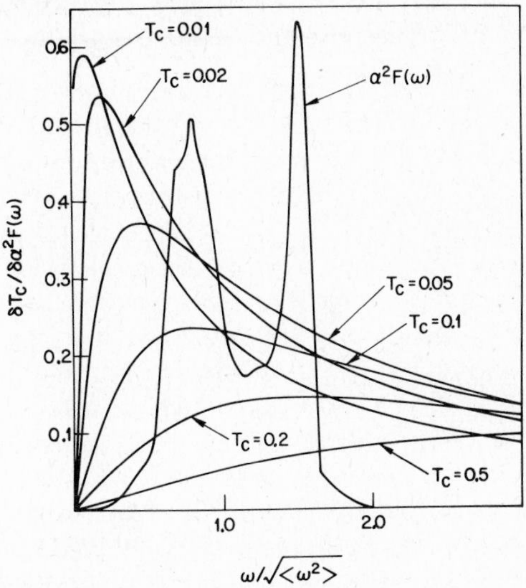

Fig. 18. Functional derivative $\delta T_c/\delta\alpha^2F(\omega)$ and $\alpha^2F(\omega)$ for Pb, plotted versus phonon energy ω. The curve labelled $T_c=0.1$ (in units of $\sqrt{\langle\omega^2\rangle}$) is correct for Pb, while the other curves correspond to fixing the shape of α^2F and μ^* but scaling the strength of α^2F and λ.

3. *Theory of* λ

This section describes how λ and thus T_c can be understood and calculated in favorable cases, and how we might go about setting up a calculation in less favorable cases where it has not yet been done. The truly important questions, such as why the A15 structure has such high T_c's, do not yet have satisfactory answers. There is empirical evidence (Vandenberg and Matthias 1977) that except for rocksalt structure (TaC, NbN, PdH, etc.) the favorable structures for high T_c are ones with anisotropic arrangements of atoms which are both metallic and covalent and have exceptionally small nearest-neighbor distances and low coordination number. Electronic band theory for metals is not yet well developed for these materials, and for the most part theorists have had to be content to study elements and systems with high coordination number where band theory is quite reliable. Hopefully the insights derived from such a study can be extended with minor modification to low-coordination-number systems, but this is not yet known. This section will *not* attempt to review all of this work, but instead will attempt a summary of the current state, emphasizing certain aspects which seem most relevant to the question of phonon anomalies.

First let us examine the order of magnitude of λ. According to (2.30), λ is density of states $N(0)$ times the Fermi surface average of $2|M_{kk'}|^2/\omega_{k-k'}$. This turns out to be a number of order 1, with remarkably large variability from metal to metal (i.e. perhaps as low as 0.1 in noble and alkali metals, and as large as 3.0 in certain amorphous thin films based on Ga, Pb, Bi, etc.) The reason why λ is $\simeq 1$ is that $M_{kk'}$ is typically the geometric mean $(E_F\omega)^{1/2}$ of an electron and a phonon energy; $N(0)$ is of course $\simeq 1/E_F$. To see why it is a geometric mean, recall that $M_{kk'}$ is the matrix element of $u\cdot\nabla V$ and thus is of order $(u/a)E_F$ since ∇V has matrix elements of order E_F/a, where a is a lattice spacing. The size of u/a can be estimated by writing $(u/a)^2$ as $\hbar/M\omega_D a^2 \simeq \hbar\omega_D/ka^2$, where the force constant k is $M\omega_D^2$. Now ka^2 is $a^2\nabla^2 V$, which must be of order E_F. Thus (u/a) is of order $(\omega_D/E_F)^{1/2} \lesssim 0.1$, and $M_{kk'}$ is of order $(\omega_D E_F)^{1/2}$.

The remarkably large variability of λ arises from several factors. The values of $N(0)$ vary considerably. Nb has $N(0) \simeq 0.73\ \mathrm{eV}^{-1}$ (Mattheiss 1970) and Cu has $N(0) \simeq 0.13\ \mathrm{eV}^{-1}$ (Janak 1969). (These are states of one spin, per atom.) The force constants k also vary a lot. Pb has $M\langle\omega^2\rangle \simeq 1.6\ \mathrm{eV/Å^2}$ while Ta, with a mass nearly as large, has $M\langle\omega^2\rangle \simeq 7\ \mathrm{eV/Å^2}$. Finally the electron–phonon matrix elements also show considerable variability. These circumstances make the study of λ quite interesting.

3.1. Factorizing λ

McMillan (1968) derived a useful simplification which is rigorous for an element,

$$\lambda = \eta / M\langle\omega^2\rangle, \tag{3.1}$$

$$\eta = N(0)\langle I^2\rangle, \tag{3.2}$$

where $\langle\omega^2\rangle$ is defined in (2.35), and η is a purely electronic parameter. The factor $\langle I^2\rangle$ is a Fermi surface average of the squared matrix element of ∇V. A generalization of (3.1) to compounds has been made by Gomersall and Gyorffy (1974a, b) and by Rietschel (1975). Following Gomersall and Gyorffy, let us calculate the product $\lambda\langle\omega^2\rangle$ which according to (2.35) is defined as

$$\begin{aligned}\lambda\langle\omega^2\rangle &\equiv 2\int_0^\infty \mathrm{d}\Omega\,\Omega\alpha^2F(\Omega)\\ &= 2N(0)\Big\langle \sum_\nu |M_{kk'}^\nu|^2\omega_{k-k'\nu}\Big\rangle_{\mathrm{FS}}.\end{aligned} \tag{3.3}$$

The second equality comes from using (2.28) with Fermi surface averaging defined by (2.15). The crucial point is that the sum over branches ν of $|M|^2\omega$ is independent of the phonon frequencies and polarizations. Using (2.8), we can write

$$2\sum_{\nu}|M^{\nu}_{kk'}|^2\omega_{k-k'\nu}=\sum_{\substack{ab\\ \alpha\beta}}(M_aM_b)^{-1/2}\left[\sum_{\nu}(\hat{\varepsilon}^{a}_{k-k'\nu})_{\alpha}(\hat{\varepsilon}^{b}_{k-k'\nu})_{\beta}\right]$$

$$\times\langle k|\partial V_a(\boldsymbol{r}-\boldsymbol{\tau}_a)/\partial\tau_{a\alpha}|k'\rangle\langle k'|\partial V_b(\boldsymbol{r}-\boldsymbol{\tau}_b)/\partial\tau_{b\beta}/k\rangle$$

$$=\sum_{a\alpha}\frac{1}{M_a}|\langle k'|\partial V_a(\boldsymbol{r}-\tau_a)/\partial\tau_{a\alpha}|k\rangle|^2, \tag{3.4}$$

where the second equality follows because according to the orthogonality relations for polarization vectors, the part of (3.4) in square brackets is $\delta_{ab}\delta_{\alpha\beta}$. Finally, taking the Fermi surface average, (3.3) becomes

$$\lambda=\sum_a \eta_a/M_a\langle\omega^2\rangle \tag{3.5}$$

$$\eta_a=N(0)\langle I_a^2\rangle=N(0)\langle|\langle k'|\partial V_a(\boldsymbol{r}-\tau_a)/\partial\tau_a|k\rangle|^2\rangle_{\mathrm{FS}}. \tag{3.6}$$

This reduces trivially to (3.1) when there is a single atom per cell. The quantity $\langle I_a^2\rangle$ is much easier to compute theoretically than is λ, and there have by now been many calculations which will be mentioned briefly in §3.3. These calculations rely on the hope that $\langle\omega^2\rangle$ can be estimated rather than calculated directly.

For compounds, (3.5) is not the only way of factorizing λ, nor is it necessarily the best. For example, (3.5) does not lead readily to the result, proved first by Taylor and Vashishta (1972), and emphasized by Rietschel, that λ is independent of isotopic mass. This is obvious for elements since $M\langle\omega^2\rangle$ is clearly mass-independent in harmonic approximation. To show the mass-independence of (3.5) for compounds, one needs a detailed theory of $\langle\omega^2\rangle$. It is simpler to go back to the formula for λ. Contained inside the Fermi surface average is the factor

$$\Phi^{\alpha\beta}_{k-k'}(a,b)=(M_aM_b)^{-1/2}\sum_{\nu}(\hat{\varepsilon}^{a}_{k-k'\nu})_{\alpha}(\hat{\varepsilon}^{b}_{k-k'\nu})/\omega^2_{k-k'\nu}. \tag{3.7}$$

Because of the additional factor ω^{-2} not found in (3.4), the orthogonality rules for $\hat{\varepsilon}$ are no help. But Φ has been shown to be related to the inverse of the $3n$ dimensional force constant matrix $K_{\alpha\beta}$ $(la,l'b)$, where n is the number of atoms per cell:

$$\Phi^{\alpha\beta}_{k-k'}(ab)=\sum_{l}K^{-1}_{\alpha\beta}(la,l'b)\exp[\mathrm{i}(k-k')\cdot(l-l')]. \tag{3.8}$$

Force constants will be discussed in the next section. For now, it suffices to note that they are mass-independent. When the complete equation for λ is written out using (3.7), there is no longer a simple exact separation into contributions from each atom. Unlike the case for $\lambda\langle\omega^2\rangle$ which does separate rigorously, λ will contain cross terms between atoms as well as terms from individual atoms. Rietschel argues that the cross terms are important, and that (3.5), although rigorous, can be misleading.

3.2. Non-transition metals

A complete calculation of λ (or what is equally difficult but more informative, $\alpha^2F(\Omega)$), requires knowing the energy bands and wavefunctions at the Fermi surface, the phonon spectrum and polarization vectors, and having a model for ∇V. In principal we have all this for most of the superconducting elements, but in practice putting it all together and performing the calculation is very tedious and time consuming. Complete calculations for d-band elements are only beginning to emerge. Calculations for sp-metals (or "simple" metals) have been available for nearly ten years, because models based on pseudopotentials simplify the situation. Only near Brillouin zone boundaries do the wavefunctions of simple metals contain significantly more than one orthogonalized or augmented plane wave (OPW or APW), and only in these regions is the Fermi surface significantly distorted from spherical. In a first approximation these effects can be neglected. It is also reasonable to assume that ∇V is the gradient of the same screened electron atom pseudopotential which enters the pseudopotential band theory. This result fails to be justified if V is a strong potential as in d-band metals, as is explained in §4.1. However, weak potentials can be regarded as linearly screened bare potentials, and the change of the potential upon displacement is screened in the same way (Heine et al. 1966). An early model was Ziman's (1962) correlation between T_c and the resistivity of the metal in the liquid phase just above the melting point. The first single-OPW pseudopotential calculations were by Swihart et al. (1964), followed closely by calculations of Ashcroft and Wilkins (1965) which contained 2-OPW corrections. Calculations for nineteen simple metals were done by Allen and Cohen (1969) and eighteen by Kakitani (1969). Both papers predict Mg and Li to be superconducting at low temperatures; as yet, this has not been confirmed. The former paper contains an extensive review of work up to 1969. Since that time many other calculations have been made. An exhaustive list of references is given by Grimvall (1976). Generally good agreement between theory and experiment is obtained. As an illustration, fig. 19 shows α^2F for Hg as measured by Hubin and Ginsberg (1969) by tunneling. There is good agreement with

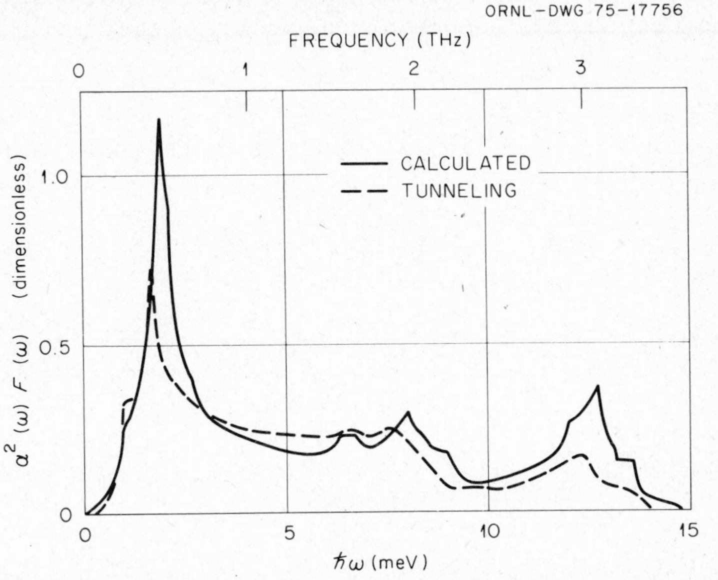

Fig. 19. Tunneling result for α^2F of Hg (Hubin and Ginsberg 1969, dashed line) compared with pseudopotential calculation based on neutron measurements (Kamitakalara et al. 1977, solid line).

a pseudopotential calculation based on neutron measurements, both done by Kamitakahara et al. (1977). The calculated value of λ is 2.0, while the experimental value is 1.6. This discrepancy is somewhat larger than usual for pseudopotential calculations, and might have been reduced somewhat by using empirical pseudopotentials (Cohen and Heine 1970). It can be seen from fig. 12 that for Pb, α^2F and F are astonishingly similar, indicating that the ratio, called $\alpha^2(\omega)$, is approximately constant. In Hg this is somewhat less true – there is a tendency for $\alpha^2(\omega)$ to decrease as ω increases, which is moderately well accounted for by the pseudopotential calculation. The overall agreement is a testimony to the care and accuracy of both tunneling and neutron experiments.

An important feature of pseudopotential calculations is the role of Umklapp scattering as first pointed out by Morel (1958). In the single OPW approximation, the matrix element $\langle \boldsymbol{k}+\boldsymbol{Q}|\nabla V|\boldsymbol{k}\rangle$ equals $\mathrm{i}\boldsymbol{Q}V(\boldsymbol{Q})$. For a pure transverse phonon, $\hat{\varepsilon}_Q\cdot\boldsymbol{Q}$ vanishes if $\boldsymbol{Q}$ is small enough to lie in the first Brillouin zone (BZ). When $\boldsymbol{Q}$ lies outside the first BZ (Umklapp process), $\hat{\varepsilon}_Q\cdot\boldsymbol{Q}$ can be written as $\hat{\varepsilon}_{Q'}\cdot(\boldsymbol{Q}'+G)$, where $\boldsymbol{Q}'$ is Q reduced to the first BZ. Transverse phonons are strongly suppressed in materials like alkalis for which $\boldsymbol{Q}$ is mostly inside the BZ. The maximum Q for a spherical surface is $2k_F$. Polyvalent metals have larger values of $2k_F$ relative to BZ dimensions, and thus an increasing fraction of the scattering events are Umklapps. This enables the transverse peak to have at least as much weight in α^2F as it has in F. This in turn is crucial to having a large value of λ or T_c. Unfortunately the values of T_c are never very large. Pb, with $T_c = 7.2$ K, seems to be fairly close to an upper limit for this class of materials.

The reason for an upper limit on T_c in sp-metals ought to be available since pseudopotentials give a fairly complete theoretical treatment. Unfortunately the situation isn't completely clear. One clue is the occurrence in the higher-T_c metals like Pb of anomalies in ω_Q which do not appear in closely related but lower-T_c metals. Fig. 20 shows neutron scattering work by Ng and Brockhouse (1968) comparing phonons in Pb with $Pb_{0.4}Tl_{0.6}$. The latter alloy has a reduced value $T_c = 4.6$ K, and a much more ordinary phonon dispersion. The theory of these phonons is less good than for most simple metals, as will be described in §4. Curiously, there is almost no hint of these anomalies in either the measured α^2F (Dynes and Rowell 1975) or in the calculated $F(\omega)$ (Allen and Dynes 1975b), indicating that the dips occupy a relatively small amount of phase space. It is commonly supposed that the dips in the Pb spectrum are a sign of incipient lattice instability. It is also commonly supposed that the occurrence of lattice instabilities is the primary factor limiting T_c. On the other hand, the softening of phonon frequencies associated with an incipient instability might raise T_c. The system of Pb-based alloys illustrates that the truth is sometimes simpler than we would like. On increasing the electron density by adding Bi, T_c continues to rise monotonically to a maximum of $\simeq 9$ K at $Pb_{0.65}Bi_{0.35}$. This concentration lies right at a phase boundary beyond which Bi is no longer soluble in Pb. Higher concentrations of Bi result in phase separation. Thus T_c is limited to values $\lesssim 9$ K by a first-order lattice instability. There is no evidence for a soft phonon, even though the dips in the spectrum deepen somewhat with the addition of Bi. There is also no evidence that the increase in T_c is especially closely related to the dips; rather the amplitude of α^2F or λ increases monotonically with electron to atom ratio, partly because of increased phase space for Umklapps. The interpretation of dips as *incipient* soft modes cannot be proved or disproved. However, dips in ω_Q in this system appear neither to cause nor to limit high T_c's, but rather to be separate effects caused (like the high T_c) by strong electron–phonon coupling. The first-order instability at $Pb_{0.65}Bi_{0.35}$ can be called a "covalent instability". Allen and Dynes (1975a) have suggested that there is a maximum value of η (approximately 2.4 eV/Å^2 in simple metals) beyond which a transition occurs to a covalent structure (as in Bi for example). Qualitative reasons for associating large η with covalent tendencies will be given in §3.3.1.

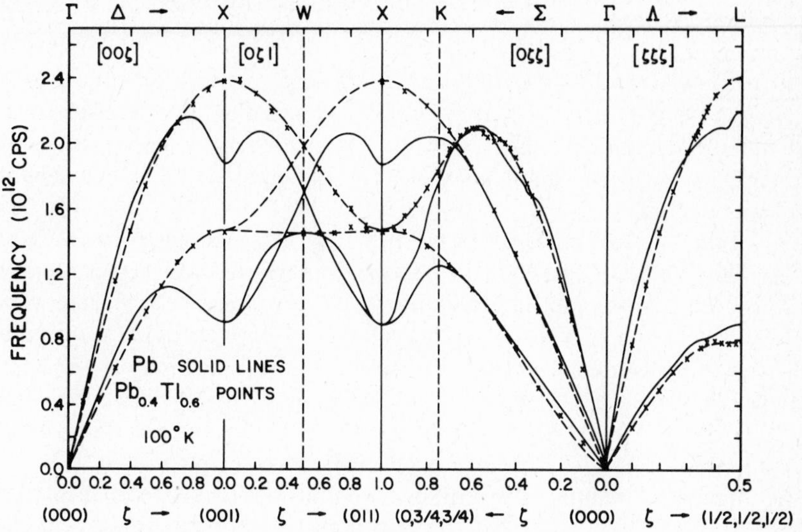

Fig. 20. Phonon dispersion curves in Pb (solid line) and of a f.c.c. alloy $Pb_{0.4}Tl_{0.6}$ (dashed line and data points).

3.3. d-band elements

The calculation of λ for d-band elements is much more difficult: there is the technical difficulty of performing the necessary four-dimensional sums with adequately precise electron and phonon eigenstates, and there is the fundamental difficulty of not being able to construct ∇V in any simple way. Various plausible models can be made for ∇V, but until very recently the technical difficulties prevented a calculation sufficiently precise to test these ideas. Therefore, the early history of this subject was primarily concerned with finding approximations and shortcuts which would give a qualitative picture without extraordinary labor on a computer. The tight-binding models of Barisić (1972), Bennemann and Garland (1972) and numerous other authors, and the early rigid-muffin-tin work of Gaspari and Gyorffy (1972) and Evans et al. (1973) were seeking such a shortcut. Unfortunately none of this early work has proved to be as accurate as is needed to get a true understanding. However, these papers played a crucial role in stimulating later, more detailed work.

3.3.1 Rigid muffin tins

The correct calculation of ∇V is a formidable problem, as will be shown in §4.2. One needs the change in the self-consistent one-electron potential (including exchange and correlation effects) when a single atom is moved. In simple metals there is a prescription for dividing up the total potential into a part (the pseudopotential) associated with each atom. Because this is a weak potential, a linear theory can be made which gives the result that the pseudopotential moves rigidly. In d-band metals, there is no known prescription which uniquely separates the potential into a part associated with each atom. However, for practical purposes of APW or KKR band calculations, the potential is forced into a muffin-tin configuration. It is a fairly obvious first guess that the part of the potential which moves rigidly with an atom is the muffin tin around that atom. Sinha (1968) has shown in detail how one may in principle (with great labor) proceed to calculate the further corrections. The "rigid-muffin-tin approximation" (RMTA) consists of ignoring these corrections. There is no known justification for this, but it is certainly possible that it gives a good estimate of $\langle k'|\nabla V|k\rangle$ even though it surely fails to give a totally correct picture for ∇V. The motivations for making this "approximation" (which should instead be called a "guess") are: (a) it will give the right order of magnitude and most of the right selection rules for $\langle k'|\nabla V|k\rangle$; (b) it avoids self-consistent calculations which are almost prohibitively difficult; and (c) the elastic matrix elements are especially simple and may be expressed in terms of phase shifts. This last point was first made by Sinha and by Golibersuch (1969), and further algebraic simplifications were made by Gaspari and Gyorffy (1972). The RMTA is sometimes ascribed to the latter authors, but it could as fairly be attributed to Sinha (1968), who has been its most persistent critic. The first complete calculations of λ by this method were by Nowak (1972) for Cu, and by Yamashita and Asano (1974) for Nb and Mo; Harmon and Sinha (1977) have recently reported the first full calculation of $\alpha^2F(\omega)$ (for Nb) by this method. The results for Cu were in good agreement with experiment, whereas Nb has so far always come out larger than experiment. However, recent calculations by Butler for Nb (to be discussed in §4.4) may restore faith in the RMTA for Nb.

Much more work has been done on calculating $\eta = N(0)\langle I^2\rangle$ than λ. McMillan (1968) stimulated interest in η by observing an unexpected lack of variation in η for the b.c.c. transition metals, which have values around 6 eV/Å^2. Hopfield (1969) gave the argument that the $\Delta l = \pm 1$ selection rule for matrix elements of a vector operator like ∇V could explain the slow variation of η. Subsequent work has shown that η does vary more than originally thought. Hopfield's arguments can be made rigorous when the wavefunction is expanded in spherical harmonics centered on the atom that moves. Confusion results if this is carelessly interpreted in conventional band-theory language, especially tight-binding language. A tight-binding d-wavefunction will only be strictly $l=2$ near the origin. Farther out, one encounters the tails of d-wavefunctions centered on *other* atoms, which can have any l-character about the origin. In the RMTA, the potential ∇V vanishes outside the muffin-tin radius, and inside there is an unambiguous assignment of l-character. This was exploited by Gaspari and Gyorffy (GG) (1972) who showed how to make a rigorous and fairly simple calculation of η in the RMTA. The GG formula is (in a form given by Pettifor 1977)

$$\frac{\eta}{N(0)} = \langle I^2 \rangle = \sum_l \frac{(2l+2)}{(2l+1)(2l+3)} M_{l,l+1}^2 f_l f_{l+1} + \text{corrections}, \tag{3.9}$$

$$M_{l,l+1} = \int_0^{\bar{r}} R_l(\partial V/\partial r) R_{l+1} r^2 \, \mathrm{d}r \Big/ \left[\int_0^{\bar{r}} R_l^2 r^2 \, \mathrm{d}r \int_0^{\bar{r}} R_{l+1}^2 r^2 \, \mathrm{d}r \right]^{1/2}, \tag{3.10}$$

where f_l is the fraction of the density of states $N(0)$ which has l-character, R_l is the radial solution of the Schrödinger equation at the Fermi energy, and $\bar{r}$ is the muffin-tin radius. The corrections to (3.9) vanish insofar as the crystal point group symmetry is approximately spherical. For cubic systems (John 1973) this is exact though $l=2$, and the correction terms in the 4d series have in fact been found small (Butler et al. 1976, Butler 1977).

The early calculations using the Gaspari–Gyorffy formalism involved simple estimates of the partial densities of states. The numerator of (3.10) can be rigorously set equal to $\sin(\delta_l - \delta_{l+1})$, where δ_l is the phase shift at the Fermi energy, provided R_l is properly normalized. Experience has shown that it is surprisingly difficult to make simple estimates of the terms in (3.9), the most recent and best effort being Pettifor's. However, direct calculation of the terms in (3.9) from band theory is not too difficult, as was demonstrated by Klein and Papaconstantapoulos (1976) and Papaconstantopoulas and Klein (1975). The most complete calculations for d-band elements, by Butler (1977), show good agreement with experiment, and give considerable insight into the trends. Contrary to earlier wisdom dating from Hopfield's (1969) paper, the $l=2$ to 3 scattering terms dominate (3.9), rather than $l=1$ to 2. Butler finds $\langle I^2 \rangle$ proportional to $f_2 f_3 / V^2$ for the 4d series, where V is the volume per atom. Both ends of the 4d series have small values of f_3 which yields small λ in spite of large $N(0)$, and accounts for the surprising absence of superconductivity. Papaconstantopoulos et al. (1977) have done the 3d series as well as 4d series. Their conclusions are similar to Butler's. However, they believe (Boyer et al. 1977) that f_3 is smaller by a factor of two when non-muffin-tin effects are included. There seems to be no intrinsic reason why elements have values of η seldom exceeding 8 eV/Å^2; values at least twice as large could in principal be obtained if the density of states and phase shifts were favorable. In fact, Butler (1977) found that technetium would have $\eta \simeq 11$ eV/Å^2 if the crystal structure were b.c.c. instead of h.c.p. Nature appears to put extrinsic restraints on the magnitude of η. Several closely related explanations have been offered. Gomersall and Gyorffy (1974b) have argued semiquantitatively that the average lattice stiffness, $M\overline{\omega^2}$, should behave as $M\overline{\Omega_0^2} - \frac{4}{5}\varepsilon_F N(0)\eta$, where $\overline{\Omega_0^2}$ is approximately the same for all transition metals, and ε_F is measured from the bottom of the valence bands. Thus a large value of η will eventually cause a second-order lattice instability with a vanishing phonon frequency at some Q. In practice, first-order instabilities usually occur first. The positions of phase boundaries of the body-centered d-band alloys are nicely predicted by this model. A more qualitative argument, given by Allen and Dynes (1975a, b) is that large η is favorable for a covalent instability. According to (3.9), large η requires large $f_l f_{l+1}$ for some l ($l=2$ for 4d elements apparently.) The existence of large fractional densities of states for two different angular symmetries at the same energy means that mixed-l directional orbitals can be formed with a minimum of promotion energy. The material can be expected to transform to a more open crystal structure where better advantage can be taken of the covalency. This in turn will surely reduce $N(0)$, η, λ and T_c. This was phrased somewhat differently by Butler (1977) who observed that large $f_2 f_3$ occurs when d-orbitals are delocalized. The size of f_3 is related to the importance of tails of d-orbitals centered on other atoms. This implies strong d-bonding and usually reduced $N(0)$.

3.3.2. *Tight binding*

Calculations of matrix elements $M_{kk'}$ from tight-binding formalism are numerous. The first model calculations of λ by this procedure were by Barisić (1972) and Bennemann and Garland (1972). Unfortunately the procedures are less clear-cut than in RMTA, and conflicting schemes have been suggested. A further element of confusion is that calculations have seldom been performed with true wavefunctions. The work of Das (1973) in Cu is a notable exception and represents one of the first complete calculations of λ. Among the models which avoid using true wavefunctions and Fermi surface averages, perhaps the best is by Birnboim (1976). Peter et al. (1977) have done a very complete tight-binding calculation for Nb, which begins with a six-function Slater–Koster fit to the energy bands of Mattheiss (1970). The matrix elements are evaluated in rigid-ion approximation. A complete treatment of anisotropy effects is given. The resulting coupling constants are too large by 30%. The anisotropy enhancement of T_c is found to be 0.24 K, consistent with experimental observation that anisotropy is small. The interpolated energy bands give a Fermi surface somewhat different in shape from the one of Mattheiss (1970) which has been fairly well confirmed by de Haas–van Alphen experiments (Karim et al. 1978).

A significant innovation has been made by Varma et al. (1979) who use an accurate non-orthogonal tight-binding parameterization of the energy bands of Nb and a method for $M_{kk'}$ which allows wavefunctions to distort while the atoms move. They have calculated $\langle I^2 \rangle$ in rigid-band approximations for b.c.c. alloys near Nb, and obtain good agreement with experiment. It may be (Varma and Dynes 1976) that non-orthogonal tight binding gives a more natural and accurate method for understanding electron–phonon coupling than the more common orthogonal method. The consequences of abandoning Bloch's method and allowing wavefunctions to distort needs further study. Perhaps the most significant aspect is that phonon dynamics have also been calculated by this method (Varma and Weber (1977); see §4.3.)

3.4. Compounds and alloys

Progress in microscopic theory of λ in compounds is hampered by two factors: band theory is less well developed, and lattice dynamics is more complicated and less well known. There are no complete calculations for d-band compounds yet. What does exist are calculations of η_a for the various atoms in certain compounds, mostly by Klein et al. (1976), with a few calculations by other groups. There is uncertainty about how to estimate λ, given η_a. Eq. (3.5) is rigorous, but there is no confidence about how to calculate $\langle \omega^2 \rangle$ because $\alpha^2 F(\omega)$ is known only in very few d-band compounds (see §5.3). Klein and Papaconstantopoulos (1976) have argued that for a diatomic binary with a large mass ratio, λ can be approximated by

$$\lambda \simeq \frac{\eta_{\mathrm{L}}}{M_{\mathrm{L}}\,\omega_{\mathrm{opt}}^2} + \frac{\eta_{\mathrm{H}}}{M_{\mathrm{H}}\,\omega_{\mathrm{ac}}^2}, \tag{3.11}$$

where the light (L) and heavy (H) atoms couple selectively through optical and acoustic phonons. This formula is in some doubt (Gomersall and Gyorffy 1974a, Rietschel 1975). The problem is that eq. (3.5) seems to suggest the parameter η_a / M_a as a measure of the importance of atom species "a" in causing T_c. This over emphasizes the importance of small M_a. If M_a is decreased with constant η_a (as in going from PdD to PdH), there is a corresponding increase in $\langle \omega^2 \rangle$ which cancels exactly, giving λ unaffected, in harmonic approximation. The version (3.11) explicitly accounts for this effect but not in a rigorous way. In the absence of calculations of α^2F or λ, I feel the safest procedure is simply to regard η_a as a reasonable measure of the importance of species "a". Klein et al. (1976) find for NbC and TaC (rocksalt structure, T_c around 10 K) that the transition element has $\eta \simeq 5$ eV/Å^2 while the C atom has $\eta \simeq 3$ eV/Å^2. Thus the carbon atom is significantly coupled. The high optical mode frequencies should help enhance $\omega_{\log}$ (eqs. (2.34), (2.37)), giving a high T_c without a particularly large λ. There is little in the way of experimental confirmation or denial of this important conclusion, except that the phonon anomalies seen by neutron scattering (Smith and Glaser 1970) occur in acoustic more than optic branches, which suggests the unimportance of optic branches. Calculations for V_3Si showed the opposite conclusion, namely that η for Si is negligible compared to 9.5 eV/Å^2 for the sum of the three V atoms. This is consistent with the common conception that the d-atom chains of the A15 structure are the T_c-determining feature.

Perhaps the most interesting compound yet studied is PdH_x which is superconducting at temperatures up to 8 K (Skoskiewicz 1972, Schirber and Northrup 1974) as x is increased to 1.0. A particularly important feature is that the substitution of D for H gives an increase (anomalous isotope effect) of the maximum T_c up to 9.8 K (Miller and Satterthwaite 1975). Alloying Pd with noble metals causes a further increase up to 16.6 K (Stritzker 1974). This is the only material for which there is good evidence of the importance of optic modes for T_c. The best evidence is from neutron scattering (Rowe et al. 1974, Rahman et al. 1976) which reveals (a) that optic modes are anomalously low in frequency, and (b) that PdD_x has optic mode frequencies still further reduced below the values expected by scaling PdH_x with force constants fixed. Result (a) strongly suggests that large electron–phonon coupling has driven down the optic mode frequencies and thus that T_c should be affected by optic modes. This observation, plus result (b), gives a very satisfying explanation of the anomalous isotope effect, which was anticipated theoretically by Ganguly (1973). The presumed cause of effect (b) is that the H atom has such a large zero-point motion that it explores outer regions of its potential well which rise more rapidly than quadratic, giving an upward anharmonic renormalization of ω_Q relative to the D atom which has smaller zero-point motions. There is reason to believe (Hui and Allen 1974) that the effect of the anharmonic phonons on T_c is adequately accounted for by putting the observed renormalized phonon frequency into the standard harmonic theories for T_c. Similar conclusions about the low optic phonon frequency and large coupling to electrons have been derived by other experiments. Superconducting tunneling (Eichler et al. 1975, Silverman and Briscoe 1975) is in principle the most powerful method, and has confirmed the neutron results, but the various experiments are not consistent with each other. The experiment of Dynes and Garno (1975) purported to see a very large coupling to the optic branch, but has never been properly published. The temperature dependence of electrical resistivity (MacLaughlan et al. 1975) has been measured and interpreted as showing strong effects due to scattering by the optic phonons.

Several theoretical formulations have been made (Burger and MacLaughlan 1976, Ganguly 1976). The calculations of Papaconstantopoulos et al. (1978) yield values of η_{Pd} and $\eta_{\mathrm{H(D)}}$ based on self-consistent APW band theory. They find $\eta = 0.9\ \mathrm{eV/\AA^2}$ for Pd and 0.4 for D. When combined with the small value of $M\omega_{\mathrm{opt}}^2$ in eq. (3.11), the large T_c (and its variation with isotope and concentration) is given a satisfactory explanation in terms of a large contribution from optic modes.

One significant aspect of the work of Papaconstantopoulos et al. (1978) on PdH_x is that they are able to take some account of the disorder due to substoichiometry, by using the CPA calculation of Faulkner (1976) for the electron states of PdH_x. There is still no theory or calculation of λ or α^2F bold enough to face the full difficulty presented by disorder, whether substitutional or positional. Until now the motivation hasn't been high since the highest T_c's are found in ordered materials. However, there is much current interest in the degradation of T_c by disorder. Also, Nb_3Ge films with $T_c \approx 23\,\mathrm{K}$ have significant inherent disorder. The existence of tunneling data for many disordered metals should make this area an excellent testing ground for theory, when it becomes capable of dealing with this degree of complexity. So far most theories have not attempted to deal with the lattice dynamics of disordered metals. An exception is the work of Grünewald and Scharnberg (1976). By making a CPA-type theory they are able to reproduce the experimentally observed shift of spectral weight to lower frequencies in disordered metals.

4. *Theory of phonons*

The quantitative microscopic theory of phonons in d-band metals is still in its infancy. The need for such a theory in order to understand T_c has encouraged the rapid development of the subject. A detailed review is given by Sinha in this volume. Excellent reviews on related topics are also available, for example the article by Brovman and Kagan (1974) on simple metals, by Sham (1974) on covalent materials, and by Bilz, et al. (1974) on ionic materials in vol. 1 of this series. At the time of this writing, rapid progress is occurring, and the first quantitative calculations are starting to appear. This section gives some background which particularly pertains to T_c, and lists the principal difficulties which are perhaps being overcome even now.

4.1. Basic concepts

In harmonic approximation, the basic microscopic concept is the force constant matrix, $K_{\alpha\beta}(l-l')$, whose negative gives the force in the α direction on the atom at $\boldsymbol{l}$, per unit displacement in the β direction of the atom at $\boldsymbol{l}'$, *all other atoms being held fixed at their equilibrium positions*. For convenience in this section we exhibit formulas only for the case of a single atom species, one atom per unit cell. Newton's law gives the acceleration of atom $\boldsymbol{l}$ in response to the displacements of all other atoms:

$$M\ddot{u}_{l\alpha}(t) = -\sum_{l'}\int_{-\infty}^{t} \mathrm{d}t'\, K_{\alpha\beta}(ll', t-t')u_{l'\beta}(t'), \tag{4.1a}$$

$$M\omega^2 u_{l\alpha} = \sum_{l'} K_{\alpha\beta}(ll',\omega)u_{l'\beta}. \tag{4.1b}$$

The second line is just the time Fourier transform of the first. Explicit allowance is made for the fact that the force on atom l is in principle slightly retarded, and depends on the positions of other atoms l' not precisely at time t, but rather at slightly earlier times. In practice it is usually sufficient to assume an instantaneous interaction (adiabatic approximation, $\mathbf{K}(ll',t-t') = \mathbf{K}(ll')\delta(t-t')$), which implies $K_{\alpha\beta}(ll'\omega)$ is independent of frequency and purely real. However, in this section we shall find it convenient to recognize that there is a weak frequency dependence, and a corresponding small imaginary part, to $\mathbf{K}$. The phonon Green's function $D_{\alpha\beta}(ll',\omega)$ is defined by

$$\sum_{l'}\left[M\omega^2\delta_{ll'}\delta_{\alpha\beta} - K_{\alpha\beta}(ll',\omega)\right]D_{\beta\gamma}(l'l'',\omega) = \delta_{ll''}\delta_{\alpha\gamma}. \tag{4.2}$$

These relations are valid in disordered as well as ordered solids. The equations are readily solved by Fourier transforming from $\boldsymbol{l}$ to $\boldsymbol{Q}$, but only if the system is a perfectly ordered crystal (where $\mathbf{K}$ depends only on $\boldsymbol{l}-\boldsymbol{l}'$ rather than l and l' individually.) In the ordered case we have

$$K_{\alpha\beta}(Q\omega) = \sum_{l'} K_{\alpha\beta}(l-l',\omega)\,\mathrm{e}^{-\mathrm{i}Q\cdot(l-l')}, \tag{4.3a}$$

$$u_{Q\alpha} = N^{-1/2}\sum_{l} u_{l\alpha}\,\mathrm{e}^{\mathrm{i}Q\cdot l}. \tag{4.3b}$$

Eq. (4.1b) then becomes

$$\left[K_{\alpha\beta}(Q\omega) - M\omega^2\delta_{\alpha\beta}\right]u_{Q\beta} = 0. \tag{4.4}$$

We shall see later that in the frequency range of interest (i.e. $\omega \lesssim \omega_D$), the real part of **K** is effectively independent of ω, and the imaginary part is linear in ω:

$$K_{\alpha\beta}(ll'\omega) = K_{\alpha\beta}(ll') - 2i\omega M\Gamma_{\alpha\beta}(ll'), \tag{4.5a}$$

$$K_{\alpha\beta}(Q,\omega) = K_{\alpha\beta}(Q) - 2i\omega M\Gamma_{\alpha\beta}(Q), \tag{4.5b}$$

where the lower line is the result for a perfect crystal. The eigenvalues $M\omega_{Q\nu}^2$ of the real part, $\mathbf{K}(Q)$, give the phonon frequencies, and the normalized eigenvectors, denoted $\hat{\varepsilon}_{Q\nu}$, are the polarization vectors. The Green's function of the perfect crystal can now be written explicitly by transforming (4.2) to $\boldsymbol{Q}$ space:

$$\left[M\omega^2\mathbf{1} - \mathbf{K}(Q) + 2Mi\omega\boldsymbol{\Gamma}(Q)\right]\mathbf{D}(Q,\omega) = \mathbf{1}. \tag{4.2b}$$

Finally, ignoring the small off-diagonal elements of Γ in the frame of the polarization vectors $\hat{\varepsilon}_{Qv}$, the solution is

$$M\mathbf{D}(Q\omega) = \sum_{\nu} \hat{\varepsilon}_{Q\nu}\left[\omega^2 - \omega_{Q\nu}^2 + 2i\omega\gamma_{Q\nu}\right]^{-1}\hat{\varepsilon}_{Q\nu}, \tag{4.6}$$

$$\gamma_{Q\nu} = \hat{\varepsilon}_{Q\nu}\cdot\boldsymbol{\Gamma}(Q)\cdot\hat{\varepsilon}_{Q\nu}. \tag{4.7}$$

Near the resonance ($\omega \approx \omega_{Q\nu}$), the Green's function (4.6) has the form $(\omega - \omega_{Q\nu} + i\gamma_{Q\nu})^{-1}$. Here $\gamma_{Q\nu}$ clearly is the intrinsic half-width at half-maximum of the experimental phonon lineshape. We will come back to this in §4.3 because of its interesting significance for superconductivity.

4.2. Electronic contribution

In this section the microscopic theory of force constants is worked out. The basic concepts are all standard and date back at least to a paper by Baym (1961). A particularly simple form of this theory emerges when a nearly-free-electron approximation is made (Toya 1958, Vosko et al. 1965). This theory is reviewed by Brovman and Kagan (1974). Very good results are obtained for alkali metals, but as the valency increases, and metals become more interesting from a superconducting point of view, the theory gets worse. The importance of correcting the theory for the higher-order effects of the lattice potential has been emphasized especially by Bertoni et al. (1974). For d-band metals it seems clear that the theory must make a clean break away from nearly-free-electron procedures. The language of nearly-free-electron theory may obscure more than it illuminates for d-band metals, and for this reason nothing more will be said here on the subject.

It is convenient to divide atoms into cores of positive charge Ze (assumed to move rigidly with the displacement $\boldsymbol{u}_l$) and valence electrons which respond in a more complicated way. For example, Nb would have $Z=5$; the core having the atomic configuration of Xe. Then the force constant can be written as two terms,

$$K_{\alpha\beta}(l-l')=K^{\mathrm{c}}_{\alpha\beta}(\boldsymbol{l}-\boldsymbol{l}')+K^{\mathrm{v}}_{\alpha\beta}(l-l'), \tag{4.8}$$

the first being the (negative) force on atom l due to the motion of the core $\boldsymbol{l}'$ in the β direction, and the second being an "indirect" force on atom $\boldsymbol{l}$ due to the valence electron charge density distortions. We will denote by $\delta\rho_\beta(\boldsymbol{r}-\boldsymbol{l}')$ the valence charge distortion which arises in response to the motion of the $\boldsymbol{l}'$ atom a unit distance in the $\boldsymbol{\beta}$ direction:

$$K^{\mathrm{c}}_{\alpha\beta}=\frac{\partial^2}{\partial R_{l\alpha}\partial R_{l'\beta}}U^{\mathrm{c}}, \tag{4.9}$$

$$K^{v}_{\alpha\beta}=\frac{\partial}{\partial R_{l\alpha}}\int \mathrm{d}r\, V^{\mathrm{b}}(\boldsymbol{r}-R_l)\delta\rho_\beta(\boldsymbol{r}-\boldsymbol{l}'). \tag{4.10}$$

Core–core interactions are given by a potential function U^{c} which is a sum of pair potentials, consisting of Coulomb force $V_{\mathrm{c}}=(Ze)^2/|\boldsymbol{R}_l-\boldsymbol{R}_{l'}|$, and possibly a short-range exchange repulsion when core electrons overlap. The potential V^{b} in (4.10) is the bare electron–core interaction, which equals $-Ze^2/|\boldsymbol{r}-\boldsymbol{R}_l|$ at long distances, but is more complicated when $\boldsymbol{r}$ is close to $\boldsymbol{R}_l$.

The central difficulty of all theories of phonons is calculating the charge density distortion $\delta\rho_\beta(\boldsymbol{r}-\boldsymbol{l}')$. In principle this can be directly calculated by solving the Schrödinger equation with the atom at $\boldsymbol{l}'$ displaced in the β direction. This calculation is very difficult because translational symmetry is lost. The closest approach to such a calculation is probably the recent work of Balderschi and Mashke (1976), who calculate $\delta\rho$ for a Si lattice with a frozen-in optic phonon. This calculation is simplified by retaining translational symmetry, so it does not permit the computation of the general force constant. A related approach, the direct calculations of energy shifts of lattices with frozen-in phonons, has recently been done by Gale and Pettifer (1977). The result for NbC are encouragingly good considering that the calculation was not self-consistent and ignored the s–p bands. One reason why direct calculation of $\delta\rho$ may be especially difficult in metals is that there are long-range oscillatory components going as $\cos(q^*\cdot(r-l'))/|\boldsymbol{r}-\boldsymbol{l}'|^3$. These Friedal oscillations are caused by extremal dimensions q^* of the Fermi surface, and give rise to long-range force constants scaling as $\cos(q^*\cdot(\boldsymbol{l}-\boldsymbol{l}))/|l-l'|^3$, which in turn cause Kohn anomalies at wavevector q^* in the phonon dispersion. d-band metals often have particularly strong Kohn anomalies (e.g. Cr, Mo, W near the H point in the [100] direction). It is not clearly understood how closely the phonon anomalies shown in §1 relate to Kohn anomalies. The Friedel tail of $\delta\rho$ might be important not just for the singular behavior at Kohn anomalies, but also for the anomalous dips in ω_Q.

A formal solution for $\delta\rho$ is easy to construct using linear response theory:

$$\delta\rho_\beta(r-l')=\int \mathrm{d}r'\,\chi(r,r',\omega)\partial V^{\mathrm{b}}(r'-R_{l'})/\partial R_{l'\beta}, \tag{4.11}$$

$$\chi(r,r'\omega)=-\frac{\mathrm{i}}{\hbar}\int_{-\infty}^{t}\mathrm{d}t'\,\mathrm{e}^{-\mathrm{i}\omega t'}\langle[\rho_{\mathrm{op}}(rt),\rho_{\mathrm{op}}(r't')]\rangle. \tag{4.12}$$

Here again, V^{b} is the bare potential and $\rho_{\mathrm{op}}(rt)$ the Heisenberg operator $\exp(\mathrm{i}Ht)\rho_{\mathrm{op}}(r)\exp(-\mathrm{i}Ht)$, where $\rho_{\mathrm{op}}(r)$ is the electron density operator $\Sigma_i\delta(r_i-r)$. Balderschi and Maschke (1976) have shown in Si that $\delta\rho$ calculated using (4.11) (with ω set to 0) agrees extremely well with their direct calculation of $\delta\rho$ using the Schrödinger equation. The calculation of χ for d-band elements currently seems more challenging. The frequency ω in (4.11–4.12) allows for retardation as described by (4.1). The Coulomb force K^{c} is for all purposes instantaneous. Much of the electronic response is also driven by instantaneous long-range Coulomb forces. However, charge conservation requires that currents flow, and these propagate at the Fermi velocity which is not quite infinitely fast compared to atomic velocities. The principal result of the non-zero electronic response time is that a small out-of-phase response exists, the phase lag ϕ of χ going as $\tan^{-1}(|\boldsymbol{r}-\boldsymbol{r}'|\omega/v_{\mathrm{F}})$. For atomic distances and phonon frequencies, this angle is small, of order $(m/M)^{1/2}$. Nevertheless, the existence of a small, long-range imaginary part (i.e. out-of-phase part) of χ, vanishing linearly with ω, has interesting consequences. As is usual in linear response theory, the out-of-phase response is dissipative – it gives the decay rate $\gamma_{Q\nu}$ (eq. 4.7) of phonons into electron–hole pairs. This decay rate is small: the ratio $1/\omega_Q\tau_Q$ is of order $(m/M)^{1/2}$, and for most purposes the phonons can be considered infinitely long-lived, or infinitely sharp in frequency.

The valence electron part of the force constant can now be written using (4.10–4.12):

$$K^{\mathrm{v}}_{\alpha\beta}(\boldsymbol{l}-\boldsymbol{l}')=\int \mathrm{d}r\,\mathrm{d}r'\,\underset{\alpha}{\nabla}V^{\mathrm{b}}(r-l)\chi(r,r')\nabla_\beta V^{\mathrm{b}}(r'-l'). \tag{4.13}$$

The result is closely parallel to the formula (2.22) for the Eliashberg interaction. The differences are: (a) bare potentials enter (4.13) while screened potentials occur in (2.22); and (b) in the force constant expression, it is the electron density propagator χ which propagates the signal from atom $\boldsymbol{l}'$ to $\boldsymbol{l}$, while in the Eliashberg interaction this role is played by the lattice displacement propagator **D**. Let us now consider the relation between the bare (V^{b}) and screened (V) electron–ion potential.

By the screened potential ∇V, we mean the total change in the self-consistent crystal potential felt by an electron when an "external" potential ∇V_{b} is applied. Both of these quantities are multiplied by the displacement $\boldsymbol{u}$ which we can assume infinitesimal, in line with the harmonic approximation. Thus we can linearize the problem and find a linear relation between ∇V and ∇V_{b}, namely

$$\nabla V(\boldsymbol{r}-\boldsymbol{l})=\int \mathrm{d}\boldsymbol{r}'\,\varepsilon^{-1}(\boldsymbol{r},\boldsymbol{r}')\nabla V_{\mathrm{b}}(\boldsymbol{r}'\boldsymbol{l}), \tag{4.14}$$

where ε is the dielectric function of the true metal, which cannot be approximated as a free electron gas.

Note that relation (4.14) is *not* true if the gradient operators are removed! The true self-consistent crystal potential V and the core potential V_{b} are not linearly related to each other because in general they are not small. The electronic response to V_{b} is not linear in V_{b}. (For simple metals it is common to assume that the bare pseudopotential is weak enough to use linear approximations.) Also, note that the dielectric function ε^{-1} in (4.14) differs from the more common usage in that ∇V is *not* the potential which a classical test charge would measure, but also contains an "exchange potential" because the "test charge" is an electron. The ε used here has been called the "proton–electron" dielectric function by Ballentine (1967) and is commonly called the "electron" dielectric function (Kugler 1975). It should also be noted that the notation ∇V is used to denote a vector field which is in general not entirely longitudinal, i.e. in spite of the notation ∇V is not necessarily the gradient of a scalar potential, unlike ∇V^{b} (Ball 1975).

To make things more explicit, the screened potential ∇V is

$$\nabla_\alpha V=\nabla_\alpha V^{\mathrm{b}}+\int \mathrm{d}r'\left[v(r-r')+v_{\mathrm{xc}}(r,r')\right]\delta\rho_\alpha(\boldsymbol{r}'-\boldsymbol{l}), \tag{4.15}$$

where $v(r-r')$ is the Coulomb interaction $e^2/|r-r'|$ and v_{xc} is an effective exchange and correlation interaction $\delta^2E_{\mathrm{xc}}/\delta\rho(r)\delta\rho(r')$ (Sham 1973). The second term gives the Coulomb field of the induced charge density. The third term is absent if ∇V is measured by a test charge, but since ∇V is measured by an electron it takes into account the depletion of charge density near the test electron at r due to exchange and correlation. The first functional derivative $\delta E_{xc}/\delta\rho(r)$ gives the exchange-correlation potential which occurs in the one-electron Schrödinger equation. Frequently this is taken to be a local function like $C\rho(r)^{1/3}$. Then v_{xc} can be written as $\frac{1}{3}C\rho(r)^{-2/3}\delta(r-r')$ (Singhal and Callaway 1976). Now we can combine (4.11) and (4.15) to get an expression for ε^{-1} as defined in (4.14):

$$\varepsilon^{-1}(r,r') = \delta(r-r') + \int \mathrm{d}r''[v(r-r'') + v_{xc}(r,r'')]\chi(r'',r'). \qquad (4.16)$$

Although χ is explicitly given by the expression (4.12), in practice this can not be evaluated for an interacting system. For a non-interacting system of electrons, e.g. (4.12) readily leads to the following expression in terms of wavefunctions $\psi_k(r)$, energies ε_k and Fermi factors f_k,

$$\chi_0(rr'\omega) = \sum_{kk'} \psi_k^{*\prime}(r')\psi_k(r')\psi_k^*(r)\psi_k'(r),$$
$$\times (f_k' - f_k)/(\varepsilon_k' - \varepsilon_k + \omega + is), \qquad (4.17a)$$

where s is a positive infinitesimal. This can be laboriously evaluated if the band structure is known. We now make the mean field hypothesis, namely that electrons respond to the bare potential ∇V^{b} in the same way that non-interacting electrons respond to the screened potential, ∇V. This allows us to replace (4.11) by

$$\delta\rho_\alpha(r-l) = \int \mathrm{d}r' \chi_0(rr'\omega)\nabla_\alpha V(r'-l). \qquad (4.18)$$

Finally we can combine (4.15) and (4.18) to get an explicit expression, not for ε^{-1} as defined in (4.14), but its inverse, namely $\varepsilon(r,r')$:

$$\varepsilon(r,r') = \delta(r-r') - \int \mathrm{d}r''[v(r-r'') + v_{xc}(r,r'')]\chi_0(r''r'). \qquad (4.19)$$

It is convenient to use a shorthand notation

$$\varepsilon = 1 - (v + v_{xc})\chi_0 \qquad (4.19a)$$

which is understood to mean the same as (4.19), i.e. 1 means the unit matrix in coordinate space, which has elements $\delta(r-r')$, while products of $\boldsymbol{r}$-matrices are constructed by doing spatial integrals.

It should now be clear why calculating the screened interaction (4.14) is so difficult. Not only is χ_0 tedious to evaluate, but also ε^{-1} involves the inversion of a densely-dimensional $\boldsymbol{r}$-matrix (4.19). Procedures for doing the inversion have been described (Sinha et al. 1971, Pick 1971, Hanke 1973, Sham 1974.) There are only just beginning to appear calculations of ε^{-1} for d-band metals, and these are so far extremely difficult, and of questionable convergence. There is a preliminary report of a solution of an integral equation for $\delta\rho_\alpha(r-l)$ by Winter (1977) using a cluster method on Th. Knowledge of $\delta\rho$ is, for our purposes, equivalent to knowing χ or ε^{-1}. It is also clear why quantitative theory of phonons in d-band metals is still so difficult. We need to know χ to get the force constant K^{v}, and χ is as difficult as ε^{-1} to calculate. Furthermore, χ needs to be known to high accuracy because considerable cancellation occurs between K^{v} and K^{c}; the result, K, can be an order of magnitude smaller than K^{v}.

Some recent progress has occurred which reduces the difficulty caused by the cancellation. First note that, in the symbolic notation of (4.19a), χ can be written as

$$\chi=\chi_0\varepsilon^{-1}=\chi_0(1-\bar{v}\chi_0)^{-1}, \tag{4.20}$$

where $\bar{v}$ is short for $v+v_{\mathrm{xc}}$. The first of these relations follows from (4.11, 4.14 and 4.18). Now multiply by 1 on the left

$$\chi=(\varepsilon^{\dagger})^{-1}(1-\chi_0^{*}\bar{v})\chi_0\varepsilon^{-1}, \tag{4.21}$$

where $(1-\chi_0^{*}\bar{v})$ is $\varepsilon^{\dagger}$, the hermitean conjugate of ε. This last expression separates into two terms, each of which is amenable to some interpretation:

$$\chi=(\varepsilon^{\dagger})^{-1}\chi_0\varepsilon^{-1}-\chi^{\dagger}\bar{v}\chi. \tag{4.22}$$

If we now insert this into eq. (4.13) for the force constant, the result is

$$\begin{aligned}K^{\mathrm{v}}_{\alpha\beta}(l,l')=&\int \mathrm{d}r\,\mathrm{d}r'\,\nabla_\alpha V(r-l)^{*}\chi_0(rr'\omega)\nabla_\beta V(r-l')\\&-\int \mathrm{d}r\,\mathrm{d}r'\,\delta\rho_\alpha(r-l)^{*}\left[\frac{e^2}{|r-r'|}+v_{\mathrm{xc}}(r,r')\right]\delta\rho_\beta(r'-l').\end{aligned} \tag{4.23}$$

Here, eq. (4.14) has been used to equate $\varepsilon^{-1}\nabla V^{\mathrm{b}}$ with ∇V, and eq. (4.11) to equate $\chi\nabla V^{\mathrm{b}}$ with $\delta\rho$. The first term of (4.22) now has screened potentials and the non-interacting susceptibility. The second term looks like the Coulomb plus exchange interaction energy of two localized charge distortions at atoms $\boldsymbol{l}$ and $\boldsymbol{l}'$. It was argued by Pickett and Gyorffy (1976) that the second term of (4.22) should cancel the long-range part of $K^{c}_{\alpha\beta}$, yielding a net short-range force constant. We prove this using some arguments due to Ball (1975). The charge density displacement $\delta\rho_\alpha(r-l)$ is a vector field and can be written as the gradient of a scalar field $\bar{\rho}$ plus the curl of a vector field B:

$$\delta\rho_\alpha(\boldsymbol{r}-\boldsymbol{l})=\nabla_\alpha\bar{\rho}(\boldsymbol{r}-\boldsymbol{l})+[\nabla\times\boldsymbol{B}(\boldsymbol{r}-\boldsymbol{l})]_\alpha. \tag{4.24}$$

We can choose $\bar{\rho}$ to vanish as $|r-l|\to\infty$, and $\boldsymbol{B}$ to be pure transverse. Translational invariance says that when the whole crystal is moved, the change in valence charge density is a rigid translation of the total valence charge $\rho(r)$,

$$\sum_l \delta\rho_\alpha(r-l)=\nabla_\alpha\rho(r). \tag{4.25}$$

Since the decomposition (4.24) of $\delta\rho_\alpha$ into longitudinal and transverse parts is unique, and since longitudinal and transverse fields retain their character under superposition, we can use (4.24) and (4.25) to write

$$\sum_l \bar{\rho}(r-l)=\rho(r), \tag{4.25a}$$

$$\sum_l \boldsymbol{B}(\boldsymbol{r}-\boldsymbol{l})=0. \tag{4.25b}$$

These relations then imply

$$\int \mathrm{d}r\,\rho(r-l)=Z, \tag{4.26a}$$

$$\int \mathrm{d}r\,\boldsymbol{B}(r-l)=0. \tag{4.26b}$$

These enable us to evaluate the second term in (4.23) when $|\boldsymbol{l}-\boldsymbol{l}'|$ is long compared to the range of $\delta\rho_\alpha(\boldsymbol{r}-\boldsymbol{l})$. First it is clear that v_{xc} is short-range and ignorable for large $|\boldsymbol{l}-\boldsymbol{l}'|$. Consider the part of $\delta\rho_\alpha$ which is pure longitudinal, i.e. written as $\nabla_\alpha\bar{\rho}$. We can then integrate the second term in (4.23) by parts to give

$$\int \mathrm{d}r\,\mathrm{d}r'\bar{\rho}(r-l)\left[\frac{\partial^2}{\partial r_\alpha \partial r'_\beta}\frac{e^2}{|r-r'|}\right]\bar{\rho}(r'-l').$$

For large $\boldsymbol{l}-\boldsymbol{l}'$, using (4.19a), this becomes approximately

$$\frac{\partial^2}{\partial r_\alpha \partial r_\beta} \frac{(Ze)^2}{|r-r'|}\bigg|_{r=l,\, r'=l'},$$

which exactly cancels the long-range part of K^c. This cancellation is *not* just a trivial result of charge neutrality, because core–core and charge–charge interactions are both repulsive and add with the *same* sign. The opposite sign in (4.23) represents the correction term for electron–electron interactions which are overcounted in the first term of (4.23). This first term in the band-structure energy, and apparently contains the dominant long-range interactions which give rise to phonon anomalies. It is no longer necessary to calculate the first term to extremely high accuracy, as there is nothing left (for larger $|l-l'|$) to cause cancellation. Thus the first term can perhaps be calculated by a careful calculation of χ_0 and a carefully educated guess about ∇V. This program has very recently been carried out by Varma and Weber (1977), using the non-orthogonal tight-binding scheme for ∇V already described in §3.3.2. The results for Nb, Mo and $Mb_{1-x}Mo_x$ alloys seem astonishingly good. The occurrence of anomalies in Nb and not in Mo in this calculation is clearly associated with contributions to χ from states near the Fermi energy. This is strongly indicated by the experiments of Powell et al. (1968) which show that relatively small amounts of Mo (corresponding to relatively small changes of the Fermi level in a rigid-band picture) give large changes in ω_Q.

A fierce debate is now occurring between two schools of thought. It is quite difficult to do justice to this debate as neither school has its ideas completely in order, and not all participants fall entirely into one school. School I believes with Pickett and Gyorffy (1976) and Varma and Weber (1977) that band structure near E_F is crucial to the explanation of ω_Q, and requires a careful calculation of χ_0. School II believes with Sinha and Harmon (1975, 1976) and Hanke et al. (1976) that local bond orbital symmetry is more important. This point of view derives originally from phenomenological theories such as the shell model, and especially the extended shell model of Weber et al. (1972) and Weber (1973a), for TaC and related metals. Attempts to generalize phenomenological theory to Nb have been made by Weber (1973b), Wakabayashi (1977) and Allen (1977). In band theory, the chemical bonds come from valence electron charge density which can seldom be adequately described by examining only a narrow set of states near E_F. Thus there seems to be a severe contradiction between these schools of thought, although it is possible that a reconciliation would result if the relevant bond orbitals were carefully specified in band-theoretic terms. At the moment, school I appears to be winning by virtue of the calculation of Varma and Weber. The program of school II is quite a lot harder to implement and only qualitative results have been achieved on the microscopic level.

In detail, the program of school II requires a complete, self-consistent calculation of $\chi(\boldsymbol{r},\boldsymbol{r}',\omega)$. They are skeptical of the accuracy of the rigid-muffin-tin or any other "guess" for ∇V in (4.16), because they see in phenomenological theory that bond orbitals readjust when atoms move. To implement the calculation of χ, it is suggested that χ_0 first be calculated in Fourier space:

$$\chi_0(\boldsymbol{Q}+\boldsymbol{G},\boldsymbol{Q}+\boldsymbol{G}',\omega) \equiv \int \mathrm{d}\boldsymbol{r}\,\mathrm{d}\boldsymbol{r}' \exp[\mathrm{i}(\boldsymbol{Q}+\boldsymbol{G})\cdot\boldsymbol{r}] \exp[-\mathrm{i}(\boldsymbol{Q}+\boldsymbol{G}')\cdot\boldsymbol{r}'] \chi_0(\boldsymbol{r}\boldsymbol{r}',\omega). \qquad (4.17\mathrm{b})$$

This is a matrix in $(\boldsymbol{G},\boldsymbol{G}')$ space which can be inverted to get χ, with the help of tricks using Wannier functions. The non-diagonal nature of χ_0 in $\boldsymbol{G}\boldsymbol{G}'$ space is crucial to the description of localized valence bonds, and is often considered to be synonymous with the "local field problem". One of the key ideas of this school is "resonant susceptibility", which means that certain types of valence bond deformation have rather weak electronic restoring forces. This "resonance" of χ is not expected to show up strongly in χ_0, which omits Coulomb energy, but it is expected to enhance $\nabla_\alpha V$ and at special wavevectors, in a way that would require a self-consistent calculation (rather than a guess) to account for it. According to this picture, anomalous phonons (occurring at special wavevectors where χ is "resonant") would contribute more than their share to superconductivity because their electron–phonon coupling matrix elements are enhanced by the resonance in $\nabla_\alpha V$.

Evidence on these questions is at the moment highly confused. There have now been many approximate calculations of χ_0 for d-band metals. The first, by Cooke et al. (1974), was a calculation of the diagonal part $\chi_0(q,q;\omega=0)$ for Nb. This calculation was done both "with" and "without" matrix elements, meaning that the factor $\langle\psi_{k+q}|\exp(\mathrm{i}q\cdot r)|\psi_k\rangle$ which enters (4.17b) was either evaluated correctly, or set to 1. It is not clear which (if either) of these two should give better insight into phonon frequencies. The matrix element which enters the first term of (4.23) is actually $\langle\psi_{k+q}|\nabla V|\psi_k\rangle$, which does not resemble closely either choice. Cooke et al. find that the choice of matrix element does make a difference, but in neither case do the results bear even qualitative relation to the experimental ω_Q. These results were extended and confirmed by Pickett and Allen (1977). A similar conclusion was reached by Gupta and Freeman (1976) who calculated $\chi_0(q,q,\omega=0)$ "without matrix elements" for

NbC and TaC. However, an important observation was made in this paper: if only those few bands which intersect E_F are included in the χ_0 calculation, then dramatic structure appears in χ_0 as a function of q. This structure has a strong correlation with the experimental ω_Q. This conclusion was also reached by Klein et al. (1976), who obtained a very similar result for $\chi_0(q)$ in TaC. Myron et al. (1977) have studied χ_0 of the layer compounds 1T-TaS_2 and $TaSe_2$, and found again that rich structure arises from bands near E_F. Much but not all of this structure can be associated with wavevectors q connecting parallel pieces of Fermi surface. There are peaks in χ at those values of q where charge-density-wave instabilities are known to occur (see §5). These results add much weight to the argument of school I that details of energy bands near E_F cause phonon anomalies. However, many points are still unclear. When many bands are summed in calculating χ_0, the dramatic structure gets overwhelmed and becomes small ripples on a featureless background. It seems as if some mechanism strongly enhances the contribution of states near E_F to the phonon self-energy relative to the contribution they make to χ_0. A second uncertainty is how much these effects are altered when the electron–phonon matrix element is used instead of $\exp(i\mathbf{q}\cdot\mathbf{r})$ or 1. Varma and Weber (1977) find in Nb that the correct matrix elements are a crucial factor, contrary to the inferences drawn from much of the χ_0 work. One wishes that the calculations of Cooke et al. (1974) on Nb could be repeated keeping only bands near E_F, to see whether the observations of Gupta and Freeman (1976) for TaC apply also to Nb.

4.3. Phonon linewidths and T_c

The theory of λ was discussed in §3. In the approach used there, the challenging part of a λ theory is calculating the electronic wavefunction, electron–phonon matrix element, and carrying out the complicated Fermi surface averages. The phonon properties ω_Q and $\hat{\varepsilon}_Q$ play an important role, but are preferably derived from experiment, and present no challenge. In this section an alternate formulation is given which puts the emphasis on phonon properties. This formulation is entirely equivalent to the one in §3, but leads to a different way of thinking about the problem. This is accomplished by the observation (to be proved later in this section) that all the complicated Fermi surface averages of matrix elements which enter λ also enter the decay rate $\gamma_{Q\nu}$ (eqs. (4.5), (4.7)) of the phonon $\boldsymbol{Q}\nu$ into electron–hole pairs. The final result is that both λ and α^2F can be conveniently expressed as simple Brillouin zone sums

$$\alpha^2F(\Omega)=N^{-1}\sum_{Q\nu}\alpha^2_{Q\nu}\delta(\Omega-\omega_{Q\nu}), \tag{4.27a}$$

$$\lambda = 2N^{-1} \sum_{Q\nu} \alpha_{Q\nu}^2 / \omega_{Q\nu}, \tag{4.27b}$$

$$\alpha_{Q\nu}^2 = \gamma_{Q\nu} / \pi N(0) \hbar \omega_{Q\nu}. \tag{4.27c}$$

A factor of two error in the original derivations (Allen 1972) is corrected here. In these formulas $\gamma_{Q\nu}$ is the electron–phonon part of the half-width at half-maximum of the phonon spectral function $S(Q,\omega)$ (eq. (4.7)), and N is the number of $\boldsymbol{Q}$-points in the Brillouin zone.

If there are other sources of broadening such as defects or anharmonic effects, these must be subtracted out of $\gamma_{Q\nu}$ before eqs. (4.27) are used. The significance of eqs. (4.27) is that they express λ and $\alpha^2 F$ as sums of contributions from individual phonons. The coupling strength for the $Q\nu$ phonon, $\alpha_{Q\nu}^2$, is given by the ratio $\gamma_{Q\nu}/\omega_{Q\nu}$ of linewidth to frequency. Except for $N(0)$, no explicitly electronic factors appear in these equations. The derivation of these equations will be done in $\boldsymbol{l}$-space instead of $\boldsymbol{Q}$-space, and as a by-product, a new and more general version of eqs. (4.27) will emerge.

As a starting point we need the formula for $\alpha^2 F(\Omega)$ in $\boldsymbol{l}$-space notation. The Eliashberg interaction, (2.22), when Fourier transformed to the frequency domain, is

$$V_{\mathrm{E}}(x,x',\omega) = \sum_{ll'} \nabla_\alpha V(x-l) D_{\alpha\beta}(ll',\omega) \nabla_\beta V(x'-l'), \tag{4.28}$$

where the phonon Green's function $D_{\alpha\beta}(ll'\omega)$ is defined by (4.2). The Fermi energy average of $N(0)V_{\mathrm{E}}(xx'\omega)$ is $\lambda(\omega)$, the Hilbert transform of $\alpha^2 F(\omega)$. The Fermi energy average was constructed in §2 by taking Bloch-wave matrix elements of $V(xx'\omega)$ and averaging over the Fermi surface. Here a more general procedure is used which is applicable, both to perfect crystals and disordered solids where Bloch's theorem and Fermi surfaces cease to be meaningful. The Fermi energy density matrix (FEDM) is defined as $\rho(xx',\varepsilon_{\mathrm{F}})$, where

$$\rho(xx',\varepsilon) = \sum_i \psi_i(x)\psi_i^*(x')\delta(\varepsilon_i - \varepsilon), \tag{4.29}$$

and $\psi_i(x)$ is an eigenfunction of the stationary solid one-electron problem of energy ε_i. This function is real by time reversal invariance. The label i runs over the quantum states; in a perfect crystal these are Bloch states and (4.29) is a Fermi surface average. Then $\lambda(\omega)$ can be written

$$\lambda(\omega) = N(0)^{-1} \int \mathrm{d}x\,\mathrm{d}x' \rho(x,x',\varepsilon_{\mathrm{F}})^2 V_{\mathrm{E}}(xx'\omega). \tag{4.30}$$

Now the Green's function $D(ll'\omega)$ satisfies a Kramers–Kronig relation which says that it is the Hilbert transform of $\operatorname{Im} D(ll',\omega)$:

$$D(ll'\omega)=\frac{1}{\pi}\int_0^\infty \mathrm{d}\Omega \operatorname{Im} D(ll'\Omega)\frac{2\Omega}{\Omega^2-\omega^2}. \tag{4.31}$$

Finally, combining (4.28, 4.30, 4.31) and (2.27), we obtain a formula for α^2F:

$$\alpha^2F(\Omega)=(\pi N(0))^{-1}\sum_{ll'}\int \mathrm{d}x\,\mathrm{d}x'\nabla_\alpha V(x-l)\rho(xx',\varepsilon_{\mathrm{F}})^2\nabla_\beta V(x'-l') \times \operatorname{Im} D_{\alpha\beta}(ll'\omega). \tag{4.32}$$

This is valid for a disordered system and reduces to (2.28) for a perfect crystal. We can relate this to the imaginary part of the force constant (4.23),

$$\operatorname{Im} K_{\alpha\beta}(l,l')=-\int \mathrm{d}r\,\mathrm{d}r'\nabla_\alpha V(r-l)^*\operatorname{Im}\chi_0(rr',\omega)\nabla_\beta V(r-l'). \tag{4.33}$$

The imaginary part of K comes only from the imaginary part of χ_0 in the first term of (4.23). Although $\nabla_\alpha V$ and $\delta\rho_\alpha$ are in principle also complex, the symmetry $K_{\alpha\beta}(l'l)=K_{\beta\alpha}(ll')$ requires that the imaginary parts of $\nabla_\alpha V$ and $\delta\rho_\alpha$ not contribute to $\operatorname{Im} K$. Using (4.17) and (4.29), χ_0 can be written

$$\chi_0(rr'\omega)=\int \mathrm{d}\varepsilon\,\mathrm{d}\varepsilon'\rho(rr'\varepsilon)\rho(rr'\varepsilon')\frac{f(\varepsilon')-f(\varepsilon)}{\varepsilon'-\varepsilon+\omega+\mathrm{i}s}. \tag{4.34}$$

The imaginary part of this expression contains the factor $[f(\varepsilon')-f(\varepsilon)]\times \delta(\varepsilon'-\varepsilon+\omega)$. For very small values of ω, such as phonon frequencies, this factor can be accurately replaced by $\omega\delta(\varepsilon)\delta(\varepsilon')$. Thus we get for low frequencies

$$\operatorname{Im}\chi_0(rr',\omega)\cong-\pi\omega\rho(rr',\varepsilon_{\mathrm{F}})^2. \tag{4.35}$$

Inserting this into (4.33), we have an essentially exact formula for the imaginary part, $-2\mathrm{i}M\omega\Gamma_{\alpha\beta}$, of the force constant $K_{\alpha\beta}$:

$$\Gamma_{\alpha\beta}(l,l')=\frac{\pi}{2M}\int \mathrm{d}r\,\mathrm{d}r'\nabla_\alpha V(r-l)\rho(rr'\varepsilon_{\mathrm{F}})^2\nabla_\beta V(r'-l'). \tag{4.36}$$

This finally enables (4.32) to be written

$$\alpha^2F(\Omega)=(2M/\pi^2N(0))\sum_{ll'}\Gamma_{\alpha\beta}(ll')\operatorname{Im} D_{\alpha\beta}(ll'\Omega). \tag{4.37}$$

This is an exact formula for an arbitrarily disordered system. No specifically electronic entities except for $N(o)$ appear. It is not necessary to subtract out anharmonic corrections from (4.37) because Γ is by definition the imaginary part of the harmonic force constant. Impurities and other defects will disorder the ll' dependence of $K-2\mathrm{i}M\omega\Gamma$, but they will not introduce a new source of damping Γ. The only interaction which makes K complex is the electron–phonon interaction. Finally to derive (4.27) we must assume the crystal is perfect. Fourier transforming, (4.37) becomes

$$\alpha^2F(\Omega)=(2M/\pi^2N(0)N)\sum_{Q}\Gamma_{\alpha\beta}(Q)\,\mathrm{Im}\,D_{\alpha\beta}(Q,\Omega), \tag{4.38}$$

$$\mathrm{Im}\,\mathbf{D}(Q\Omega)=-(\pi/2M)\sum_{\nu}\frac{\hat{\varepsilon}_{Q\nu}\hat{\varepsilon}_{Q\nu}}{\omega_{Q\nu}}\delta(\Omega-\omega_{Q\nu}). \tag{4.39}$$

Eq. (4.39) follows from (4.6) using the approximation $\gamma_{Q\nu}\ll\omega_{Q\nu}$ to convert the lorentzian lineshape to a delta function. Combining (4.7), (4.38) and (4.39), the result (4.27) is finally obtained. If the lineshape (4.39) deviates significantly from a delta function, it is (4.38) and *not* (4.27) which correctly gives $\alpha^2F(\Omega)$ (Allen and Silberglitt 1974). In imperfect crystals, (4.38) and (4.27) both break down, but (4.37) remains valid.

4.4. Measurements and calculations of phonon widths

Equations (4.27) show that measurements of $\gamma_{Q\nu}$ in superconducting metals would give ideally detailed information about the relation between T_c and phonons. The importance to T_c of *individual* phonons would be learned, whereas a tunneling measurement of $\alpha^2F(\Omega)$ tells only the integrated importance of all phonons at frequency Ω. Unfortunately, present neutron sources and spectrometers do not normally yield sufficient resolution to measure intrinsic electron–phonon linewidths of phonons. The problem is compounded by the fact that anharmonic and defect broadening are likely to be present and must be subtracted out. An ingenious way around this second problem was found by Axe and Shirane (1973a, b). In the superconducting state, at $T=0$, $\mathrm{Im}\,\chi(rr'\omega)$ and thus $\gamma_{Q\nu}$ vanish for $\omega<2\Delta$ because of the energy gap. (Formulas for $\alpha^2F(\Omega)$ are to be interpreted as containing the *normal* state χ or γ.) For phonons with $\omega_{Q\nu}<2\Delta$, it is thus possible to extract $\gamma_{Q\nu}$ by comparing observed neutron lineshapes at $T=0$ and T just above T_c. Fig. 21 shows the results of a measurement of this type by Shapiro et al. (1975) on Nb. From these data a value $\gamma_{Q\nu}=0.03\pm0.005$ meV has been extracted. The ratio γ_Q/ω_Q is 0.01. This is perhaps the sharpest resolution ever yet obtained by neutron scattering. It is easy to show from eqs. (4.27) (Allen 1972) that the *average* value of

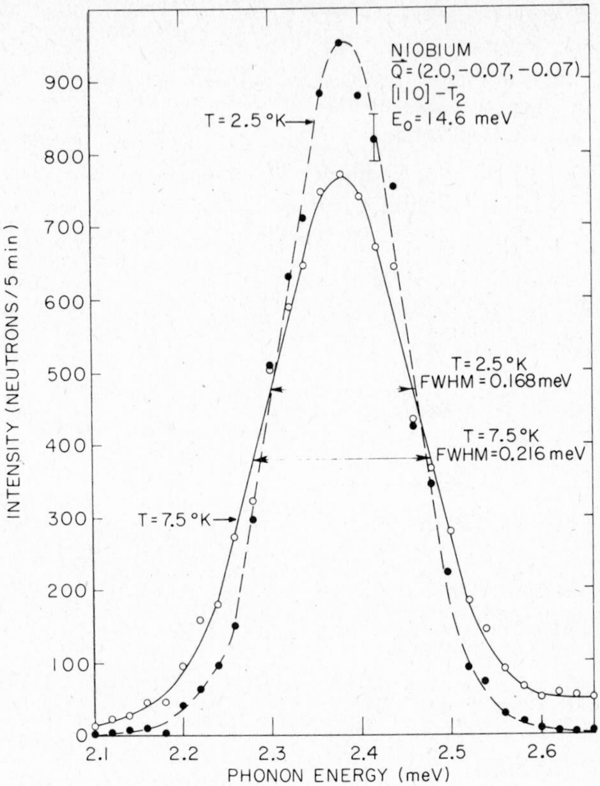

Fig. 21. Lineshape of a [110] phonon in Nb polarized [001]. At 2.5 K, ω_Q is less than $2\Delta(T)$ and the linewidth is instrumental. At 7.5 K, ω_Q exceeds $2\Delta(T)$ and the extra linewidth is caused by breaking Cooper pairs ($T_c = 9.2$ K) (Shapiro et al. 1975).

(γ_Q/ω_Q) for the complete set of $(Q\nu)$'s for an element is given by

$$(\gamma/\omega)_{\text{ave}} = \frac{\pi}{6} N(0)\hbar\langle\omega\rangle\lambda = \frac{1}{3N}\sum_{Q\nu} \gamma_{Q\nu}/\omega_{Q\nu}. \qquad (4.40)$$

Using the values $N(0) = 0.73$ eV^{-1} (spin atom)$^{-1}$ (Mattheiss 1970) and $\hbar\langle\omega\rangle = 14$ meV (Bostock et al. 1976, Robinson et al. 1976) and $\lambda \simeq 1$ for Nb, one finds $(\gamma/\omega)_{\text{ave}} \simeq 0.01$. In other words, the particular phonon shown in fig. 21 has about an average value of the coupling constant α^2. It would be more interesting to look at phonons with larger Q values, but unfortunately the method breaks down because $\omega_Q > 2\Delta$. This is illustrated in fig. 22 which shows the variation of measured linewidths as the temperature is increased through T_c, for phonons of various frequencies both less

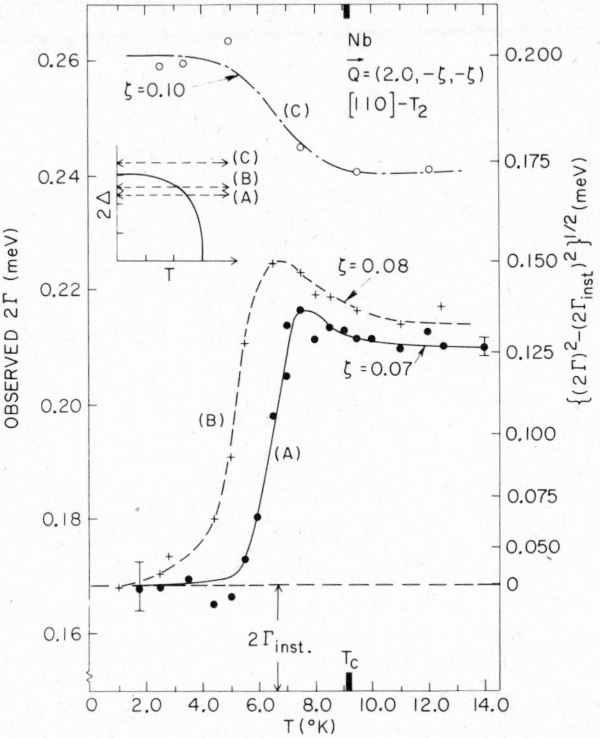

Fig. 22. Linewidth versus temperature for three different [110] phonons polarized [001] in Nb. The relation between ω_Q and $2\Delta(T)$ is shown in the inset (Shapiro et al. 1975).

than and greater than 2Δ. When $\omega_Q \gtrsim 2\Delta$, there is a slight enhancement of γ_Q below T_c which is predicted by BCS theory (Bobetic 1964).

An alternative to the Axe and Shirane method is to use very pure crystals and low temperatures to minimize defect and anharmonic scattering rates, and make absolute linewidth measurements. The size of the linewidth, according to (4.40), increases with $N(0)\hbar\langle\omega\rangle$. This indicates that d-band metals are favorable cases, while sp-band materials with smaller densities of states and phonon frequencies, are unfavorable. This strategy has been employed by Butler et al. (1977). As part of a program to calculate $\alpha^2F(\Omega)$ for Nb using eq. (4.27), Butler calculated $\gamma_{Q\nu}$ on a fairly dense mesh in $\boldsymbol{Q}$-space. The calculation predicted some fairly sharp peaks in the dispersion of γ with $\boldsymbol{Q}$. Within the error bars of the neutron experiment, there is excellent agreement as shown in fig. 23. It is worth emphasizing that the calculations were done *before* the experiment (although published simultaneously). This is probably the first time that

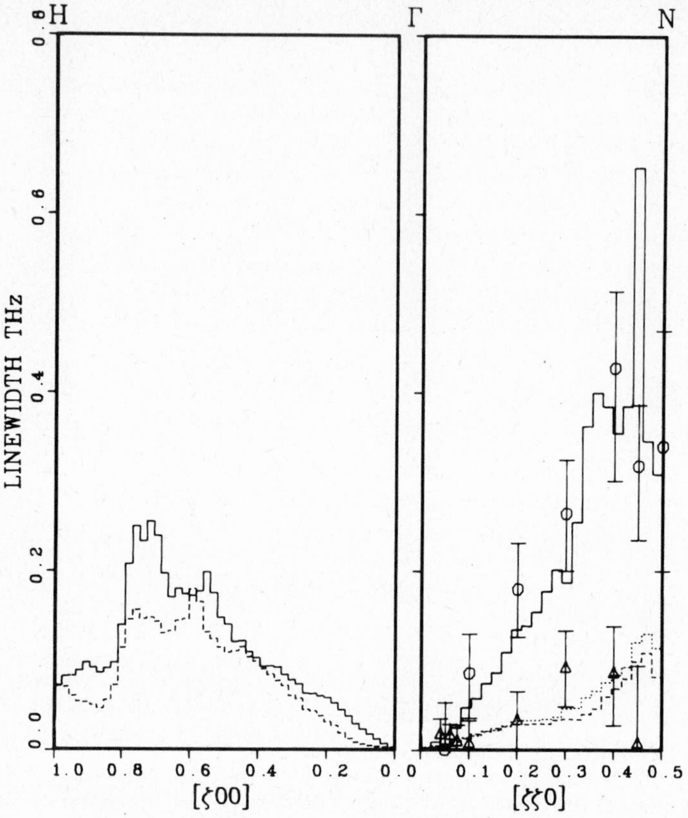

Fig. 23. The solid lines are calculated linewidths for LA (solid lines) and TA (dashed lines) phonons in Nb. The data points are obtained by comparing neutron measurements of lineshapes with the known instrumental resolution (Butler et al. 1977).

microscopic theoretical predictions for an electron–phonon phenomenon have successfully compared with subsequent experiment for a d-band metal. The theory depends sensitively on the band structure and wavefunctions at the Fermi level, and on the model for ∇V. It is hard to escape the conclusion that the rigid-muffin-tin model used in the theory is giving accurate Fermi surface matrix elements of ∇V. It is also interesting (but not surprising) that phonons which lie in dips in ω_Q have enhanced values of γ_Q. However, no support is given to the argument of school II that a "resonant" enhancement should occur. The idea of the resonant enhancement is that the ∇V operator should be enhanced at special values of Q *beyond* what is predicted in a model like rigid muffin tin which doesn't attempt self-consistent screening. The success of the rigid-muffin-tin calculation seems to be contradicting this idea. It should also be said that the experiment is very difficult, and the data of fig. 23 should perhaps be regarded as preliminary.

On intuitive grounds one might expect that since $\gamma_{Q\nu}$ and $\omega_{Q\nu}$ are real and imaginary parts of the same analytic function, there should be some way of estimating $\gamma_{Q\nu}$ from knowledge of $\omega_{Q\nu}$ alone. Rigorous calculation of $\gamma_{Q\nu}$ requires more than knowledge of $\omega_{Q\nu}$ because the latter is the real part of a self-energy at $\omega=0$ and the complete ω-dependence is necessary to Kramers–Kronig transform. Nevertheless the hope persists that an approximate scheme can be found. The author has written a series of papers on this subject (Allen and Cohen 1972, Allen and Dynes 1975b, Pickett and Allen 1977). Early optimism has not been confirmed by the most recent results. The paper of Pickett and Allen (1977) suggests that a method is available which should work in principle for Nb, but implementing this method is not practical because it requires extremely accurate knowledge of ω_Q for a series of alloys. Fortunately, the calculations of Butler have shown that $\gamma_{Q\nu}$ can be calculated; hopefully techniques for doing the measurement will improve.

5. *Lattice instability*

Consider a crystal which is perturbed by impressed lattice distortions $\boldsymbol{u}_l$ and an impressed charge density distortion $\delta\rho(\boldsymbol{r})$, both of which represent arbitrary small changes which we imagine to be imposed externally and frozen in place. The energy of the crystal will increase to second order by an amount

$$
\begin{aligned}
U = {} & \frac{1}{2}\sum_{ll'}\left[\partial^2 V_{\mathrm{c}}(\boldsymbol{R}_l-\boldsymbol{R}_{l'})/\partial\boldsymbol{R}_l\partial\boldsymbol{R}_{l'}\right]:\boldsymbol{u}_l\boldsymbol{u}_{l'} \\
& +\sum_l \int \mathrm{d}\boldsymbol{r}\,\delta\rho(\boldsymbol{r})\left[\partial V^{\mathrm{b}}(\boldsymbol{r}-\boldsymbol{R}_l)/\partial\boldsymbol{R}_l\right]\cdot\boldsymbol{u}_l \\
& +\frac{1}{2}\int \mathrm{d}\boldsymbol{r}\,\mathrm{d}\boldsymbol{r}'\,\delta\rho(\boldsymbol{r})\chi^{-1}(\boldsymbol{r}\boldsymbol{r}')\,\delta\rho(\boldsymbol{r}'). \qquad (5.1)
\end{aligned}
$$

There can be no first-order terms in the expansion (5.1) because $\boldsymbol{u}_l=0$ and $\delta\rho(\boldsymbol{r})=0$ represents equilibrium. The first term of (5.1) is ion–ion Coulomb energy as in (4.9). The second term is the interaction energy of the charge distortion and lattice displacement, with V^{b} being the bare potential. The third term represents the total change in band energy (with corrections for exchange and Coulomb double counting). A derivation of this is given by Sham (1969). Rather than repeat Sham's somewhat tricky proof, we can demonstrate the plausibility of (5.1) by showing that it leads easily to (4.13). This is done by observing that the adiabatic approximation says that if we unfreeze the electron distortion $\delta\rho$, it will rapidly readjust to minimize the functional (5.1). Algebraically, this means that the functional derivative of U by $\rho(r)$ vanishes:

$$\delta U/\delta\rho(r)=0$$

$$=\sum_{l}\partial V^{\mathrm{b}}(\boldsymbol{r}-\boldsymbol{R}_l)/\partial \boldsymbol{R}_l\cdot\boldsymbol{u}_l+\int \mathrm{d}r'\chi^{-1}(r,r')\delta\rho(r'). \qquad (5.2)$$

When this is solved for $\delta\rho$, which is then substituted back into (5.1), the result is

$$U=\frac{1}{2}\sum_{ll'}\left[\mathbf{K}^{\mathrm{c}}(l-l')+\mathbf{K}^{\mathrm{v}}(l-l')\right]:\boldsymbol{u}_l\boldsymbol{u}_{l'}, \qquad (5.3)$$

with $\mathbf{K}^{\mathrm{c}}$ and $\mathbf{K}^{\mathrm{v}}$ given by (4.9) and (4.13). Thus χ^{-1} plays the role of an "electronic restoring force" which drives charge density distortions back to equilibrium.

The occurrence of phonon anomalies is clearly associated with large values of χ which according to (4.13) and (5.3), will tend to cause cancellation of K^{c} and K^{v}, making the phonon frequencies low. But (5.1) shows that large χ means small χ^{-1} or small electronic restoring forces. In §5.1 we focus on the electronic distortions, which are more fundamental, and of which the phonon anomalies can be regarded as a secondary manifestation. In §5.2 a connection is made with the theory used for non-metals. In §5.3, the Batterman–Barrett instability of A15 metals is discussed in some detail.

5.1. Electronically driven instabilities

In d-band metals, the charge distortion $\delta\rho(r)$ is moderately well localizable near the atoms (or possibly in the bonds). Let us assume the existence of a convenient set of orthonormal localized functions $f_L(\boldsymbol{r}-\boldsymbol{l})$ where L summarizes the labels of the functions.

$$\delta\rho(r)=\sum_{L,\boldsymbol{l}}\delta\rho_{lL}f_L(\boldsymbol{r}-\boldsymbol{l}). \qquad (5.4)$$

In actual fact, choosing an appropriate set of localized functions is by no means trivial. The Wannier functions are complete and orthonormal, but poorly localized and not at all simple. The procedure used here is valid conceptually but may not be satisfactory as a computational procedure.

The assumption (5.4) allows us to rewrite (5.1) in the form

$$U=\frac{1}{2}\sum_{ll'}K^{c}_{\alpha\beta}(l-l')u_{l\alpha}u_{l'\beta}$$
$$+\sum_{ll'}W_{\alpha L}(l-l')u_{l\alpha}\rho_{l'L}+\frac{1}{2}\sum_{ll'}\chi_{LL'}(l-l')\rho_{lL}\rho_{l'L'}, \tag{5.5}$$

where repeated Greek and upper case Roman subscripts are summed. Or, Fourier transforming

$$U=\frac{1}{2}\sum_{Q}K^{c}_{\alpha\beta}(Q)u_{Q\alpha}u_{-Q\beta}$$
$$+\sum_{Q}W_{\alpha L}(Q)u_{Q\alpha}\rho_{-QL}+\frac{1}{2}\sum_{Q}\chi^{-1}_{LL'}(Q)\rho_{QL}\rho_{-QL'}. \tag{5.6}$$

Here, $W_{\alpha L}$ and $\chi^{-1}_{LL'}$ are the matrix elements of $\partial V_{\mathrm{b}}/\partial R$ and χ^{-1} in the localized basis set:

$$W_{\alpha L}(l-l')=\int \mathrm{d}r f_{L}(r-l')\partial V^{\mathrm{b}}(r-R_{l})/\partial R_{l\alpha},$$
$$\chi^{-1}_{LL'}(l-l')=\int \mathrm{d}r\,\mathrm{d}r' f_{L'}(r'-l')\chi^{-1}(r',r)f_{L}(r-l). \tag{5.7}$$

In this language the phonon frequencies are found by solving Newton's law after adiabatically eliminating ρ_{QL}

$$\partial U/\partial\rho_{QL}=0, \tag{5.8a}$$
$$\partial U/\partial u_{Q\alpha}=-M\ddot{u}_{Q\alpha}. \tag{5.8b}$$

The result is that the eigenvalues $M\omega^2$ of the following dynamical matrix $K_{\alpha\beta}(Q)$ determine the frequencies $\omega_{Q\nu}$:

$$K_{\alpha\beta}(Q)=K^{c}_{\alpha\beta}(Q)-W_{\alpha L}(Q)\chi_{LL'}(Q)W_{\beta L'}(Q), \tag{5.9}$$

where $\chi_{LL'}$ is the inverse in LL' space of $\chi^{-1}_{LL'}(Q)$ in (5.6). Equations very similar to this have been suggested by Sinha and Harmon (1975, 1976) and Hanke et al. (1976), the principal difference being that in these works it is the wavefunctions rather than the charge distortion which is expanded in localized functions.

Let us focus on the charge distortion ρ_{QL} rather than the displacement $u_{Q\alpha}$. Suppose temporarily that the electron–phonon coupling $W_{\alpha L}(Q)$ in (5.6) is negligible, and that the charge distortion can be described by a single function $f_{L_0}(\boldsymbol{r}-\boldsymbol{l})$. Suppose further that the electronic restoring force, $\chi^{-1}_{L_0L_0}(Q)$ becomes zero at a temperature T_0 and wave vector Q_0:

$$\chi^{-1}_{L_0L_0}(Q_0)=\alpha(T-T_0)/T_0. \tag{5.10}$$

Then at temperature near and below T_0, large amplitude charge distortions $\rho_{Q_0L_0}$ would occur, and it would be essential to consider higher-order terms in $\rho_{Q_0L_0}$ in the expansion (5.6). A particularly simple form might be

$$U^{\text{el}}(Q_0)=\frac{1}{2}\alpha(T/T_0-1)\rho_{Q_0}\rho_{-Q_0}+\frac{1}{4}B(\rho_{Q_0}\rho_{-Q_0})^2+\text{H.C.} \tag{5.11}$$

The subscript L_0 is omitted from ρ_{Q_0} for convenience; the symbol +H.C. (Hermitean conjugate) reminds us to add on terms with $Q_0\to -Q_0$. The adiabatic approximation (5.8a) then gives

$$\langle\rho_{Q_0}\rangle=\begin{cases}0 & T>T_0\\ \left[\alpha(1-T/T_0)/B\right]^{1/2} & T<T_0\end{cases}. \tag{5.12}$$

This describes an electronic charge density wave (CDW) transition in which a charge distortion with Fourier components $\pm Q_0$ becomes stable below T_0. There are several mechanisms which in principle can cause such an effect in a metal. The simplest was discussed by Peierls (1955) who neglected electron–electron Coulomb interactions and showed explicitly that one-dimensional metals would always have instabilities with Q_0 near $2k_{\text{F}}$. The Peierls transition is *not* exactly what is described here because the electron–lattice displacement interaction plays an indispensable role in his analysis. The next simplest was discussed by Overhauser (1968) who argued that even delocalized electrons with no lattice to pin them could have CDW instabilities at $Q_0=2k_{\text{F}}$ driven by Coulomb and exchange interactions. Both Peierls and Overhauser derive the stability of the distorted structure from the occurrence of gaps near the Fermi surface which lower the energy of electrons near the Fermi level. For the free-electron case, Overhauser's claims are still controversial, but for metals with fairly narrow bands and reduced dimensionality, all doubt was erased by the observation by Wilson et al. (1974) and Williams et al. (1974) of incommensurate lattice distortions in layer structure transition metal dichalcogenide metals. The connection between Q_0 and the Fermi surface dimensions has been established both by band theory (Mattheiss 1973, Freeman et al. 1977, Myron et al. 1977), and by rigid-band interpretation of alloying experiments (Di Salvo et al. 1975). Landau-type theories much more sophisticated than (4.30) have been given by McMillan (1975) and by

Moncton et al. (1975, 1977). The hexagonal symmetry and the location of $\boldsymbol{Q}_0$ together require formulas more complicated than (5.11) containing third-order terms. These help explain a rich experimental phenomenology (Wilson et al. 1975).

The source of the instability is probably an abnormally sharp peak in $\chi_0(\boldsymbol{Q}, \omega=0, T=0)$ (see eq. (4.17)) which is strongly affected by the Fermi surface shape. The calculation of Myron et al. (1977) for 1T-$TaSe_2$ (which neglect possible $\boldsymbol{Q}$-variation caused by matrix elements) shows a strong peak of $\chi_0(Q)$ at exactly the wavevector Q_0 of the observed CDW. These peaks arise from bands at and very near to the Fermi level. If the Fermi surface were simple enough, simple geometrical "nesting" of the surface with a copy of itself displaced by Q_0 should suffice to locate Q_0. However, the surfaces of d-band metals are sufficiently complicated that geometrical arguments are risky. A similar calculation (Myron et al. 1977) for TaS_2 shows less good correlation between χ_0 and Q_0.

The source of the temperature dependence of χ^{-1} in (5.10) is generally thought to be the Fermi factors in the non-interacting susceptibility (4.10). Usually such effects are small (of order kT/E_F) because of Fermi degeneracy. However, in one dimension the susceptibility χ_0 of the non-interacting electrons for $Q_0 = 2k_F$ (i.e. exactly spanning the Fermi surface) has a term of the form $N(0)\ln(E_F/k_B T)$, diverging at low temperatures. Rice and Scott (1975) have recently shown that a $\log(E_F/k_B T)$ singularity in χ_0 can also occur in two-dimensional band structures when Q_0 connects two saddle points. If the saddle points do not happen to lie at the Fermi energy, the form for χ_0 at $Q=2k_F$ in two dimensions has a weaker singularity. Using the result that χ^{-1} equals $\chi_0^{-1} - \bar{v}$ (from 4.13), one expects a CDW transition at a temperature $E_F \exp(-1/N(0)\bar{v})$. A similar logarithmic divergence occurs in three dimensions for the "Cooper pair susceptibility" (with a cutoff θ_D instead of E_F, unfortunately). This divergence locates the superconducting transition temperature $\theta_D \exp(-1/N(0)V)$. However, the charge susceptibility (4.10) has only a very weak temperature dependence in three dimensions. Thus the behavior described by (5.10–5.12) is usually considered to be unlikely except in metals with reduced dimensionality.

Now let us couple the lattice distortions back into the problem. That is, we add onto (5.11) the lattice energy $\frac{1}{2}Ku^2$ and the coupling energy $Wu\rho$ from (5.6).

$$U = \frac{1}{2} K u_{Q_0} u_{-Q_0} + W u_{Q_0} \rho_{-Q_0} + \frac{1}{2}\alpha(T/T_0 - 1)\rho_{Q_0}\rho_{-Q_0} + \frac{1}{4}B(\rho_{Q_0}\rho_{-Q_0})^2 + \text{H.C.} \tag{5.13}$$

There are various ways to think about this equation. First consider what happens if we minimize (5.13) with respect to u_{Q_0}. This is the reverse of the usual adiabatic approximation – it assumes that ions quickly readjust to the electron displacements rather than the other way. This would be nonsense for the dynamic properties of the model, but for the static properties it is fine – it doesn't matter which variable is eliminated first. The minimum occurs when u_{Q_0} equals $-(W/K)\rho_{Q_0}$. Substituting this into (5.13), the result is identical to (5.11) with a renormalized transition temperature T_0^*:

$$T_0^* = T_0(1 + W^2/\alpha K). \tag{5.14}$$

Thus coupling the CDW to the lattice *raises* the temperature of the instability, by an amount which can be expected to be of order T_0, since W^2 and αK should be of similar size, a few (eV/Å)^2. In fact, T_0 can be negative and still yield a positive T_0^* (notice that (5.10) requires αT_0 to be positive).

Let us now think about eq. (5.13) in a different way which emphasizes the dynamical aspects. The $B\rho^4$ term plays a crucial role for $T \leqslant T_0^*$, but can be neglected at temperatures high enough that the electronic restoring forces are large and $\langle \rho_Q \rho_{-Q} \rangle$ is small. The remaining terms of (5.14) are a quadratic form which can be diagonalized to give the restoring forces for new modes which are mixtures of phonon and charge density. It is interesting to relax the adiabatic approximation (5.8a), replacing it by $m\omega^2 \rho_Q = \partial U / \partial \rho_{-Q}$, where m is comparable to the electron mass. This is parallel to (5.8b) for phonons, and should be thought of as a crude classical model for the heavily damped plasma oscillations which describe the actual time-dependent motions of electrons in real metals. The secular equation is

$$\begin{pmatrix} K - M\omega^2 & W \\ W & \chi^{-1} - m\omega^2 \end{pmatrix} \begin{pmatrix} u \\ \rho \end{pmatrix} = 0, \tag{5.15}$$

where χ^{-1} is $\alpha(T/T_0 - 1)$. We must assume T significantly larger than T_0 in order to neglect the $B\rho^4$ term. Solving these equations using $m \ll M$, there is a low-frequency mode ω_{L} and a high one ω_{H}, given by

$$M\omega_{\mathrm{L}}^2 = K - W^2\chi; \qquad \rho_{\mathrm{L}} = -\chi W u_{\mathrm{L}}, \tag{5.16a}$$

$$m\omega_{\mathrm{H}}^2 = \chi^{-1} + (m/M) W^2 \chi; \qquad u_{\mathrm{H}} = (m/M)\chi W \rho_{\mathrm{H}}. \tag{5.15b}$$

The interpretation is that the low-frequency mode is the phonon whose frequency has the customary renormalization because the charge follows essentially adiabatically; ρ_L is χ times the potential, $-Wu_{\mathrm{L}}$, felt by the electrons. As T is reduced, χ increases, as $1/\alpha(T/T_0-1)$ and ω_L^2 decreases. At the same temperature T_0^* where the CDW transition occurs (5.14), the phonon frequency ω_L goes to zero as $(T-T_0^*)^{1/2}$. This behavior is typical of a second-order structural phase transition, and indeed, once the quartic term $B\rho^4$ is included we can see that the lattice displacement u will acquire a non-zero average amplitude $\langle u\rangle = -\chi^{-1}W^{-1}\langle\rho\rangle$, where $\langle\rho\rangle$ behaves as $(T_0^*-T)^{1/2}$ for T below T_0^* as in (5.12). Thus as the temperature is lowered, a coupled lattice distortion and CDW is induced at a temperature T_0^*, higher than the instability temperature T_0 of the uncoupled electron system. If there are no third-order terms (like ρ^3) in the expansion (5.13), the transition may well be second order, accompanied by a soft phonon. Although slightly over-simplified, this is close to the usual picture of an electronically driven lattice instability. The high-frequency mode, ω_H, is unaffected by phonons to order m/M. Roughly speaking, this is a plasmon, although of course real plasma modes in metals must be described by the complete Schroedinger equation rather than the *ad hoc* version of Newton's law used here. Nevertheless, it is an interesting prediction of the theory that an electronically driven lattice instability should be accompanied by softening of the plasma response of the metal at the same wavevector Q_0 as the distortion. The softening is not completed at $T=T_0^*$; as can be seen from (5.16b), ω_H is predicted to behave as $(T-T_0)^{1/2}$, the prediction being valid only for $T>T_0^*>T_0$. This effect has not yet been observed, but ought to be observable in principle by inelastic electron or X-ray scattering experiments.

5.2. Anharmonic effects

Testardi (1972) pointed out that there is ample evidence of anharmonic behavior in A15 metals, and gave a qualitative explanation of how highly anharmonic interatomic forces can account for anomalous elastic behavior, without any need for a pathological electronic spectrum. Smith and Finlayson (1978) argued that their anomalous thermal expansion data provide evidence in favor of Testardi's approach for V_3Si. Unfortunately, anharmonic effects have not so far been easy to interpret in detail. Simple soluble models are scarce, and quantitative models capable of fitting the rich data are unavailable.

The previous section described structural phase transitions specifically driven by electronic effects. The temperature behavior is associated with k_BT/E_F corrections coming from departures from Fermi degeneracy. In order to build models which agree with data, it is often necessary to choose the degeneracy parameter E_F to be extremely small, of order 100 K. Current band calculations show peaks of width 1000 K (Mattheiss 1975), and although band theory cannot be trusted to the scale of 100K, it nevertheless seems unlikely that such fine structure is a common occurrence or essential to high-T_c superconductivity. This is especially true because all but the very best samples of Nb_3Sn are likely to have a defect-limited energy uncertainty $\hbar/\tau$ exceeding 100K. On the other hand, low-temperature structural transitions also occur in non-metals, where the relevant degeneracy parameter is a band gap of several volts. It is often plausible to call these "electronically driven", in spite of the discrepancy in temperature scales. The way out of the seeming contradiction is to assign the occurrence of negative quadratic restoring forces to electronic renormalization, but the stabilization of the symmetric phase at high temperatures to anharmonic renormalization (rather than reduction of the electronic renormalization). In other words, $M\omega^2 = K - W^2\chi$ is negative at all temperatures in harmonic approximation, and only anharmonic terms make $\omega^2 > 0$ when $T > T_0$. A simplified picture can be made as follows:

$$U(Q_0) = \frac{1}{2} K u_{Q_0} u_{-Q_0} + W u_{Q_0} \rho_{-Q_0} + \frac{1}{2} \chi^{-1} \rho_{Q_0} \rho_{-Q_0} + \frac{1}{4} b (u_{Q_0} u_{-Q_0})^2 + \text{H.C.} \tag{5.17}$$

There are two differences between this and eq. (5.13): χ^{-1} is *not* significantly temperature dependent here, and the quartic term is bu^4 rather than $B\rho^4$. Of course, whenever a structural transition is close by, both u_{Q_0} and ρ_{Q_0} have sufficiently large amplitudes that higher-order terms of *both* types (and cross-terms as well) should be kept. However, higher powers of u (which are often neglected in theories proposed for metals) have a fundamentally stronger role in altering the temperature dependence than do higher powers of ρ. The amplitude of thermal fluctuation of both u and ρ is determined by Bose–Einstein factors $2N+1$. In the case of lattice displacements, $2N+1$ becomes $2kT/\hbar\omega_{Q_0}$, whereas in the case of charge density fluctuations there is no significant thermal excitation and $2N+1$ is close to 1 and independent of temperature. Thus in (5.17) we keep the higher powers only of u, in the belief that higher powers of ρ will have no significant qualitive effect.

We now use the adiabatic approximation to eliminate ρ_{Q_0} from (5.17), that is we minimize by ρ_{Q_0} and evaluate (5.17) at the minimum:

$$U(Q_0) = -\frac{1}{2} K' u_{Q_0} u_{-Q_0} + \frac{1}{4} b (u_{Q_0} u_{-Q_0})^2 + \text{H.C.} \tag{5.18}$$

Here $-K' = K - \chi W^2$ is the renormalized harmonic force constant which is to good approximation temperature independent. The minus sign is introduced because we are interested in the case where the harmonic force constant is negative – i.e. the lattice is unstable at all temperatures in harmonic approximation. This is a double well potential with minima at $u = \pm(K'/b)^{1/2}$ and a local maximum at $u = 0$. The anharmonic term bu^4 can now stabilize the lattice if the fluctuations of u are large enough, and if $b > 0$. Realistic problems of this type become rapidly too difficult to solve exactly, but good qualitative insights can be derived from a mean field treatment. A sophisticated form of mean field theory, the self-consistent phonon approximation, can solve quantum mechanical problems like (5.18) to good accuracy, provided there are no important odd powers of u_Q in the potential, and provided anharmonic phonon decay is not important. This method is reviewed by Werthamer (1969), and in the article by Götze and Michel (1974) in vol. 1 of this series. For the present purposes we need not worry about numerical accuracy, and the essential physics is more easily seen in a less sophisticated mean field treatment. We replace u^4 by $(\langle u^2\rangle + \delta u^2)^2$, where $\delta u^2 = u_{Q_0}u_{-Q_0} - \langle u_{Q_0}u_{-Q_0}\rangle$ is the fluctuation of u^2 about its mean value $\langle u^2\rangle$. Assuming that a stable oscillation frequency ω_0 exists for mode Q_0, $\langle u^2\rangle$ can be replaced by $k_B T/M\omega_0^2$. This assumes that $k_B T$ is large compared to $\hbar\omega_0$, a convenient simplification. The mean field (MF) approximation then neglects $(\delta u^2)^2$, giving

$$U_{MF} = \frac{1}{2}\frac{bk_B}{M\omega_0^2}(T - T_0)u_{Q_0}u_{-Q_0}, \tag{5.19}$$

where $k_B T_0$ is $M\omega_0^2 K'/b$. The final result is that there are stable oscillations of the lattice above a temperature T_0. Below this temperature, the mean field theory must be modified: u^2 should not be replaced by $\langle u^2\rangle + \delta u^2$, but rather by $\langle(\langle u\rangle + \delta u)^2\rangle + \delta(\langle u\rangle + \delta u)^2$, where $\langle u\rangle$ is the mean value and δu is the fluctuation about that value of the lattice displacement u. Then the mean value $\langle u\rangle$ is fixed by requiring that there should be stable oscillations of δu about $\langle u\rangle$; i.e. terms in U_{MF} linear in δu should vanish. One finds the standard mean field result that $\langle u\rangle$ behaves as $(T_0 - T)^{1/2}$. Strictly speaking these results would become modified by completing the self-consistency loop and solving for ω_0 as the stable oscillation frequency. The justification for not doing so is that the higher-order terms in (5.17–5.18) should contain $bu_{Q_0}u_{-Q_0}u_{Q_1}u_{-Q_1}$ for all Q_1, as well as other more complicated terms. The factor ω_0 in $k_B T/M\omega_0^2$ should be chosen as an average frequency for all the modes Q_1 which are coupled to Q_0; the bulk of these will be temperature independent and unaffected by self-consistency.

Many models have been presented to explain the structural instability of A15 metals. These will be briefly described in §5.3. They all have the feature that the temperature dependence enters via an electronic susceptibility $\chi(T)$ as in eq. (5.13), rather than via anharmonic renormalization as in eqs. (5.17 – 5.19)

The point of the discussion given here is that anharmonic effects provide an adequate alternative source for the temperature-dependent parameters associated with structural phase transitions. Furthermore, anharmonic effects are inescapable; whenever a phonon frequency is abnormally low, the corresponding amplitude u is sufficiently large that anharmonic terms will modify the temperature behavior. Thus the neglect of anharmonic effects in theories of structural transitions in metals is ultimately unjustifiable. It is possible that anharmonic effects could leave the physical consequences of an electronic model qualitatively unchanged, simply causing a renormalization of, for example, the effective degeneracy parameter. However, it seems equally possible that the majority of the models proposed so far assign the temperature dependence to the wrong mechanism. The first microscopic discussion of the question in the literature, to my knowledge, is McMillan's (1977) suggestion that the CDW transition in 2H-$TaSe_2$, although surely arising from electronic susceptibility, is driven away at high T by lattice fluctuations (anharmonic renormalization). As evidence, he cites three discrepancies between traditional CDW theory and experiment. First, the specific heat jump at T_0 is anomalous. Second, the phonon softening extends over a fairly wide range in Q space, implying a shorter coherence length than anticipated. Third, and most direct, infrared measurements (Barker et al. 1975) show a possible CDW-induced electronic gap which is seven times larger than the value $1.76k_BT_0$ found if electronic promotion energy is responsible for the transition to the normal phase. Not everyone agrees with this interpretation of the infrared data.

An additional experimental test ought now to be possible. A direct measure of $\chi(Q,\omega=0)$ can be made in principle by inelastic scattering experiments. The cross section for an incident particle to scatter with energy loss ω and momentum transfer $\boldsymbol{q}$ is proportional to $\operatorname{Im}\varepsilon^{-1}(q,\omega)$. This in turn is proportional to $\operatorname{Im}\chi(q,\omega)$. By Kramers–Kronig transforming, this gives $\chi(q)$:

$$\chi_1(q,\omega=0)=\frac{2}{\pi}\int_0^\infty \frac{d\omega}{\omega}\operatorname{Im}\chi(q,\omega). \tag{5.20}$$

If the left-hand side is to peak at some particular value $q=Q_0$, this should be experimentally as a shift in the spectral weight of the cross section $\operatorname{Im}\chi(q,\omega)$ as $\boldsymbol{q}$ varies through Q_0. The classical model of eqs. (5.15–5.16) says $\mathrm{Im}\,\chi(q,\omega) \propto \delta(\omega - \omega_H(q))$. A peak in χ_1 at Q_0 is then associated with a dip in the plasmon dispersion relation at $q = Q_0$. In reality the energy loss spectrum is surely much more complicated, but an anomalously strong low ω response should be expected at $q = Q_0$. More dramatic, the T-dependence of χ should be directly visible as a T-dependence of the cross section. Electron energy loss or inelastic X-ray scattering experiments might be able to see this effect, and thus discover from experiment whether χ carries the T-dependence which stabilizes the undistorted phase.

5.3. Martensitic transition in A15 structure

As described in §1, the A15 structure not only exhibits the highest known values of T_c, but also exhibits the Batterman–Barrett transition (in V_3Si and Nb_3Sn). This transition is often called "martensitic", probably because many samples are only partially transforming. The instability is a prototype soft-mode, almost-second-order transition. There is not much relation to the transition in martensite (Anderson and Blount 1965).

There are now four well-developed microscopic theories which all describe quite successfully not only parameters like $C_{11}(T) - C_{12}(T)$ which are closely related to the transition, but also the anomalous T-dependence of the magnetic susceptibility, $\chi_H(T)$ (see fig. 24). These theories all begin with more or less idealized models for the band structure. An honest first-principles theory is clearly beyond theoretical abilities at this time, given the inability of theory to calculate C_{11} or χ_H even at $T = 0$ in a d-band element. Thus the microscopic theories clearly aim for a metaphorical rather than literal truth. The theories, in historical order, are: (1) Labbé and Friedel (LF) (1966a, b); (2) Gor'kov (1973), Gor'kov and Dorokhov (1976); (3) Bhatt (1977); and (4) Lee, Birman and Williamson (LBW) (1977a, b), Lee and Birman (1978). There are also numerous other theories which are less specific about band structure. The subject is too large to review here, and specific reviews are available (R9–R11). A partial listing of the more important papers is: Pytte (1971), Sham (1971), Noolandi and Sham (1973), Varma et al. (1974), Kragler and Thomas (1975). Theories (1) and (2) involve one-dimensional models of the energy bands, and bear no discernible relation to recent band-theoretical work (Mattheiss 1975, Klein et al. 1977, Jarlborg and Arbman 1977, Ho et al. 1978). Thus in the literal sense these theories are surely wrong. Recent Landau-type formulations (to be discussed shortly) by Bhatt and McMillan (1976) of Gor'kov theory and by Bhatt (1978), Kragler and Thomas (1975), and Kragler (1978) of LF theory, help clarify the metaphorical truths of these theories. Theories (3) and (4) consider more realistic band models. The LBW model identifies the source of the unique properties of

A15 metals with the unique occurrence of a six-fold degeneracy at the R point (R(4) representation) of the Brillouin zone. The attractive feature is that a simple answer is proposed for the vexing question, what is special about the A15 structure? The proposed answer is that no other structure is known which allows such a high degeneracy. Most band calculations do in fact show that the Fermi level lies near the R(4) point. The LBW model treats the bands near R(4) by a $\boldsymbol{k}\cdot\boldsymbol{p}$ expansion which has the virtue of being exact and model-independent near enough to the R point. Quite a large number of free parameters occur in such an expansion. After simplifying assumptions, LBW find a class of models with a logarithmic singularity of $N(\varepsilon)$, and a wider class with a broadened logarithmic singularity. Any structural change which lowers the crystal symmetry will lift the degeneracy at R. If the Fermi level is near enough to the R(4) level, this will lower the electronic energy. In spite of its attractive features, this model has the difficulty of focusing strongly on a small part of k-space. This runs counter to experience in studying superconductivity in d-band elements, where Fermi surfaces have large areas and electrons participate quite democratically. It seems probable that this experience is relevant to structural transitions in compounds.

Bhatt and Mattheiss (1979) claim that calculations done so far in the LBW model have used parameters which correspond to unrealistically flat

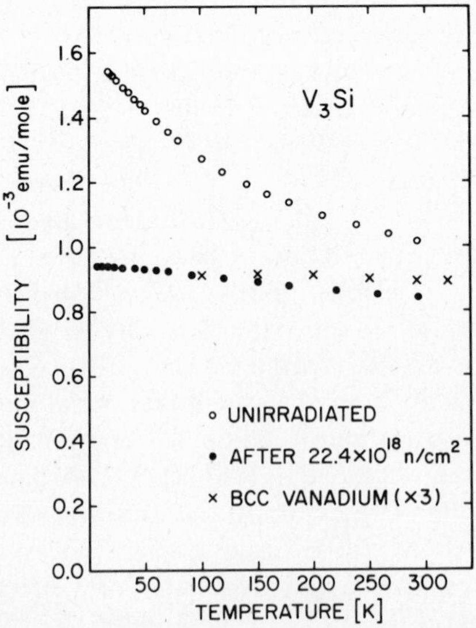

Fig. 24. Susceptibility $\chi(T)$ of V_3Si, measured by Guha et al. (1978). The unradiated sample shows a rapid thermal degradation of χ, whereas after radiation damage the susceptibility per atom is almost identical to that of b.c.c. V metal (measured by Kohlhaas and Weiss 1964).

bands (by a factor of 100) in order to fit anomalous susceptibilities. Bhatt's model is the only one which is closely tied to available band calculations (Mattheiss 1975). He takes six x^2-y^2 d-functions, one per transition atom, and builds two tight-binding bands which fit in rough approximation Mattheiss' bands near E_F. A small second-neighbor overlap is chosen to produce a peaked density of states coming from several distinct saddle points lying close in energy. These saddle points are sensitive to distortion, which causes the peak to split into subpeaks. Provided the Fermi level lies in the peak, the energy is lowered by the distortion. Both models (3) and (4) give fits to $\chi_H(T)$ and $C_{11}(T)-C_{12}(T)$.

None of these theories shed much light on T_c. They all have the common feature of a sharp peak in $N(\varepsilon)$ in which the Fermi level is more or less accidentally located. The large values of $N(0)$ obviously help increase λ and T_c. Band theory does not indicate larger values of $N(0)$ in A15 structure than are found in other comparably complicated d-band compounds, so the special status of A15 structure remains a matter of conjecture. The lack of insight into T_c perhaps reflects our relatively more sophisticated level of understanding of T_c than of ω_Q in d-band elements. We know that T_c is a complicated property, so we do not expect simplified model theories to be much help.

Several of the model theories have recently been recast as Landau theories by Bhatt and McMillan (1976) and Bhatt (1978). This important step clarifies the physics by reducing the reliance on specific band models. The Landau procedure allows many deductions about properties related to the structural transition, based on minimum assumptions. Correspondingly, Landau theory is unable to say much about $\chi_H(T)$ or $C_{44}(T)$ which are not directly coupled to the order parameter. I do not think this is any disadvantage. Consider the behavior of $\rho(T)$, which is always anomalously flat at high T in A15 metals (see fig. 25). Fisk and Webb (1976) attribute the anomaly to a breakdown of Boltzmann theory due to very short mean free paths. This contrasts sharply with earlier thinking (Cohen et al. 1967, Bader and Fradin 1976) which attribute the behavior to Fermi smearing of sharp structure in $N(\varepsilon)$. It is perhaps premature to rule out the earlier explanations, but evidence is accumulating that the room-temperature phonon-limited mean free path is in fact extremely short (Mattheiss et al. 1978, Allen et al. 1978). It is also difficult to believe that such a widespread phenomenon as flatness of $\rho(T)$ could arise from accidental features of $N(\varepsilon)$ and the placement of Fermi levels. It is appropriate to consider $\chi(T)$ in a similar light. Nearly all d-band elements have surprisingly strong T-dependence of χ (Galoshina 1974). Anomalous increases of $\chi(T)$ at low T are found in many d- and f-band metals and compounds (layer metals with CDW instabilities, Ce and Ce compounds believed to show valence fluctuations, etc.). Anderson (1977) has speculated that there

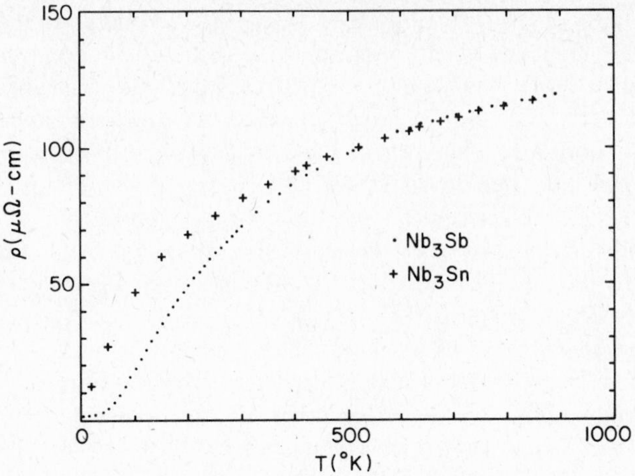

Fig. 25. Resistivity $\rho(T)$ of Nb_3Sn (measured by Woodard and Cody 1964) and Nb_3Sb (measured by Fisk and Webb 1976, source of this figure).

may be unifying underlying explanations for these $\chi(T)$ and $\rho(T)$ anomalies which are not yet appreciated. On these grounds it seems to me that the successful fit of $\chi(T)$ by Fermi smearing a model $N(\varepsilon)$ with judicious placement of Fermi level is *not* impressive. There are too many uncertainties involved, such as (1) separation of χ into spin and orbital parts, (2) choice of enhancement parameter and (3) possibility of phonon effects in $\chi(T)$ (Bhatt 1977). An interesting discussion by Herring (1965, p. 283) on this point has been widely ignored. He suggests that the T-independent quasiparticle energy picture may break down, and enhancement parameter I may become T-dependent, at the level of accuracy at which Fermi smearing becomes important. Thus there is perhaps some virtue in the restricted scope of Landau theories, being confined to phenomena not too far from T_0, and which couple to the order parameter.

The first Landau theory for A15's was by Anderson and Blount (1965). They used as order parameters the uniaxial strians ε_{xx}, ε_{yy} and ε_{zz}. These transform among themselves under the symmetry operations of the crystal, and thus form a basis for a representation of the point group. This representation is reducible, and decomposes into $\Gamma_1^+ + \Gamma_{12}^+$. It is clear that the volume dilation, $\varepsilon_0 = \varepsilon_{xx} + \varepsilon_{yy} + \varepsilon_{zz}$, transforms according to the identity representation. The two remaining strains can be written as

$$\varepsilon_1 = (2\varepsilon_{zz} - \varepsilon_{xx} - \varepsilon_{yy})/\sqrt{6}\,,$$

$$\varepsilon_2 = (\varepsilon_{xx} - \varepsilon_{yy})/\sqrt{2}\,. \tag{5.21}$$

These transforms as partners of the two dimensional (Γ_{12}^{+}) representation of the cubic point group, just as the d-functions $2z^2-x^2-y^2$ and x^2-y^2. The strain $\varepsilon_0=0=\varepsilon_2, \varepsilon_1=\varepsilon$ is a tetragonal deformation in the $\hat{z}$ direction, while $\varepsilon_0=0=\varepsilon_1, \varepsilon_2=\varepsilon$ is an orthorhombic deformation. The free energy F is an invariant function of the strains. We are interested in the behavior when $\varepsilon_0=0$ and $\varepsilon_1, \varepsilon_2$ are small. Taylor expanding about the cubic state, we get

$$F_1=\tfrac{1}{2}A_1(\varepsilon_1^2+\varepsilon_2^2)+\tfrac{1}{3}B_1(\varepsilon_1^2-3\varepsilon_2^2)\varepsilon_1 + \tfrac{1}{4}C_1(\varepsilon_1^1+\varepsilon_2^2)^2+\ldots, \tag{5.22}$$

where A_1, B_1, C_1 are parameters which could (in principle) be measured by measuring the stress–strain relation. In particular, A_1 is just the elastic constant $C_{11}-C_{12}$. There is no linear term because $\varepsilon_1=\varepsilon_2=0$ is meant to represent the stable cubic state at high temperature, and because there is no way to make a linear function of $\varepsilon_1, \varepsilon_2$ which is invariant under the group operations. It turns out that in cubic symmetry, exactly one invariant function of $\varepsilon_1, \varepsilon_2$ can be constructed which is homogeneous of order 2, 3 and 4. These functions are exhibited explicitly in (5.22). The third-order term is particularly important. Suppose one has three different pairs of functions (ψ_1, ψ_2), (ψ_1', ψ_2') and (ψ_1'', ψ_2''), where each pair transforms according to representation Γ_{12}^{+}. Then it is possible to construct exactly one invariant which is trilinear in the three species, namely

$$f=\psi_1\psi_2'\psi_2''+\psi_2\psi_1'\psi_2''+\psi_2\psi_2'\psi_1''-\psi_1\psi_1'\psi_1''. \tag{5.23}$$

It is an important restriction on (5.23) that the rotation matrices $\Gamma_{ij}^{12}(R)$ for all three sets be *identical* (i.e., not just similar) to the matrices which rotate the pair $(\varepsilon_1, \varepsilon_2)$. The third-order term of (5.22) is just (5.23) with all three pairs set equal to $(\varepsilon_1, \varepsilon_2)$.

In order for (5.22) to describe a transition to a low-T tetragonal phase, one assumes following Landau (Landau and Lifshitz 1969) that B_1 and C_1 are weak functions of T, while

$$A=\alpha_1(T/T_0-1), \quad \alpha>0. \tag{5.24}$$

If $B_1=0$ and $C_1>0$, the free energy is minimized by

$$\varepsilon=\begin{cases} 0, & T>T_0 \\ [\alpha_1(1-T/T_0)/C]^{1/2}, & T<T_0 \end{cases}, \tag{5.25}$$

where $\varepsilon=(\varepsilon_1^2+\varepsilon_2^2)^{1/2}$ is the amplitude of the distortion. The direction is undetermined – it can be tetragonal in $\hat{x}, \hat{y}$ or $\hat{z}$ directions, or orthorhombic (linear combination of tetragonal in more than one direction). The existence of a third-order invariant ($B \neq 0$) in (5.22) alters this result in two ways. First, it singles out pure tetragonal distortions. The minima of (5.22) occur when $\varepsilon_2 = 0$ or $\varepsilon_2 = \pm\sqrt{3}\,\varepsilon_1$, which correspond to tetragonal distortions in the $\hat{z}, \hat{y}$ or $\hat{x}$ directions, respectively (Bhatt and McMillan 1976). Second, it forces the transition to be first order. If B is small, (5.25) remains valid except close to T_0. The third-order term causes ε to take a discontinuous jump of magnitude ε^* at an enhanced temperature T_0^*:

$$\varepsilon^* = -2B/3C,$$
$$T_0^* = T_0(1 + 2B^2/9\alpha C). \tag{5.26}$$

The results (5.26), with $B/C \ll 1$, correspond closely to experiment in V_3Si and Nb_3Sn. The approximate form (5.25) is accurate over a fairly wide range not too close or too far away from T_0. The sign of B determines the sign of the tetragonality, which is $c > a$ for V_3Si and opposite for Nb_3Sn.

At the time when Anderson and Blount first wrote down eq. (5.22), it was not yet known that C_{11}-C_{12} behaved as $(T$-$T_0)$ (see fig. 4), but X-ray measurements showed that the transition was second order within experimental resolution. This requires a small value of B, which seems somewhat of an accident. To avoid invoking chance, they proposed that strains $(\varepsilon_1, \varepsilon_2)$ might not be the true order parameter. Instead ε might enter F only in higher order of coupling (presumably in third or fourth order, i.e. $\psi^2\varepsilon$ or $\psi^2\varepsilon^2$) to a "hidden order parameter" ψ. The most probable choice for ψ is a sublattice distortion (i.e. an internal rearrangement of the atoms such as an optic phonon, not coupled in lowest order, $\psi\varepsilon$, to macroscopic strain. Then the absence of a term of the type ψ^3 in F would permit a second-order transition, and the higher-order coupling would cause $\langle\varepsilon\rangle$ to become finite and proportional to ψ or ψ^2 below T_c. Perel et al. (1968) searched unsuccessfully for this distortion in V_3Si by X-rays. Shirane and Axe (1971b) searched in Nb_3Sn by neutrons, and found only a small (0.015 Å) sublattice distortion of symmetry Γ_{12}, indicated by arrows in fig. 26. This does not qualify as the hidden order parameter required by Blount and Anderson because it has the same symmetry as the strain, is coupled bilinearly ($\psi\varepsilon$), and has an allowed cubic term (ψ^3) in F. However, the need for a hidden order parameter had vanished because Testardi et al. (1965) had found that C_{11}-C_{12} did in fact behave as T-T_0 above T_0. This behavior requires that the order parameter have Γ_{12} symmetry, and thus it follows that the transition must be first order. The Γ_{12} sublattice distortion automatically must accompany a tetragonal strain, as was implicit in the paper of Anderson and Blount, and explicit in the appendix of Perel et al. (which contains a nice elaboration of the group theory). This is because the free energy (5.22) can be supplemented by additional terms involving the amplitude of the sublattice distortion. Following Bhatt and McMillan (1976), we define three amplitudes, Q_x, Q_y, Q_z, which measure the amplitude of the sublattice distortion separately on the three chains of atoms in the $\hat{x}$, $\hat{y}$ and $\hat{z}$ directions, respectively (atoms (1,2), (3,4), (5,6) of fig. 24). These three amplitudes transform into each other under rotations and thus are the basis for a representation $(\Gamma_2^+ + \Gamma_{12}^+)$ of the group. It is important to take into account the screw rotations (shown in fig. 26) in deriving this result (which is given by Perel et al.). The combination $Q_0 = Q_x + Q_y + Q_z$ transforms according to Γ_2^+, and any choice of two orthogonal amplitudes (Q_1', Q_2') form a basis for Γ_{12}^+. It is important to make a unitary transformation to a set (Q_1, Q_2) defined to have *identical* rotation matrices to the set $(\varepsilon_1, \varepsilon_2)$ (eq. 5.21). A correct choice is

$$Q_1 = (Q_x - Q_y)/\sqrt{2}\,,$$
$$Q_2 = -(2Q_z - Q_x - Q_y)/\sqrt{6}\,. \tag{5.27}$$

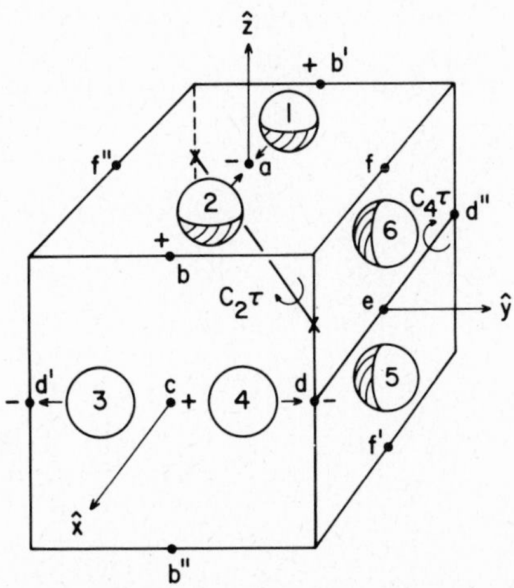

Fig. 26. A unit cell of A15 structure showing the six A atoms (chemical formula $(A_3B)_2$). The B atoms are located at the corners and center of the cube. Also shown are locations of "bond charges" a, b, …, f defined to be halfway between A atoms, along chains in the $\hat{x}$, $\hat{y}$ and $\hat{z}$ directions. The short arrows on atoms 1–4 indicate the displacement pattern of the $Q_x - Q_y$ phonon and of the low T distortion of Nb_3Sn. The ± signs on bond charges a–d indicate the $\phi_x - \phi_y$ charge density wave which drives the $Q_x - Q_y$ phonon soft in Gor'kov's model. Also shown are two of the screw rotation axes. The screw rotation $C_4\tau$ leaves the crystal invariant under a 90° rotation about the axis dd'' followed by a translation $\frac{1}{2}a(100)$. The screw rotation $C_2\tau'$ leaves the crystal invariant under a 180° rotation about the axis xx' followed by a translation $\frac{1}{2}a(110)$.

The free energy (5.22) should then be supplemented by terms for the sublattice

$$F_2 = \tfrac{1}{2}A_2(Q_1^2+Q_2^2)+\tfrac{1}{3}B_2(Q_1^2-3Q_2^2)Q_1+\tfrac{1}{4}C_1(Q_1^2+Q_2^2)^2, \tag{5.28}$$

and by coupling terms

$$F_{12} = D_{12}(Q_1\varepsilon_1+Q_2\varepsilon_2)+E_{12}[2Q_1Q_2\varepsilon_2+(Q_2^2-Q_1^2)\varepsilon_1] + G_{12}[2\varepsilon_1\varepsilon_2Q_2+(\varepsilon_2^2-\varepsilon_1^2)Q_1]+\dots. \tag{5.29}$$

The second-order term in (5.29) is invariant by the "shell sum theorem" (Lax 1974), while the two third-order mixed invariants are special cases of (5.23). Already we have at least nine free parameters, and things are getting out of hand. Only a few combinations of these parameters are relevant to the predicted behavior. After minimizing $F_1+F_2+F_{12}$ successively by the various order parameter components, one will be left with a model equivalent to either F_1 or F_2 alone, but with the parameters A_i, B_i, C_i renormalized. We can understand qualitatively what happens by examining a simplified case. A minimum model which displays the physics of two coupled order parameters is to keep A_1, A_2 and $D_{12}\neq 0$, plus one third-order and one fourth-order term. Let us arbitrarily take B_1 and C_1 as the non-vanishing third- and fourth-order coefficients:

$$F(Q,\varepsilon)=\tfrac{1}{2}A_1\varepsilon^2+\tfrac{1}{3}B_1\varepsilon^3+\tfrac{1}{4}C_1\varepsilon^4+\tfrac{1}{2}A_2Q^2+D_{12}Q\varepsilon. \tag{5.30}$$

Here we use the result that a minimum of F occurs when $Q_2=\varepsilon_2=0$, and drop the subscripts on ε_1, Q_1.

First let us eliminate the variable Q by minimizing:

$$\partial F/\partial Q|_{Q=\bar{Q}}=0,$$

$$\bar{Q}=-D_{12}\varepsilon/A_2, \tag{5.31}$$

$$F(\bar{Q},\varepsilon)=\tfrac{1}{2}(A_1-D_{12}^2/A_2)\varepsilon^2+\tfrac{1}{3}B_1\varepsilon^3+\tfrac{1}{4}C_1\varepsilon^4. \tag{5.32}$$

This describes a first-order transition at a renormalized temperature to a state of spontaneous strain. A sublattice distortion $\bar{Q}$ proportional to the strain accompanies the transition. Equivalent results can be derived by minimizing first by ε:

$$\partial F/\partial\varepsilon|_{\varepsilon=\bar{\varepsilon}}=0,$$

$$\bar{\varepsilon}=-D_{12}Q/A_1-B_1\bar{\varepsilon}^2/A_1-C_1\bar{\varepsilon}^3/A_1.$$

The second of these equations can be solved perturbatively for $\bar{\varepsilon}$ as a function of $Q' = D_{12}Q/A$:

$$\bar{\varepsilon} = -Q' - (B_1/A_1)Q'^2 - \left[2(B_1/A_1)^2 - C_1/A_1\right]Q'^3 + \dots . \tag{5.33}$$

Using this result to eliminate ε from (5.30), the free energy is

$$F(Q,\bar{\varepsilon}) = \tfrac{1}{2}(A_2 - D_{12}^2/A_1)(A_1/D_{12})^2 Q' - \tfrac{1}{3}B_1 Q'^3 + \tfrac{1}{4}(C_1 - 2B_1^2/A_1)Q'^4 + \dots . \tag{5.34}$$

This describes a first-order transition to a state of spontaneous sublattice distortion. A strain $\bar{\varepsilon}$ approximately proportional to the sublattice distortion accompanies the transition. Note that the transition temperature in both (5.32) and (5.34) is determined approximately by $A_1A_2 - D_{12}^2 = 0$ (assuming weak third-order terms). The general result is that the determinant of the matrix of coefficients of quadratic terms determines the transition temperature, if it is a second-order transition. For a first-order transition, T_0^* is influenced by higher than fourth-order coefficients which have been dropped in (5.34). Therefore when $B_1 \neq 0$ and the transition is first order, (5.32) and (5.34) make slightly different predictions, because of truncation errors.

Now we can ask an interesting question: which order parameter is more fundamental, strain or sublattice distortion? Lacking any other choice, the answer seems to be the one whose quadratic coefficient A_i contains the dominant temperature dependence. If the Landau free energy functional is derived theoretically from an underlying Hamiltonian, the primary order parameter should be clear. However, if the Landau functional is fitted to experiment, the distinction between primary and secondary order parameters is not completely meaningful. For example, many of the Landau predictions depend only weakly on which of the three parameters A_1, A_2 or D_{12} carries the T-dependence. The coefficient $A_1 - D_{12}^2/A_2$ of the quadratic term of (5.32) is physically identified with the renormalized elastic constant C_{11}-C_{12}. The elastic constant is the second derivative of F by strain. The correct procedure is to differentiate $F(\overline{Q}(\varepsilon), \varepsilon)$ rather than $F(Q, \varepsilon)$, because the sublattice displacement Q is rapid, and follows the strain adiabatically. Formulas for C_{11}-C_{12} are given below for the three different sources of T-dependence, assuming always $T > T_0^*$ (where T_0^* can be assumed to be only weakly enhanced above T_0 by third-order terms).

Case (a): $A_1 = A_1(0) + A_1'(0)T$,

$$C_{11} - C_{12} = A_1'(0)(T - T_0), \quad A_1'(0)T_0 = D_{12}^2/A_2 - A_1(0).$$

$$\textit{Case (b):}\quad A_2 = A_2(0) + A_2'(0)T,$$

$$C_{11} - C_{12} = (A_1 A_2'(0)/A_2(0))(T - T_0),$$
$$A_2'(0)T_0 = D_{12}^2/A_1 - A_2(0).$$

$$\textit{Case (c):}\quad |D_{12}| = |D_{12}(0)| - |D_{12}'(0)|T,$$

$$C_{11} - C_{12} = (2|D_{12}(0)||D_{12}'(0)|/A_2)(T - T_0),$$

$$2|D_{12}'(0)|T_0 = (D_{12}(0)^2 - A_1 A_2)/|D_{12}(0)|. \tag{5.35}$$

In all three cases, $C_{11} - C_{12}$ vanishes at T_0, which would be the transition temperature if third-order terms were absent. Note in cases (a) and (b) that it is no longer necessary to have $A_1(T)$ or $A_2(T)$ pass through zero in order to have an instability. Provided the coupling $|D_{12}|$ is large enough, and provided either A_1 or A_2 is an *increasing* function of temperature, T_0 is a positive temperature. The sign of D_{12} is irrelevant, but note in case (c) that $|D_{12}|$ must be a decreasing function of temperature. In both (b) and (c), terms higher than linear in T have been dropped, on the grounds that only first-order variations near T_0 are kept in the simplest version of Landau theory.

The observed vanishing of $C_{11} - C_{12}$ thus only indicates that the order parameter has Γ_{12} symmetry, but does not indicate whether the "primary" order parameter is strain, sublattice displacement or some other Γ_{12} quantity. A measurement of the temperature dependence of the optic mode Γ_{12} (where eigenvector is the same as the sublattice distortion) would be more revealing in this regard. The restoring force $\overline{M}\omega_0^2$, where $\overline{M}$ is the effective reduced mass and ω_0 is the Γ_{12} optic phonon frequency, is the second derivative $\partial^2 F/\partial Q^2$. It is clear that the derivative should be taken with strain $\varepsilon = 0$, because strains respond slowly compared to the frequency ω_0. Thus $\overline{M}\omega_0^2$ is a direct measure of the unrenormalized parameter A_2. In cases (a) and (c), A_2 is independent of T (to lowest order), while in case (b), A_2 is assumed to be a linear function of T, but finite and positive at T_0:

$$M\overline{\omega}_0^2 = D_{12}^2/A_1 + A_2'(0)(T - T_0)$$
$$(T > T_0^*, \quad \text{case (b)}). \tag{5.36}$$

Thus no model predicts a vanishing optic mode frequency (unless the coupling D_{12} accidentally vanishes), but case (b) predicts that ω_0 should have a measurable T-dependence. Unfortunately, ω_0 has not yet been measured at any temperature. In practice one may expect to find all three parameters A_1, A_2, D_{12} contributing some T-dependence. It seems likely, but it is surely not guaranteed, that one source will dominate.

Microscopic theory goes beyond the level of the Landau theory discussed so far by seeking in electronic theory the source of the T-dependence and of the negative value of $C_{11}-C_{12}$ at $T=0$ in the undistorted structure. It has been traditional to attribute both effects to the same mechanism, but this is not at all necessary. A deeper insight can be achieved by expanding the scope of Landau theory to include an electronic order parameter. This was first done by Kragler and Thomas (1975) and in a different way by Bhatt and McMillan (1976). Let us consider as possible electronic order parameters, the charges ρ_i' on each of the six transition atoms in the unit cell. The numbering scheme is shown in fig. 26. This is identical in spirit to eq. (5.4), except we are only considering a single type of local charge density distortion, interpreted as total charge. (Any other scalar distortion of the local charge density, such as a "breathing" distortion, would be an equally valid, though less probable, interpretation). The six charge parameters ρ_i' transform among themselves under rotations, forming a basis for the permutation group of the transition atoms as well as the point group of the crystal. This representation reduces to $\Gamma_1^+ + \Gamma_{12}^+ + \Gamma_{25}^-$ (Perel et al. 1968). The odd combinations $(\rho_1'-\rho_2', \rho_3'-\rho_4', \rho_5'-\rho_6')$ are a basis for Γ_{25}^-, while the symmetric total charge, $\rho_1'+\rho_2'+\ldots+\rho_6'$ transforms as Γ_1^+. Neither of these couples in low order to tetragonal strain, whereas the remaining two components do. Picking the phase so that the rotation matrices are identical to those for $(\varepsilon_1, \varepsilon_2)$ (eq. 5.21), these are (Kragler and Thomas 1975)

$$\rho_1 = (2\rho_z - \rho_x - \rho_y)/\sqrt{6}\,,$$

$$\rho_2 = (\rho_x - \rho_y)/\sqrt{2}\,, \tag{5.37}$$

where ρ_x, ρ_y, ρ_z are the total charges on each of the three chains $\rho_x = \rho_1' + \rho_2'$, etc. A generalized Landau theory will now have additional terms like

$$F_3 = \tfrac{1}{2}A_3(\rho_1^2+\rho_2^2) + \tfrac{1}{3}B_3(\rho_1^2 - 3\rho_2^2)\rho_1 + \tfrac{1}{4}C_3(\rho_1^2+\rho_2^2)^2, \tag{5.38}$$

as well as coupling terms such as

$$F_{13} = D_{13}(\varepsilon_1\rho_1 + \varepsilon_2\rho_2) + \ldots,$$

$$F_{23} = D_{23}(Q_1\rho_1 + Q_2\rho_2) + \ldots\,. \tag{5.39}$$

There are four third-order invariants with coefficients $E_{13}, E_{23}, G_{13}, G_{23}$, analogous to (5.29), which are suppressed in (5.39), and also a third-order trilinear invariant of the type (5.23). Thus in a complete Landau theory (eqs. (5.22), (5.28), (5.29), (5.38), (5.39)) there are 10 allowed third-order invariants. The number of allowed fourth-order invariants is 21, of which (mercifully) only the three homogeneous ones are displayed. A full Landau expansion to fourth order with three different Γ_{12} order parameters thus has $6+10+21=37$ free parameters. Any of the first six can carry the dominant T-dependence.

Kragler and Thomas (1975) considered only the two order parameters ε and ρ (following Labbe–Friedel (LF) theory), and stopped at second order. In LF theory, the T-dependence enters the coefficient A_3 (because of a very sharp peak in electronic density of states). Recently Bhatt (1978) has extended this theory by including a sublattice distortion which is coupled only to strain, not to charge (i.e. $D_{23}=0$). Of the higher-order terms, only C_3 is kept in fourth order, while in third order, only the term quadratic in ρ and linear in ε. The optic mode restoring force $\overline{M}\omega_0^2$ is now calculated as $\partial^2 F(\varepsilon, Q, \bar{\rho}(Q,\varepsilon))/\partial Q^2$, because it can be assumed that charge distortion responds rapidly, and follows the optic phonon adiabatically. Because $D_{13}=0$, ω_0 is predicted to be independent of T. By adding inhomogeneous charge response (i.e. terms like $G|\nabla\rho|^2$), the model is able to fit the T_1 phonon dispersion (acoustic branch with $Q\,\|[110], \hat{\varepsilon}\,\|[1\bar{1}0]$) out to finite $\boldsymbol{Q}$, as well as predict dispersion of ω_0 (which turns out to be T-dependent when $\boldsymbol{Q}\neq 0$).

While LF theory couples interchain charge fluctuations to macroscopic strains, Gor'kov's theory couples intrachain charge fluctuations to optic phonons. In the atom-charge model just discussed, the intrachain fluctuations (i.e. $\rho_1' - \rho_2'$, etc.) have Γ_{25}^- symmetry and do not couple in second order. However, by considering bond charges $(\phi_a, \phi_b, \ldots, \phi_f)$ rather than atom charges, a different result is obtained. In fig. 26, the six independent bond locations a, b, ..., f, are shown. Under rotations the scalar variables ϕ_a etc. transform among themselves like the permutations of the locations. They form a representation $\Gamma_1^+ + \Gamma_2^+ + 2\Gamma_{12}^+$. The even combinations $(\phi_a + \phi_b, \phi_c + \phi_d, \phi_e + \phi_f)$ transform as $\Gamma_2^+ + \Gamma_{12}^+$, and are in the present context indistinguishable from the charge variables ρ_x, ρ_y, ρ_z already discussed. The odd combinations $(\phi_x, \phi_y, \phi_z$, where $\phi_x = \phi_a - \phi_b$, etc.) transform as $\Gamma_2^+ + \Gamma_{12}^+$. In their transformation properties they are identical to the optic mode amplitudes (Q_x, Q_y, Q_z). The symmetric combination $\phi_x + \phi_y + \phi_z$ has Γ_2^+ symmetry, and the other two components should be chosen as

$$\phi_1 = (\phi_x - \phi_y)/\sqrt{2}\,,$$
$$\phi_2 = -(2\phi_z - \phi_x - \phi_y)/\sqrt{6}\,, \tag{5.40}$$

following eq. (5.27). The electronic distortions (ϕ_x, ϕ_y, ϕ_z) can be regarded as independent electronic charge-density-waves (CDW's) on the three chains. One expects a CDW to accompany an optic mode displacement.

This is indicated in fig. 26 where the arrows denote the Q_1 phonon displacements and the $\pm$ signs denote the corresponding (ϕ_1) charge displacements. There should be electronic charge accumulation (covalent bonding) between atoms on the chain in the regions of atomic compression. In the Peierls (1955) model the charge displacements occur because a gap opens in a half-filled, one-dimensional electron band when the atoms dimerize. The charge distortion lowers the electron energy and thus drives the transition. This is the essence of Gor'kov's model.

One can now imagine a Landau theory with four independent types of Γ_{12}^+ order parameters. The Landau analysis of Gor'kov theory by Bhatt and McMillan keeps only three types, ε, Q, and ϕ. Following Gor'kov, the dominant T-dependence enters the term $\frac{1}{2}A_4\phi^2$ and comes from the Peierls instability. Also, the second-order coupling of ϕ to ε is omitted. This theory has a T-dependent optical phonon frequency ω_0, as in eq. (5.36). One could easily build a theory with only interchain charge fluctuations ρ (and $\phi=0$) which would also have ω_0 depend on T as in (5.36), simply by allowing $D_{23}\neq 0$. Thus a measurement of ω_0 depending on T would not distinguish between the two types of electronic distortion.

As Bhatt and McMillan have shown, Gor'kov theory also has a possible instability involving a Γ_2^+ distortion involving a $Q_0=Q_x+Q_y+Q_z$ optic mode amplitude and a $\phi_0=\phi_x+\phi_y+\phi_z$ charge distortion. The reason is that it is physically expected that if a $\langle\phi_1\rangle$ or a $\langle\phi_2\rangle$ distortion lowers electronic energy, a $\langle\phi_0\rangle$ distortion should also. The specific Gor'kov form used by Bhatt and McMillan suggests that Γ_{12}^+ should be preferred. However, there are terms allowed by symmetry and omitted in this theory which would in principle permit Γ_2^+ to become the preferred distortion. It is interesting to speculate that possibly some of the A15 metals which do not have a Batterman–Barrett instability may instead have a Γ_2^+ instability which does not couple directly to macroscopic strain.

There is one experiment which in principle permits a determination of the relative importance of intrachain (ϕ) and interchain (ρ) distortions, namely a measurement of the T-dependence of the Γ_2^+ optic phonon, which should depend on T if ϕ-distortions dominate because of a $\phi_0 Q_0$ term in the free energy, but should not depend on T if ρ distortions dominate because symmetry forbids a $\rho_0 Q_0$ term. The implicit assumption is made here that the electronic restoring force $\frac{1}{2}A_5(T)\phi_0^2$ should be almost as strong a function of T as the $\frac{1}{2}A_4(T)(\phi_1^2+\phi_2^2)$ term is.

Now I would like to raise an objection to all these theories, namely to question the assumption that the T-dependence resides in an electronic restoring force like $A_3(T)$ or $A_4(T)$. These assumptions are based on microscopic LF and Gor'kov theories which use one-dimensional bands and have pathological response functions $\chi(T)$. The theories of LBW and

Bhatt show that it is *possible* to get strongly T-dependent electronic response in three-dimensional models also. However, it is difficult to understand why this should occur except by accident, or why disorder effects and lifetime broadening don't destroy the effect more readily. A way out was already suggested in §5.2 and used by McMillan (1977) in discussing CDW transitions in 2H-$TaSe_2$. A plausible microscopic mechanism for T-dependent restoring forces is anharmonic renormalization, which is surely present whenever harmonic restoring forces are small. This provides sufficient reason for believing that the quadratic coefficients A_1 and A_2 in the *free energy* (not the restoring forces K' in the *potential* energy (5.18)) should be *increasing* functions of T. The harmonic forces are small, corresponding vibrational amplitudes are large and the potential energy wells probably rise *more rapidly* than quadratic for large displacements, owing to d-shell overlap repulsion making the anharmonic term b in (5.18) positive. This yields the T-dependent effective restoring force (5.19) which enters the free energy. In this picture the source of the instability is that electronic renormalization (such as ρ or ϕ distortions) give a renormalization of A_1 and A_2 such that at least one of them becomes negative at $T=0$, after adiabatically eliminating ρ and ϕ from F. The adiabatic renormalizations ($-D_{ij}^2/A_j$, where i is 1 or 2 and j is 3 or 4) are probably only weak functions of T. The source of the T-dependence which makes the cubic phase stable at high T can come from either $A_1(T)$ or $A_2(T)$ or both. It becomes difficult in this picture to identify one order parameter as primary. As already suggested, measurement of the electronic energy-loss function might be a helpful experiment.

One final difficulty with assigning the T-dependence to electronic parameters is that microscopic theory gives long coherence lengths ξ_0 to electronic order parameters. The "bending energy" of a CDW, $\frac{1}{2}\alpha\xi_0^2|\nabla\phi|^2$, is large. According to Bhatt and Lee (1977), Gor'kov theory predicts ξ_0 of order $(a/\pi)(E_F/k_BT_0)$, and LF theory predicts ξ_0 of order $(a/\pi)(E_B/k_BT_0)^{1/2}$. Both E_F and E_B are electronic parameters of order the electron bandwidth. Thus ξ_0 is predicted to be many times larger than the lattice constant. The predicted T_1 acoustic phonon dispersion is (for $T>T_0^*$)

$$\omega^2(T_1)=(\alpha/\rho)(T/T_0-1+\xi_0^2Q^2)Q^2,$$

where $\alpha(T/T_0-1)$ is the renormalized value of $C_{11}-C_{12}$. Neutron measurements show that the softening of the phonon dispersion extends far out into the Brillouin zone, showing that ξ_0 is actually small. It is possible according to Bhatt and Lee that this difficulty is overcome by the more realistic three-dimensional models of Bhatt (1977) and LBW. McMillan (1977) has emphasized that if the coherence length is short, many low-frequency phonons are available to contribute to the free energy, and lattice entropy is likely to dominate over electronic entropy. This is an additional argument in favor of assigning the dominant T-dependence to a lattice coordinate, in this case, strain.

6. *Conclusions*

Many interesting questions lie unresolved. This article has emphasized three: (i) whether high T_c depends on anomalously low phonon frequencies, (ii) whether phonon anomalies in d-band metals are better described in terms of energy bands near E_F or valence charge fluctuations, and (iii) whether lattice instabilities are driven away at high T by electronic or lattice fluctuations. The first has been the subject of much speculation since 1970, the second is now the subject of a noisy debate, while the third has been only recently raised. Several other questions have been touched on. Why do so many high-T_c metals show strongly T-dependent phonon spectra, persisting to high temperature with anomalous sign? What effect, if any, do highly anharmonic phonons have on electron–phonon coupling and T_c? Surely other important questions are lurking in the data, but we have not yet been keen enough to appreciate them. It is safe to predict continued progress in this field. However, these are complicated questions unlikely to be settled rapidly, unless motivation remains high, and first-rate experiments continue to be done. I am told by outsiders that this is a "fashionable" field, and fashion is notoriously fickle.

Several indications point towards high-temperature phenomena as an increasingly important frontier for superconductivity. The anomalous high-T behavior of χ and ρ, the "saturation" idea (Fisk and Webb 1976) are interesting puzzles whose understanding may or may not be relevant to T_c. The anomalous high-T behavior of phonon spectra is surely relevant to T_c. The high-T specific heat is known to be affected by anomalous T-dependent phonon spectra (Knapp et al. 1975, 1976). One of the key questions in understanding the high T_c (or lack of it) in A15 compounds is the nature of the high-T phase diagrams which determine those compounds and compositions which can be made and whether a metastable phase can be created. It seems to me likely that the underlying Gibbs energy exhibits significant and interesting new behavior related to anharmonic or other high-temperature effects.

To many, the ultimate justification for this research is to raise T_c. I see no reason for pessimism here, and we shall all be quite disappointed if this does not happen. However, a more realistic justification is that our obsession with raising T_c will greatly enrich the frontiers of physics and materials science, whether the goal is achieved or not.

Recent developments

Many developments of the past two years are reviewed in a forthcoming conference proceedings (C4). There is also a new review by Smith and Finlayson (1979) on the experimental status of the correlation between high T_c and lattice instability.

Several recent experiments are worth special comment. The Γ_{12} optic phonon of V_3Si has at last been measured in a Raman scattering experiment by Wipf et al. (1978). As discussed in §5, the temperature dependence of the frequency of this mode is an important clue as to the mechanism for the structural instability. The experiment reveals a significant increase in linewidth $\gamma_Q(T)$ as T is lowered, and a small but definite decrease in frequency $\omega_Q(T)$. The latter proves that sublattice distortion plays a role in the structural dynamics. The former is harder to understand. Normally anharmonic linewidths decrease as T decreases. Therefore the result was interpreted as an electron–phonon effect, although usually the electron–phonon broadening is not a strong function of temperature.

Another recent experiment may shed some light on this. Staudenmann and Testardi (1979) have measured an anomalous behavior of the Debye–Waller factors by X-ray scattering from V_3Si. The behavior seems to indicate that $\langle u^4\rangle - 3\langle u^2\rangle^2$ is strongly different from its harmonic value of 0 for the V atoms. This anharmonic effect is very pronounced at 77 k, but goes away at room temperature. This mysterious behavior seems similar to the anomalous width measured by Wipf et al., so perhaps both are anharmonic effects. Staudenmann and Testardi interpret their result as indicating a strong tendency for atoms to sit off-center at 77 K (above the structural transition temperature).

Far infrared transmission in V_3Si has been measured by McKnight et al. (1979). The departure from Drude behavior has been fitted by assuming a form for the electron–phonon spectral function $\alpha^2F(\omega)$. The best fit requires much more weight at small ω than is expected from neutron scattering. The data were taken above T_c where the theory is simple. Sharper structure would be expected below T_c (see fig. 11) but the theory is quite complicated.

The "tungsten bronzes" A_xWO_3 where A is usually an alkali metal, show a rich interplay of superconductivity, structural changes and metal–insulator transitions. Neutron scattering measurements have been made on the hexagonal bronzes with $x=0.33$, by Kamitakahara et al. They find nearly dispersionless low frequency optic branches whose frequency ω_A varies as $M_A^{-1/2}$ where M_A is the mass of A atom. The variation of the T_c with atomic species A can be explained by assuming that a dominant contribution comes from these flat branches.

Perhaps the most striking theoretical development has been the continued success of the Varma–Weber procedure for calculating lattice eigenfrequencies. A long paper (Varma and Weber 1979) explains the calculations for elements and compounds have also been done now (Weber 1979). The dispersion curves of NbN were predicted in advance of the experiment (Christensen et al. 1979) and show a qualitative resemblance to the data. A more quantitative fit was achieved by using a lower Fermi energy, as is perhaps appropriate since the sample was substoichiometric ($NbN_{0.93}$). Similar experimental and theoretical achievements have now been reported for VN (Weber et al. 1979).

It is still not easy to reconcile the success of these theories with alternate formulations of the problem. Harmon and Ho (1979) have calculated the self-consistent electronic charge redistribution in Nb and Mo when a phonon $Q=(2\pi/a)\left(\frac{2}{3},\frac{2}{3},\frac{2}{3}\right)$ is "frozen" in place. They report (a) avoiding self-consistency by using the rigid-muffin-tin potential works reasonably in Nb but not at all for Mo; (b) the self-consistent Nb results plus a rigid band shift of E_F fails badly to explain Mo; and (c) states well below E_F are important, but their behavior may be in some way "correlated" with the behavior of states at E_F.

Recent calculations of λ have been reviewed by Butler (1979) who argues convincingly for the success (so far) of rigid-muffin-tin theory. However, Glötzel et al. (1979) have done a series of accurate rigid-muffin-tin calculations and conclude that either rigid-muffin-tin or local-density theory is failing, on the basis of significant discrepancies between theory and experiment for T_c. Finally, it is worth noting that theory has now advanced to a point where a prediction about T_c can be taken seriously. Gyorffy et al. (1979) have made calculations of electronic structure of $Pd_{1-x}Ag_x$ alloys, using KKR band theory combined with the coherent potential approximation (CPA). Their computed values of η are the first which treat the disordered alloy problem in a realistic way. They predict that as x increases, η does not fall nearly as rapid as the density of states $N(0)$. This suggests that for a range of x values, spin fluctuations will be suppressed yet λ will remain high enough for measurably large values of T_c to be found. The predictions rests on an extrapolation of the phonon frequencies; the problem of predicting these for disordered alloys is still much too hard.

Acknowledgements

I thank the many colleagues who have assisted in the preparation of this review, especially W. H. Butler. I thank D. Siegel and L. Zarifi for an excellent job of typing the manuscript. This work was supported in part by U.S. National Science Foundation grant no. DMR76-82946.

Conference Proceedings

Three recent published proceedings have a partial emphasis on T_c and phonons. These are referred to in the references as C1, C2, C3.

C1. The First Rochester Conference, Rochester, N.Y., October, 1971. Ed. by D. H. Douglass, 1972, *Superconductivity in d- and f-band metals* (American Institute of Physics, New York).

C2. The Second Rochester Conference, Rochester, N.Y., April, 1976. Ed. by D. H. Douglass, 1976, *Superconductivity in d- and f-band metals* (Plenum, New York).

C3. The Puerto Rico Conference, San Juan, Puerto Rico, November, 1975. Ed. by M. Gomez, J. A. Gonzalo and M. I. Kay, 1977, Ferroelectrics **16** and **17**, nos. 1/2.

Review Articles

These are referred to in the references as R1, R2,...

R1. Bergmann, G. (1976), Amorphous Metals and their Superconductivity, Physics Reports **27**, 159–185.

R2. Cody, G. D. and G. W. Webb (1973), Superconductivity: Phenomena, Theory, Materials, Crit. Rev. Sol. State Sci. **4**, 27–83.

R3. Dew-Hughes, D. (1975), Superconducting A15 Compounds: A Review, Cryogenics, 435–454.

R4. Geballe, T. H. and M. R. Beasley (1976), Superconducting Materials for Energy-Related Applications, in *Energy and materials*, Ed. by G. G. Libowitz and M. S. Whittingham (Academic Press, New York).

R5. Geilikman, B. T. (1973), Problems of High Temperature Superconductivity in Three-Dimensional Systems, Usp. Fiz. Nauk. **109**, 65–90 (Sov. Phys. – Usp. **16**, 17–30).

R6. Grimvall, G. (1976), The Electron–Phonon Interaction in Normal Metals, Physica Scripta **14**, 63–78.

R7. Izyumov, Yu A. and E. Z. Kurmaev (1974), Physical Properties and Electronic Structure of Superconducting Compounds with the β-Tungsten Structure, Usp. Fiz. Nauk. **113**, 193–238 (Sov. Phys. – Usp. **17**, 356–380).

R8. Izyumov, Yu. A. and E. Z. Kurmaev (1976), Superfluidity of Compounds Based on Transition Elements, and its Connection with Lattice Instability, Usp. Fiz. Nauk. **118**, 53–100 (Sov. Phys. – Usp. **19**, 26–52).

R9. Kopaev, Yu. V. (1976), Theory of the Relationship between Electron and Structural Transformations and Superconductivity, in *Proceedings of the P. N. Lebedev physics institute*, Vol. 86 (English Translation: Consultants Bureau, New York, 1977).

R10. Maksimov, E. G. (1976), Electron–Phonon Interaction and Superconductivity, in *Proceedings of the P. N. Lebedev physics institute*, Vol. 86 (English Translation: Consultants Bureau, New York, 1977).

R11. Probst, C. and J. Wittig (1979), Superconductivity: Metals, Alloys and Compounds, in *Handbook on the physics and chemistry of rare earths*, Ed. by K. A. Gschneider, Jr. and L. Eyring (North-Holland, Amsterdam), Vol. 1, p. 749.

R12. Roberts, B. W. (1976), Survey of Superconductive Materials and Critical Evaluation of Selected Properties, J. Phys. Chem. Ref. Data **5**, 581–821. A much more complete list of reviews and books is given on pp. 792–794 of this reference.
R13. Testardi, L. R. (1973), Elastic Behavior and Structural Instability of High Temperature A15 Structure Superconductors, in *Physical acoustics*, Ed. by W. P. Mason and R. N. Thursten (Academic Press, New York), Vol. 10, p. 193.
R14. Testardi, L. R. (1975), Structural Instability and Superconductivity in A-15 Compounds, Rev. Mod. Phys. **47**, 637–648.
R15. Testardi, L. R. (1977), Structural Instability of A15 Superconductors, in *Physical acoustics*, Ed. by W. P. Mason and R. N. Thurston (Academic Press, New York), Vol. 13, p. 27.
R16. Vonsovskii, S. V., Yu. A. Izyumov and E. Z. Kurmaev (1977), Superconductivity of Transition Metals and their Alloys and Compounds (in Russian), Nauka, Moscow.
R17. Weger, M. and I. B. Goldberg (1973), Some Lattice and Electronic Properties of the β-Tungstens, in *Solid state physics*, Ed. by F. Seitz, D. Turnbull and H. Ehrenreich, Vol. 23, pp. 1–177.
R18. Wolf, E. L., Electron Tunneling Spectroscopy, Rept. Prog. Phys. **41**, 1439–1508.

References

Allen, P. B. (1972), Phys. Rev.B **6**, 2577.
Allen, P. B. (1973), Solid State Commun. **12**, 379.
Allen, P. B. (1977), Phys. Rev.B **16**, 5139.
Allen, P. B. and M. L. Cohen (1969), Phys. Rev. **187**, 525.
Allen, P. B. and M. L. Cohen (1972), Phys. Rev. Lett. **29**, 1593.
Allen, P. B. and R. C. Dynes (1975a), Phys. Rev.B **12**, 905 (referred to as AD).
Allen, P. B. and R. C. Dynes (1975b), Phys. Rev.B **11**, 1895.
Allen, P. B. and R. S. Silberglitt (1974), Phys. Rev.B **9**, 4733.
Allen, P. B. and W. E. Pickett, K. M. Ho and M. L. Cohen (1978), Phys. Rev. Lett. **40**, 1532.
Anderson, P. W. (1958), Phys. Rev. **112**, 900.
Anderson, P. W. (1959), J. Phys. Chem. Solids **11**, 26.
Anderson, P. W. (1961), Some Recent Developments in the Theory of Superconductivity, in *Proceedings of the seventh international conference on low temperature physics*, Toronto, 1961, Ed. by G. M. Graham (Plenum, New York) pp. 298–310.
Anderson, P. W. (1977), unpublished remarks. Also published remarks at the International Conference on Physics of Transition Metals, Toronto, 1977.
Anderson, P. W. and E. I. Blount (1965), Phys. Rev. Lett. **14**, 217.
Appel, J. (1968), Phys. Rev. Lett. **21**, 1164.
Arnold, G. B. (1978), Phys. Rev.B **18**, 1076.
Arnold, G. B., J. Zasadzinski and E. L. Wolf (1978), Phys. Lett. **69A**, 136.
Aschermann, Friedrich, Justi and Kramer (1941), Physik. Z. **42**, 349.
Ashcroft, N. W. and J. W. Wilkins (1965), Phys. Lett. **14**, 285.
Axe, J. D. and G. Shirane (1973a), Phys. Rev.B **8**, 1965.
Axe, J. D. and G. Shirane (1973b), Phys. Rev. Lett. **30**, 214.
Bader, S. D. and F. Y. Fradin (1976), in C2, pp. 567–582.
Bader, S. D., S. K. Sinha and R. N. Shelton (1976), in C2, pp. 209–221.
Baldereschi, A. and K. Maschke (1976), in *Proceedings of the international conference on lattice dynamics*, Ed. by M. Balkanski (Flammarion, Paris) pp. 36–38.

Ball, M. A. (1975), J. Phys.C **8**, 3328.
Ballentine, L. E. (1967), Phys. Rev. **158**, 670.
Bardeen, J. and D. Pines (1955), Phys. Rev. **99**, 1140.
Bardeen, J., L. N. Cooper and J. R. Schrieffer (1957), Phys. Rev. **108**, 1175.
Barisic, S. (1972), Phys. Rev.B **5**, 932, 941.
Barker, A. S., J. A. Ditzenberger and F. J. DiSalvo (1975), Phys.B **12**, 2049.
Batterman, B. W. and C. S. Barrett (1964), Phys. Rev. Lett. **13**, 390.
Baym, G. (1961), Ann. Phys. (N.Y.) **14**, 1.
Bennemann, K. H. and J. W. Garland (1972), Theory for Superconductivity in d-Band Metals, in *Superconductivity in d- and f-band metals*, Ed. by D. H. Douglass (American Institute of Physics, New York), pp. 103–137.
Bergmann, G. and D. Rainer (1973), Z. Physik **263**, 59.
Bertoni, C. M., V. Bortolani, C. Calandra and F. Nizzoli (1974), J. Phys. F. **4**, 19.
Bhatt, R. N. (1977), Phys. Rev. B **16**, 1915.
Bhatt, R. N. (1978), Phys. Rev. B **17**, 2947.
Bhatt, R. N. and P. A. Lee, (1977), Phys. Rev. B **16**, 4288.
Bhatt, R. N. and L. F. Mattheiss (1979), Phys. Rev. B **20**, 2542.
Bhatt, R. N. and W. L. McMillan (1976), Phys. Rev. B **14**, 1007.
Bilz, H., B. Gliss and W. Hanke (1974), Theory of Phonons in Ionic Crystals, in *Dynamical properties of solids*, Ed. by G. K. Horton and A. A. Maradudin (North Holland, Amsterdam), Vol. 1, pp. 343–390.
Birnboim, A. (1976), Phys. Rev. B **14**, 2857.
Bobetic, V. M. (1964), Phys. Rev. **136**, A1535.
Bogoliubov, N. N., V. V. Tolmachev and D. V. Shirkov (1958), *A new method in the theory of superconductivity*, Academy of Science, Moscow, (Consultants Bureau, New York, 1959).
Bostock, J. and M. L. A. MacVicar (1978), Phys. Lett. **71A**, 373.
Bostock, J., K. H. Lo, W. N. Cheung, V. Diadiuk and M. L. A. MacVicar (1976), in C2, p. 367.
Boyer, L. L., B. M. Klein and D. A. Papaconstantopoulos (1977), Ferroelectrics **16**, 291.
Brovman, E. G. and Yu. M. Kagan (1974), Phonons in Non-Transition Metals, in *Dynamical properties of solids*, Ed. by G. K. Horton and A. A. Maradudin (North-Holland, Amsterdam), Vol. 1, pp. 191–300.
Burger, J. P. and D. S. MacLachlan (1976), J. de Physique **37**, 1227.
Butler, W. H. (1977), Phys. Rev.B **15**, 5267.
Butler, W. H. and P. B. Allen (1976) in C2, p. 73.
Butler, W. H., J. J. Olson, J. S. Faulkner and B. L. Gyorffy (1976), Phys. Rev.B **14**, 3823.
Butler, W. H., H. G. Smith and N. Wakabayashi (1977), Phys. Rev. Lett. **39**, 1004.
Butler, W. H., F. J. Pinski and P. B. Allen (1979), Phys. Rev. B **19**, 3708.
Chaikin, P. M., G. Arnold and P. K. Hansma (1977), J. Low Temp. Phys. **26**, 229.
Chu, C. W., V. Diatschenko, C. Y. Huang and F. J. DiSalvo (1977), Phys. Rev.B **15**, 1340.
Cohen, M. L. and V. Heine (1970), in *Solid state physics*, Ed. by F. Seitz, D. Turnbull and H. Ehrenreich (Academic Press, New York), Vol. 24, pp. 249–463.
Cohen, R. W., G. D. Cody and J. J. Halloran (1967), Phys. Rev. Lett. **19**, 840.
Cooke, J. F., H. L. Davis and M. Mostoller (1974), Phys. Rev.B **9**, 2485.
Das, S. G. (1973), Phys. Rev.B **7**, 2238.
DiSalvo, F. J., J. A. Wilson, B. G. Bagley and J. V. Waszczak (1975), Phys. Rev.B **12**, 2220.
Dynes, R. C. (1972), Solid State Commun. **10**, 615.
Dynes, R. C. and J. P. Garno (1975), Bull. Am. Phys. Soc. **20**, 422.
Dynes, R. C. and J. M. Rowell (1975), Phys. Rev.B **11**, 1884.
Eichler, A., H. Wühl and B. Stritzker (1975), Solid State Commun. **17**, 213.

Eliashberg, G. M. (1960), Zh. Eksp. Teor. Fiz. **38**, 966; **39**, 1437 (Sov. Phys. JETP **11**, 696 (1960); **12**, 1000 (1961)).
Evans, R., G. D. Gaspari and B. L. Gyorffy (1973), J. Phys.F **3**, 39.
Farrell, D. F. and B. S. Chandrasekhar (1977), Phys. Rev. Lett **38**, 788.
Faulkner, J. S. (1976), Phys. Rev.B **13**, 2391.
Fisk, Z. and G. W. Webb (1976), Phys. Rev. Lett. **36**, 1084.
Freeman, A. J., M. Gupta, H. W. Myron, J. Rath and T. Watson-Yang (1977), in *Proceedings of the international conference on lattice dynamics*, Ed. by M. Balkanski (Flammarion, Paris), p. 204.
Friday, B. R. (1973), Physics Today **26** #6, 11.
Fröhlich, H. (1952), Proc. Roy. Soc. (London) **A215**, 291.
Gale, S. G. and D. G. Pettifor (1977), Sol. State Commun. **24**, 175.
Galoshina, E. V. (1974), Sov. Phys. - Usp. **17**, 345 (Usp. Fiz. Nauk. **113**, 105 (1974)).
Ganguly, B. N. (1973), Z. Phys. **265**, 433.
Ganguly, B. N. (1976), Phys. Rev.B **14**, 3848.
Gärtner, K. and A. Hahn (1976), Z. Naturforsch. **31a**, 361.
Gaspari, G. D. and B. L. Gyorffy (1972), Phys. Rev. Lett. **28**, 801.
Gavaler, J. R. (1973), Appl. Phys. Lett. **23**, 480.
Giaever, I. (1960), Phys. Rev. Lett. **5**, 147.
Giaever, I., H. R. Hart, Jr. and K. Megerle (1962), Phys. Rev. **126**, 941.
Golibersuch, D. C. (1969), Phys. Rev. **157**, 532.
Gomersall, I. R. and B. L. Gyorffy (1974a), J. Phys.F **4**, 1204.
Gomersall, I. R. and B. L. Gyorffy (1974b), Phys. Rev. Lett **33**, 1286.
Gompf, F., H. Lau, W. Reichardt and J. Salgardo (1972), in *Neutron inelastic scattering, Proceedings of a symposium in Grenoble* (IAEA, Vienna), pp. 137–148.
Gor'kov, L. P. (1973), Zh. Eksp. Teor. Fiz. **65**, 1658 (Soviet Phys. - JETP **38**, 830 (1974)).
Gor'kov, L. P. and O. N. Dorokhov (1976), J. Low Temp. Phys. **22**, 1.
Götze, W. and K. H. Michel (1974), Self-Consistent Phonons, in *Dynamical properties of solids*, Ed. by G. K. Horton and A. A. Maradudin (North-Holland, Amsterdam), Vol. 1, pp. 499–540.
Grimvall, G. (1976), R6.
Grünewald, G. and K. Scharnberg (1976), Phys. Rev. Lett. **37**, 361.
Guha, A., M. P. Sarachik, F. W. Smith and L. R. Testardi (1978), Phys. Rev.B **18**, 9.
Gupta, M. and A. J. Freeman (1976), Phys. Rev.B **14**, 5205.
Gurvitch, M., A. K. Ghosh, C. L. Snead, Jr. and M. Strongin (1977), Phys. Rev. Lett. **39**, 1102.
Hanke, W. (1973), Phys. Rev.B **8**, 4585, 4591.
Hanke, W., J. Hafner and H. Bilz (1976), Phys. Rev. Lett. **37**, 1560.
Hardy, G. F. and J. K. Hulm (1953), Phys. Rev. **89**, 884.
Harmon, B. N. and S. K. Sinha (1977), Phys. Rev.B **16**, 3919.
Heine, V., P. Nozieres and J. W. Wilkins (1966), Phil. Mag. **13**, 741.
Herring, C. (1965), Exchange Interactions Among Itinerant Electrons, in *Magnetism*, Ed. by G. T. Rado and H. Suhl (Academic Press, New York), Vol. 5.
Ho, K. M., W. E. Pickett and M. L. Cohen (1978), Phys. Rev. Lett. **61**, 580.
Hopfield, J. J. (1969), Phys. Rev. **186**, 443.
Horsch, P. and H. Rietschel (1977), Z. Phys.B **27**, 153.
Hubin, W. and D. M. Ginsberg (1969), Phys. Rev. **188**, 716.
Hui, J. C. K. and P. B. Allen (1974), J. Phys.F **4**, L42.
Janak, J. F. (1969), Phys. Lett. **28A**, 570.
Jansen, A. G. M., F. M. Mueller and P. Wyder (1977), Phys. Rev.B **16**, 1325.

Jarlborg, T. and G. Arbman (1977), J. Phys. F **7**, 1635.
John, W. (1973), J. Phys. F **3**, L231.
Joyce, R. R. and P. L. Richards (1970), Phys. Rev. Lett. **24**, 1007.
Kakitani, T. (1969), Prog. Theor. Phys. **42**, 1238.
Kamerlingh Onnes, H. (1911), Akad. van Wetenschappen (Amsterdam) **14**, 113.
Kamitakahara, W. A., H. G. Smith and N. Wakabayashi (1977), Ferroelectrics **16**, 111.
Karakozov, A. E., E. G. Maksimov and S. A. Mashkov (1975), Zh. Eksp. Teor. Fiz. **68**, 1937 (Sov. Phys.-JETP **41**, 971 (1976)).
Karim, D. P., J. B. Ketterson and G. W. Crabtree (1978), J. Low Temp. Phys. **30**, 389.
Keller, K. R. and J. J. Hanak (1967), Phys. Rev. **154**, 628.
Kimball, C. W., L. W. Weber and F. Y. Fradin (1976a), Phys. Rev.B **14**, 2769.
Kimball, C. W., L. W. Weber, G. Van Landuyt, F. Y. Fradin, B. D. Dunlap and G. K. Shenoy (1976b), Phys. Rev. Lett. **36**, 412.
Kittel, C. (1963), Quantum Theory of Solids, (Wiley, N.Y.).
Kittel, C. (1976), *Introduction to solid state physics*, fifth ed. (Wiley, New York).
Kirzhnits, D. A., E. G. Maksimov and D. I. Khomskii (1973), J. Low Temp. Phys. **10**, 79.
Klein, B. M. and D. A. Papaconstantopoulos (1976), J. Phys. F **6**, 1135.
Klein, B. M., D. A. Papaconstantopoulos and L. L. Boyer (1976), Calculations of the Superconducting Properties of Compounds, in C2, pp. 339–360.
Klein, B. M., D. A. Papaconstantopoulos and L. L. Boyer (1977), Ferroelectrics **16**, 299.
Knapp, G. S., S. D. Bader, H. V. Culbert, F. Y. Fradin and T. E. Klippert (1975), Phys. Rev.B **11**, 4331.
Knapp, G. S., S. D. Bader and Z. Fisk (1976), Phys. Rev.B **13**, 3783.
Kohlhaas, R. and W. D. Weiss (1964), Z. Naturforsch, **29a**, 1227.
Kragler, R. (1978), Physica **93B**, 314.
Kragler, R. and H. Thomas (1975), J. de Physique Lett. **36**, 153.
Kugler, A. A. (1975), J. Stat. Phys. **12**, 35.
Labbé, J. and J. Friedel (1966a), J. Phys. (Paris) **27**, 153.
Labbé, J. and J. Friedel (1966b), J. Phys. (Paris) **27**, 303.
Labbé, J., S. Barisić and J. Friedel (1967), Phys. Rev. Lett **19**, 1039.
Landau, L. D. and E. M. Lifshitz (1969), *Statistical physics*, second ed. (Addison-Wesley, Reading, Mass.), ch. XIV.
Lax, M. (1974), *Symmetry principles in solid-state and molecular physics* (Wiley, New York).
Leavens, C. R. (1975), Solid State Commun. **17**, 1499.
Leavens, C. R. (1976), Solid State Commun. **19**, 395.
Leavens, C. R. (1977), J. Phys. F, **7**, 1911.
Leavens, C. R. and J. P. Carbotte (1974), J. Low Temp. Phys. **14**, 195.
Lee, T. K. and J. L. Birman (1978), Phys. Rev.B **17**, 4391.
Lee, T. K., J. L. Birman and S. J. Williamson (1977a), Phys. Rev. Lett. **39**, 839.
Lee, T. K., J. L. Birman and S. J. Williamson (1977b), Phys. Lett. **64A**, 89.
Lie, S. G. and J. P. Carbotte (1978), Solid State Commun. **26**, 511.
Lou, L. F. and W. J. Tomasch (1972), Phys. Rev. Lett. **29**, 858.
Louie, S. G. and M. L. Cohen (1977), Solid State Commun. **22**, 466.
MacLachlan, D. S., R. Mailfert, J. P. Burger and B. Souffaché (1975), Solid State Commun. **17**, 281.
Mailfert, R., B. W. Batterman and J. J. Hanak (1967), Phys. Lett. **24A**, 315.
Mailfert, R., B. W. Batterman and J. J. Hanak (1969), Phys. Stat. Sol. **32**, K67.
Markowitz, D. and L. P. Kadanoff (1963), Phys. Rev. **131**, 563.
Mattheiss, L. F. (1970), Phys. Rev.B **1**, 373.
Mattheiss, L. F. (1973), Phys. Rev.B **8**, 3719.
Mattheiss, L. F. (1975), Phys. Rev.B **12**, 2161.

Mattheiss, L. F., L. R. Testardi and W. W. Yao (1978), Phys. Rev.B **17**, 4640.
Matthias, B. T., T. H. Geballe, S. Geller and E. Corenzwit (1954), Phys. Rev. **95**, 1435.
Matthias, B. T., T. H. Geballe, B. D. Longinotti, E. Corenzwit, G. W. Hull and J. P. Maita (1967), Science **156**, 645.
McMillan, W. L. (1968), Phys. Rev. **167**, 331.
McMillan, W. L. (1975), Phys. Rev.B **12**, 1187.
McMillan, W. L. (1977), Phys. Rev.B **16**, 643.
McMillan, W. L. and J. M. Rowell (1965), Phys. Rev. Lett. **14**, 108.
McMillan, W. L. and J. M. Rowell (1969), Tunneling and Strong-Coupling Superconductivity, in *Superconductivity*, Ed. by R. D. Parks (M. Dekker, New York), pp. 561–613.
Medvedev, M. V., E. A. Pashitskii and Yu. S. Pyatiletov (1973), Zh. Eksp. Teor. Fiz. **65**, 1186 (Sov. Phys.-JETP **38**, 587 (1974)).
Meissner, W. and H. Franz (1930), Z. Phys. **65**, 30.
Migdal, A. B. (1958), Zh. Eksp. Teor. Fiz. **34**, 1438 (Sov. Phys.-JETP **7**, 996 (1958)).
Miller, R. J. and C. B. Satterthwaite (1975), Phys. Rev. Lett. **34**, 144.
Moncton, D. E., J. D. Axe and F. J. DiSalvo (1975), Phys. Rev. Lett. **34**, 734.
Moncton, D. E., J. D. Axe and F. J. DiSalvo (1977), Phys. Rev.B **16**, 801.
Morel, P. (1958), Phys. Rev. Lett **1**, 244.
Myron, H. W., J. Rath and A. J. Freeman (1977), Phys. Rev.B **15**, 855.
Nakajima, S. and M. Watabe (1963), Prog. Theor. Phys. **29**, 341.
Ng, S. C. and B. N. Brockhouse (1968), Atomic Vibrations in Face-Centered Cubic Alloys of Bi, Pb, and Tl, in *Neutron inelastic scattering, Proceedings of the IAEA symposium*, Copenhagen, 1968 (IAEA, Vienna), p. 253.
Noolandi, J. and L. J. Sham (1973), Phys. Rev.B **8**, 2468.
Nowak, D. (1972), Phys. Rev.B **6**, 3691.
Overhauser, A. W. (1968), Phys. Rev. **167**, 691.
Owen, C. S. and D. J. Scalapino (1971), Physica (Utr.) **55**, 691.
Papaconstantopoulos, D. A. and B. M. Klein (1975)
Papaconstantopoulos, D. A., L. L. Boyer, B. M. Klein, A. R. Williams, V. L. Morruzzi and J. F. Janak (1977), Phys. Rev.B **15**, 4221.
Papaconstantopoulos, D. A., B. M. Klein, E. N. Economou and L. L. Boyer (1978), Phys. Rev.B **17**, 141.
Peierls, R. E. (1955), *Quantum theory of solids* (Oxford University Press, Oxford) p. 108ff.
Perel, J., B. W. Batterman and E. I. Blount (1968), Phys. Rev. **166**, 616.
Peter, M., J. Ashkenazi and M. Dacorogna (1977), Helv. Phys. Acta **50**, 267.
Pettifor, D. G. (1977), J. Phys. F **7**, 1009.
Pick, R. M. (1971), Microscopic theory and shell model in insulators, in *Phonons, Proceedings of the international conference*, Rennes, 1971, Ed. by M. A. Nusimovici (Flammarion, Paris), p. 20.
Pickett, W. E. and P. B. Allen (1977), Phys. Rev.B **16**, 3127.
Pickett, W. E. and B. L. Gyorffy (1976), Theory of Lattice Dynamics of Strong Coupling Systems in the Rigid Muffin Tin Approximation, in C3, p. 251.
Powell, B. M., P. Martel and A. D. B. Woods (1968), Phys. Rev. **171**, 727.
Powell, B. M., A. D. B. Woods and P. Martel (1972), In *Neutron inelastic scattering*, (IAEA, Vienna), pp. 43–51.
Powell, B. M., P. Martel and A. D. B. Woods (1977), Can. J. Phys. **55**, 1601.
Pytte, E. (1971), Phys. Rev.B **4**, 1094.
Rahman, A., K. Skold, C. Pelizzari and S. K. Sinha (1976), Phys. Rev.B **14**, 3630.
Rice, T. M. and G. K. Scott (1975), Phys. Rev. Lett. **35**, 120.
Rietschel, H. (1975), Z. Phys. B **22**, 133.
Roberts, B. W. (1976), R12.

Robinson, B., T. H. Geballe and J. M. Rowell (1976), in C2, p. 381.

Rowe, J. M., J. J. Rush, H. G. Smith, M. Mostoller and H. E. Flotow (1974), Phys. Rev. Lett. **33**, 1297.

Rowell, J. M. (1976), Solid State Commun. **19**, 1131.

Rowell, J. M., A. G. Chynoweth and J. C. Phillips (1962), Phys. Rev. Lett. **9**, 59.

Rowell, J. M., P. W. Anderson and D. E. Thomas (1963), Phys. Rev. Lett. **10**, 334.

Scalapino, D. J. and P. W. Anderson (1964), Phys. Rev. **133**, A921.

Scalapino, D. J., J. R. Schrieffer and J. W. Wilkins (1966), Phys. Rev. **148**, 263.

Schirber, J. E. and C. J. M. Northrup, Jr. (1974), Phys. Rev.B **10**, 3818.

Schrieffer, J. R., D. J. Scalapino and J. W. Wilkins (1963), Phys. Rev. Lett. **10**, 336.

Schweiss, B. P., B. Renker, E. Schneider and W. Reichardt (1976), in C2,p. 189–208.

Sham, L. J. (1969), Effects of Electron Dynamics on Lattice Vibrations, in *Modern solid state physics*: Vol. 2, *Phonons and their interactions* Ed. by R. H. Enns and R. R. Haering (Gordon and Breach, New York), pp. 143–198.

Sham, L. J. (1971), Phys. Rev. Lett. **27**, 1725.

Sham, L. J. (1973), Phys. Rev.B **7**, 4357.

Sham, L. J. (1974), Theory of Lattice Dynamics of Covalent Crystals, in *Dynamical properties of solids*, Ed. by G. K. Horton and A. A. Maradudin (North-Holland, Amsterdam), Vol. 1, pp. 301–342.

Shapiro, S. M., G. Shirane and J. D. Axe (1975), Phys. Rev.B **12**, 4899.

Shen, L. Y. L. (1970), Phys. Rev. Lett. **24**, 1104.

Shen, L. Y. L. (1972a), Phys. Rev. Lett. **29**, 1082.

Shen, L. Y. L. (1972b), in C1, pp. 31–44.

Shirane, G. and J. D. Axe (1971a), Phys. Rev. Lett. **27**, 1803.

Shirane, G. and J. D. Axe (1971b), Phys. Rev.B **4**, 2957.

Silverman, P. J. and C. V. Briscoe (1975), Phys. Lett A **53**, 221.

Singhal, S. P. and J. Callaway (1976), Phys. Rev.B **14**, 2347.

Sinha, S. K. (1968), Phys. Rev. **169**, 477.

Sinha, S. K. and B. N. Harmon (1975), Phys. Rev. Lett. **35**, 1515.

Sinha, S. K. and B. N. Harmon (1976), Phonon Anomalies in the d-Band Metals, in C2, pp. 269–296.

Sinha, S. K., R. P. Gupta and D. L. Price (1971), Phys. Rev. Lett. **26**, 1324.

Skoskiewicz, T. (1972), Phys. Stat. Sol. **A11**, K123.

Smith, H. G. (1972a), Phonon Spectra and Superconductivity in some Transition- and Actinide-Metal Carbides, in C1.

Smith, H. G. (1972b), Phys. Rev. Lett. **29**, 353.

Smith, H. G. and W. Gläser (1970), Phys. Rev. Lett. **25**, 1611.

Smith, T. F. and T. R. Finlayson (1978), *High pressure and low temperature physics*, Ed. by C. W. Chu and J. A. Woollam (Plenum, New York), p. 315.

Spengler, W., R. Kaiser, A. N. Christiansen and G. Müller-Vogt (1978), Phys. Rev. B **17**, 1095.

Stedman, R., L. Almquist and G. Nilsson (1967), Phys. **162**, 549.

Stritzker, B. (1974), Z. Phys. **268**, 261.

Swihart, J. C., D. J. Scalapino and Y. Wada (1964), in *Proceedings of the IXth international conference on low temperature physics*, Columbus, Ohio 1964, Ed. by J. G. Daunt (Plenum, New York 1965).

Talbot, E. and C. R. Leavens (1978), private communication.

Taylor, D. W. and P. Vashishta (1972), Phys. Rev.B **5**, 4410.

Testardi, L. R. (1971), Phys. Rev.B **3**, 95.

Testardi, L. R. (1972), Phys. Rev.B **5**, 4342.

Testardi, L. R., T. B. Batterman, W. A. Reed and V. G. Chirba (1965), Phys. Rev. Lett. **15**, 250.

Testardi, L. R., J. H. Wernick and W. A. Royer (1974), Solid State Commun. **15**, 1.
Toya, T. (1958), J. Res. Inst. Catal. Hokkaido Univ. **6**, 161, 183.
Vandenberg, J. M and B. T. Matthias (1977), Science **198**, 194.
Varma, C. M. and R. C. Dynes (1976), Empirical Relations in Transition Metal Superconductivity, in C2, pp. 507–533.
Varma, C. M. and W. Weber (1977), Phys. Rev. Lett. **39**, 1094.
Varma, C. M., J. C. Phillips and S. T. Chui (1974), Phys. Rev. Lett. **33**, 1223.
Varma, C. M., E. I. Blount, P. Vashishta and W. Weber (1979), Phys. Rev. B **19**, 6130.
Vosko, S. H., R. Taylor and G. H. Keech (1965), Can. J. Phys. **43**, 1187.
Wakabayashi, N. (1977), Solid State Commun. **23**, 737.
Weber, W. (1973a), Phys. Rev.B **8**, 5082.
Weber, W. (1973b), Phys. Rev.B **8**, 5090.
Weber, W., H. Bilz and U. Schroder (1972), Phys. Rev. Lett. **28**, 600.
Werthamer, N. R. (1969), Am. J. Phys. **37**, 763.
Williams, P. M., G. S. Parry and C. B. Scruby (1974), Phil. Mag. **29**, 695.
Wilson, J. A., F. J. DiSalvo and S. Mahajan (1974), Phys. Rev. Lett. **32**, 882.
Wilson, J. A., F. J. DiSalvo and S. Mahajan (1975), Advan. Phys. **24**, 117.
Winter, H. (1977), Calculation of the Superconducting Transition Temperature of Thorium, in *Proceedings of the international conference on physics of transition metals*, Toronto, 1977, Ed. by M. J. G. Lee (to be published).
Woodard, D. W. and G. D. Cody (1964), Phys. Rev. **136**, 166A.
Wu, H. -S., C. -H. Tsai, C. -T. Kung, K. T. Chi and C. -T. Tsai (1977), Scientia Sinica XX, 583.
Wühl, H., A. Eichler and J. Wittig (1973), Phys. Rev. Lett. **31**, 1393.
Yamashita, J. and S. Asano (1974), Prog. Theor. Phys. **51**, 317.
Yanson, I. K. and Yu. N. Shalov (1976), Zh. Eksp. Teor. Fiz. **71**, 286, (Sov. Phys.–JETP **44**, 148).
Ziman, J. M. (1962), Phys. Rev. Lett. **8**, 272.

Conference Proceedings

C4. The Third Rochester Conference, San Diego, Calif., June 1979, Ed. by H. Suhl and M. B. Maple (Academic Press, New York, to be published).

Review Articles

Smith, T. F. and T. R. Finlayson (1979), Superconductivity and Lattice Instability, Contemp. Phys. (to be published).
Stritzker, B. and H. Wuhl (1978), Superconductivity in Metal-Hydrogen Systems, in G. Alefeld and J. Völkl, eds., *Hydrogen in Metals*, (Springer-Verlag, Berlin) V. II p. 243.

References

Butler, W. H. (1979), in C4.
Christensen, A. N., O. W. Dietrich, W. Kress, W. D. Teuchert, and R. Currat (1979) Solid State Commun. **31**, 795.
Glötzel, D., D. Rainer and H. R. Schober (1979) Z. Phys. B (in press).
Gyorffy, B. L., A. J. Pindor and W. Temmerman (1979) (to be published).
Harmon, B. N. and K. M. Ho (1979) in C4.

Kamitakahara, W. A., K. Scharnberg and H. R. Shanks (1979) to be published.
McKnight, S. W., S. Perkowitz, D. Tanner and L. R. Testardi (1979) Phys. Rev. B **19**, 5689.
Staudenmann, J. L. and L. R. Testardi (1979) Phys. Rev. Lett. **43**, 40.
Varma, C. M. and W. Weber (1979) Phys. Rev. B **19**, 6142.
Weber W., (1979) in C4.
Weber, W., P. Roedhammer, L. Pintschovius, W. Reichardt, F. Gompf and A. N. Christensen (1979) Phys. Rev. Letters **43**, 868.
Wipf, H., M. V. Klein, B. S. Chandrasekhar, T. H. Geballe and J. H. Wernick (1978) Phys. Rev. Letters **41**, 1752.

CHAPTER 3

Spectroscopy of Collective Pair Excitations

P. A. FLEURY

Bell Laboratories, Murray Hill
New Jersey 07974, USA

Dynamical Properties of Solids, edited by
G. K. Horton and A. A. Maradudin

Contents

1. *Introduction*

One of the most useful concepts in solid state physics has been that of the collective excitation. For example, the entire field of lattice dynamics – upon which this series of volumes is based – exists by virtue of the vast simplification to the many-body (10^{23}) problem afforded by the quantized lattice vibrations or phonons. However, this viewpoint is useful and productive of physical insight only to the extent that the "modes" so derived, characterized by a frequency $\omega_{\boldsymbol{k}}$, and wave vector, $\boldsymbol{k}$, are in some sense well defined. A particularly successful familiar example is the set of phonon dispersion curves for simple solids in the low-temperature limit. There the existence of spatial structural order permits a quantitative and consistent characterization of the dynamics, ranging from the long-wavelength, low-frequency acoustic modes, to the short-wavelength, high-frequency motions represented by optic phonons throughout the Brillouin zone. Similar successes can be cited for the collective excitations of ordered spin systems: the magnons or spin waves.

At the other extreme one may consider the small-scale, short-time dynamics of condensed matter systems which lack the underlying spatial periodicity of the crystalline solids, such as liquids and amorphous solids. In these cases collective description is considerably less useful. In fact, for classical liquids there is still no systematic dynamic theory which extends much beyond the hydrodynamic regime. The breakdown of the harmonic approximation – in which the dynamics are adequately described in terms of long-lived, essentially non-interacting collective excitations – is of considerable importance, not only in amorphous or disordered systems, but also in ordered solids at finite temperatures. Interactions among elementary excitations are obviously fundamental to this breakdown. In addition to those interactions which merely renormalize the excitation frequency and impart a finite lifetime to the single collective excitation, interactions between particular *pairs* of excitations have recently received increasing attention. Such interactions may give rise to collective pair modes and bound states in a wide variety of condensed matter systems. These pair excitations have been found to exhibit remarkably similar behavior in such diverse materials as liquids, antiferromagnets and crystalline solids. So that detailed spectroscopic studies of pair modes have provided a unifying viewpoint upon which might be based deeper understanding of the small-scale, short-time dynamics of condensed matter.

In this chapter we shall be concerned with the (primarily experimental) study of collective pair mode excitations in solids and liquids. Further, we shall concentrate upon information provided by spectroscopic techniques because of the large dynamic range in both space and time which they can probe. Scattering of both light and neutrons has been employed to examine pair excitations in crystalline solids and both classical and quantum liquids. While neutron scattering provides the opportunity to study pair modes with essentially arbitrary total wave vector, the subject of pair excitations has received less attention in neutron spectroscopy than it has in optical spectroscopy and light scattering. Accordingly we shall give more emphasis here to the optical experiments. Although optical techniques are generally restricted to studying essentially $k \approx 0$ processes, this restriction is relatively unimportant for pair excitations, since it does not restrict to small values the *wave vectors of the individual excitations* comprising the pair. In general, any process in which a pair of excitations gives rise to a dynamic polarizability can be examined by light scattering. As related in more detail below the processes involving pairs of *phonons* (whether from the same or different excitation branches) is called "second-order Raman scattering" and clearly dominates the literature in terms of the number of publications devoted to it. We shall not, however, be concerned with most of this work because it does not contain as an essential element the effects of *interaction* between members of the pair. In other words, in crystalline solids, especially at temperatures well below melting, the harmonic approximation holds sufficiently well for phonons that the second-order Raman spectrum can be adequately described in terms of joint densities of states derived from one-phonon dispersion curves. As far as phonon pairs in crystalline solids are concerned we shall consider only two kinds of situations: first, wherein the one-phonon dispersion curves are so configured that the possibility of a *bound phonon-pair* state exists; and second, wherein the effects of anharmonicity become exaggerated for a *particular* mode owing to its low frequency, as occurs in the vicinity of second-order structural phase transitions.

Because of the intrinsically more complicated excitation spectrum and the approximate nature of theories for anharmonic coefficients, phonon-pair spectra in crystals have not, however, provided the best prototypical system for the theory/experiment of collective pair excitations. Indeed, as we shall see, even the question of the simplest case phonon-pair bound state in diamond remains controversial. Rather the most thoroughly studied and theoretically best understood examples of collective pair excitations are provided by antiferromagnetic crystals, which sustain both magnon–magnon and magnon–exciton pairs. Accordingly, we shall devote a section of this chapter to optical and theoretical studies of pair modes involving magnetic excitations. Not only does one have the advantage with magnetic systems of a microscopic hamiltonian and the controlled approximations to which it gives rise, but also by varying temperature, magnetic field and composition one may follow continuously the effect upon pair modes of the loss of long-range magnetic order, without having to melt the host solid lattice.

The strongest candidate for a non-magnetic pair mode bound state appears to be the $k=0$, two-roton excitation in liquid helium. Because it is a monatomic liquid having only a single branch of excitations (the phonon–roton branch) and because it remains a liquid at such low temperatures that the single excitations are well defined for very large $\boldsymbol{k}$, liquid helium may not seem a surprising candidate for a bound pair state. Nonetheless, owing to the very small value of the roton–roton interaction and to the small but non-zero decay channel of roton pairs into phonon pairs, as well as some questionable theoretical approximations, even this situation is not completely clear.

Further the concept of pair modes in other simple liquids (hydrogen and the rare gas liquids) can be said to have only qualitative validity at best, since the single excitation (one-phonon) branch is not well defined much beyond the hydrodynamic limit. Nevertheless, spectra interpretable as arising from pairs of highly damped phonons have been obtained and can be compared suggestively with those from (a) the corresponding crystalline solids below melting and (b) liquid helium.

Following the presentation of some general theoretical ideas regarding interacting pairs of excitations and their manifestations in inelastic scattering spectra (§2) we shall discuss in turn pair excitations in magnetic systems (§3), in simple liquids (§4) and finally in crystalline solids (§5). In §6 we record some brief concluding remarks.

2. *Theoretical background*

2.1. Light scattering and correlation functions

The spectrum of inelastically scattered light or neutrons from a condensed medium contains considerable information about the elementary excitations sustained in that medium. However, there are a number of potential complications which must be appreciated if one is to relate observed spectral features quantitatively to the medium's elementary excitations. We concentrate here on the formalism of light scattering (Fleury

1972) in order to illustrate these complications. Except for the coupling between the excitations and the probe particle (which determine scattering selection rules and complications due to finite wave vector) the description of neutron scattering is quite similar. For details the reader should consult G. Dolling's chapter 10 in volume 1 of this series.

In general the spectrum of scattered light is expressed as the functional dependence of observed intensity $I(q,\Omega)$ upon frequency ω_2 and wave vector $\boldsymbol{k}_2$. As shown schematically in fig. 1, Ω is the difference in frequency between the incident photon ω_1 and the scattered photon ω_2. Similarly $\hbar \boldsymbol{q}$ is the vector difference in momenta between the scattered and incident photon $\hbar(\boldsymbol{k}_2 - \boldsymbol{k}_1)$.

The scattered intensity spectrum is determined by the autocorrelation function of the scattered field. Since, in the dipole approximation, the scattered field is proportional to the product of the incident field and the medium's polarizability $\alpha(\boldsymbol{r},t)$, it follows that the fundamental physical quantity of interest to us is the autocorrelation function of the polarizability $\alpha(\boldsymbol{r},t)$ operator. Thus

$$I(\boldsymbol{q},\omega) = \mathrm{FT}\, I(\boldsymbol{r}-\boldsymbol{r}', t-t') = \mathrm{const}\,\mathrm{FT}\langle \alpha(\boldsymbol{r},t)\alpha(\boldsymbol{r}',t')\rangle, \tag{2.1}$$

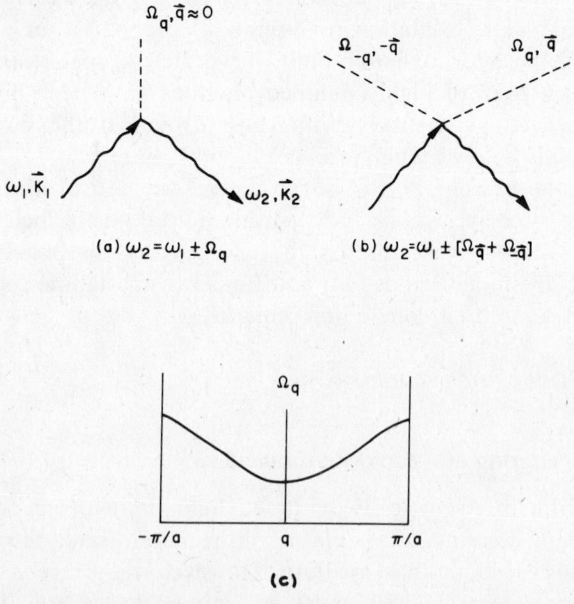

Fig. 1. Kinematics of first- and second-order light scattering. (a) First-order ($\omega_2 = \omega_1 \pm \Omega_q$), (b) second-order ($\omega_2 = \omega_1 \pm [\Omega_q + \Omega_{-q}]$), (c) Simple dispersion curve for a single-branch excitation. In second-order scattering, excitations from opposite halves of the zone are paired.

where FT represents the Fourier transform in both space and time. This correlation function is simply related to the generalized susceptibility $\chi(\boldsymbol{r}-\boldsymbol{r}', t-t')$ defined in terms of the commutator

$$\chi(\boldsymbol{r}-\boldsymbol{r}', t-t') = \mathrm{i}\theta(t-t')\langle[\alpha(\boldsymbol{r},t), \alpha(\boldsymbol{r}',t')]\rangle, \tag{2.2}$$

$$\chi(\boldsymbol{q},\omega) = \mathrm{i}\int\int \mathrm{d}t\,\mathrm{d}\boldsymbol{r}\,\theta(t)\langle[\alpha(\boldsymbol{r},t), \alpha(0,0)]\rangle \mathrm{e}^{\mathrm{i}(\omega t+\boldsymbol{q}\cdot\boldsymbol{r})}, \tag{2.3}$$

where $\theta(t)=1$ for $t \geqslant 0$, and zero otherwise. The fluctuation dissipation theorem then relates the spectrum to χ as

$$I(\boldsymbol{q},\omega) = -\frac{\hbar}{\pi}(1-\mathrm{e}^{-\beta\hbar\omega})^{-1}\,\mathrm{Im}\,\chi(\boldsymbol{q},\omega). \tag{2.4}$$

where $\beta = 1/kT$.

Eqs. (2.2) and (2.3) will be recognized as the familiar forms for the so-called retarded Green's function (Abrikosov et al. 1963) describing the evolution of the operator α. We note that α is actually a tensor, but will defer any appendage of subscripts to a later section so as to maximize clarity here. Since $\alpha(\boldsymbol{r},t)$ receives contributions from all changes in state or configuration of charges in the medium, a complete microscopic derivation of α is hopelessly complicated for a condensed medium. However, optical spectroscopy permits separation of the contributions to $\alpha(\boldsymbol{r},t)$ on the basis of time scale (frequency), length scale (wave vector) and symmetry (polarization selection rules), so that it is often possible experimentally to focus attention upon the spectroscopic manifestation of a particular type of excitation.

In general $\alpha(\boldsymbol{r},t)$ can be expanded in powers of the various dynamic variables of the medium. If only one type of elementary excitation is present we may write (Fleury 1974)

$$\alpha(\boldsymbol{r},t) = \alpha_0 + \sum_{n=1} a_n U^n(\boldsymbol{r},t). \tag{2.5}$$

Here $U(\boldsymbol{r},t)$ is linear in the creation and annihilation operators for the single elementary excitation. Symmetry considerations determine the structure of the set of coefficients $\{a_n\}$. It is seldom necessary to consider terms beyond $n=2$. The most common scattering processes are depicted in fig. 1 and are called first-order ($a_1 \neq 0, a_{n\neq 1}=0$) and second-order ($a_2 \neq 0$, $a_{n\neq 2}=0$) scattering, respectively. Crudely speaking, these correspond to the creation (or destruction) of a *single* excitation (first order) or of a *pair* excitation (second order).

2.2. First-order scattering

In the simplest case where $\alpha(\boldsymbol{r},t)$ can be expressed as

$$\alpha(\boldsymbol{r},t)=a_1 U(\boldsymbol{r},t),$$

the observed spectrum is quite directly related to the single excitation Green's function. That is

$$I(\boldsymbol{q},\omega)=-\frac{\hbar}{\pi}(1-\mathrm{e}^{-\beta\hbar\omega})^{-1}a_1^2\,\mathrm{Im}\,G_{\mathrm{I}}(\boldsymbol{q},\omega), \tag{2.6}$$

where

$$G_{\mathrm{I}}(\boldsymbol{r},t|\boldsymbol{r}',t')=G_{UU}(\boldsymbol{r},t|\boldsymbol{r}',t')=-\mathrm{i}\langle\theta(t)U(\boldsymbol{r},t)U(\boldsymbol{r}',t')\rangle, \tag{2.7}$$

$$U(\boldsymbol{r},t)=\frac{1}{V^{1/2}}\sum_q\sqrt{\frac{\omega_q}{2}}\,\{b_q\exp[\mathrm{i}(\boldsymbol{q}\cdot\boldsymbol{r}-\omega_q t)] + b_q^*\exp[-\mathrm{i}(\boldsymbol{q}\cdot\boldsymbol{r}-\omega_q t)]\}, \tag{2.8}$$

where θ represents time ordering. Usually we shall be concerned with two types of interactions; those which renormalize and broaden the single elementary excitation, and those which impart structure to the pair excitation which is not present in the single excitation. The former will be handled in this chapter phenomenologically by replacing ω_q by $\tilde{\omega}_q$ and ε by Γ, where (Abrikosov et al. 1963).

$$G_{\mathrm{I}}^0(\omega,\boldsymbol{q})=\tfrac{1}{2}\omega_q\left([\omega-\omega_q+\mathrm{i}\varepsilon]^{-1}-[\omega+\omega_q-\mathrm{i}\varepsilon]^{-1}\right), \tag{2.9}$$

with $\varepsilon=0^+$, is the propagator for a single non-interacting excitation.

2.3. Second-order scattering

In some cases a_1 in eq. (2.5) may be vanishingly small or identically zero (e.g. symmetry requirements) so that $\alpha(\boldsymbol{r},t)$ may be expressed as

$$\alpha(\boldsymbol{r},t)=a_2 U(\boldsymbol{r},t)U(\boldsymbol{r},t),$$

where the U's are defined as above. Here the spectrum is determined by the Fourier transform of the four-point correlation function (Fleury 1978)

$$\chi_2=\chi_{UUUU}=-\mathrm{i}\langle\theta(a_2U(\boldsymbol{r},t)U(\boldsymbol{r},t)a_2U(\boldsymbol{r}',t')U(\boldsymbol{r}',t'))\rangle, \tag{2.10}$$

which for $a_2=$constant is directly proportional to the two-particle Green's function, $G_{\mathrm{II}}(\boldsymbol{r},t|\boldsymbol{r}',t')$. We will see later that the spatial and symmetry characteristics of a_2 can modify the observed lineshape significantly particularly in the cases of roton pairs and magnon pairs. Ignoring this complication for the moment let us consider what to expect for spectra arising from terms like eq. (2.10). Calculations of the pair of Green's functions, G_{II}, in a condensed medium (particularly at finite temperature) are not generally possible and many approximate techniques – applicable to particular systems – have been developed to attack such problems (Abrikosov et al. 1963, Halley 1978). In most cases of experimental interest to us the interaction between elementary excitations which relate G_{II} to G_{I} have been introduced in the simplest way, through the approximate Bethe–Salpeter equation

$$G_{\text{II}}(x,x') = \mathrm{i}\left[2(G_{\text{I}}(x,x'))^2 + \mathrm{i}g_4\int \mathrm{d}x''\, G_{\text{I}}(x,x'')G_{\text{I}}(x'',x') + \dots\right], \quad (2.11)$$

where x represents $(\boldsymbol{r},t)$ and G_{I} is the single excitation Green's function. The diagrammatic representation of eq. (2.11) appears in fig. 2. The approximation here considers the interaction g_4 to be a constant in both space and time – which in general it is not. Upon the Fourier transform the integrals in eq. (2.11) reduce to a geometric series of algebraic terms whose sum can be written down immediately as

$$G_{\text{II}}(\boldsymbol{q},\omega) = \frac{G^0_{\text{II}}(\boldsymbol{q},\omega)}{1 - g_4 G^0_{\text{II}}(\boldsymbol{q},\omega)}. \quad (2.12)$$

Here G^0_{II} describes the propagator of all *pairs* of non-interacting excitations ($g_4 = 0$) whose total momentum is $\hbar\boldsymbol{q}$. But they may be dressed (i.e. $\omega_q \to \tilde{\omega}_q; \varepsilon \to \Gamma_q$) in eq. (2.9). That is (Ruvalds and Zawadowski 1970a)

$$G^0_{\text{II}}(\boldsymbol{q},\omega) = \frac{\mathrm{i}}{(2\pi)^4}\int\int \mathrm{d}\boldsymbol{k}\,\mathrm{d}\omega'\, G_{\text{I}}(\boldsymbol{k}+\boldsymbol{q},\omega-\omega')G_{\text{I}}(-\boldsymbol{k},\omega'). \quad (2.13)$$

In the special case where no interactions whatsoever are considered the single excitation Green's function is given by eq. (2.12) and the spectrum is

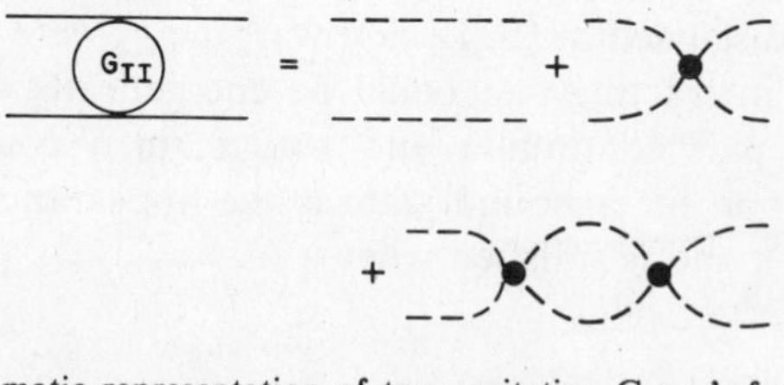

Fig. 2. Diagrammatic representation of two-excitation Green's function G_{II}; single dashed lines represent single excitation Green's function G_{I}.

just the *density* of states $\mathrm{Im}\, G_{\mathrm{II}}^{00}$:

$$I(\boldsymbol{q},\omega) = -\frac{\hbar}{\pi} a_2^2 (1 - \mathrm{e}^{-\beta\hbar\omega})^{-1} \mathrm{Im}\, G_{\mathrm{II}}^{00}(\boldsymbol{q},\omega). \tag{2.14}$$

The second-order Raman spectrum is then describable solely as overtone scattering from pairs of harmonic excitation at $\boldsymbol{k}$ and $-\boldsymbol{k}$ from the same branch. In crystalline solids, even in the harmonic limit, phonon-pair spectra from different branches can frequently occur with appreciable intensity and thus can complicate the spectrum. In principle these inter-branch combination processes can be calculated by a simple generalization of eq. (2.13):

$$G_{\mathrm{II}}^{0}[a,b] = \frac{\mathrm{i}}{(2\pi)^4} \int\int \mathrm{d}\boldsymbol{k}\, \mathrm{d}\omega'\, G_{\mathrm{I}}^{a}(\boldsymbol{k}+\boldsymbol{q},\omega-\omega') G_{\mathrm{I}}^{b}(-\boldsymbol{k},\omega'). \tag{2.15}$$

However, the scattering coefficients a_2 are generally not the same for different branches and very few situations exist for which even non-interacting phonon-pair spectra have been quantitatively calculated. The low-temperature second-order Raman spectra of the f.c.c. rare gas solids represents perhaps the best and simplest example here.

Let us now consider the effects of non-zero interactions between members of the pair $g_4 \neq 0$. Examination of eq. (2.12) reveals that the effect of increasing g_4 will be to redistribute the spectral weight in $\mathrm{Im}\,\chi_2(\omega)$ with respect to where it occurs in $\mathrm{Im}\,\chi_2^0(\omega)$.

However, for a more specific answer to this question we must examine both the single excitation dispersion curves and density of states, as well as the sign and magnitude of the interaction g_4. For example, if ω_{m} is the frequency of a *maximum* on the single excitation dispersion curve, and if the non-interacting pair density of states has a peak at $2\omega_{\mathrm{m}}$ the effect of a negative g_4 (corresponding to an attractive interaction) will shift the peak in the response to *lower* frequency and hence *into* the pair continuum. Thus the interacting pair will be able to decay into a free pair and is then properly called a *pair-resonance*. This case is realized in the antiferromagnetic magnon pairs discussed in §3. If, however, the g_4 were positive in the above example, the interacting pair could be energetically split off above the non-interacting pair continuum and may form a *bound state*. The mathematical condition for a bound state is the appearance of a pole in $G_{\mathrm{II}}(\boldsymbol{q},\omega)$ of eq. (2.12), and is satisfied when

$$g_4 G_{\mathrm{II}}^{0}(\boldsymbol{q},\omega_{\mathrm{B}}) = 1. \tag{2.16}$$

The frequency ω_B is the pair bound state frequency and the binding energy is just $\hbar(\omega_B - 2\omega_M)$. This situation has been proposed to exist in diamond, to account for the sharp feature observed near $2\omega_M$ in the second-order Raman spectrum. We shall discuss this example in more detail in §5.

All of the preceding discussion holds as well in the case of a dispersion curve *minimum*, except that an attractive interaction ($g_4 < 0$) is required (but not sufficient) for existence of a bound state. The latter situation is generally agreed to occur for roton pairs in liquid helium, as we shall discuss in §4.

Finally we note that interactions and bound states may also exist between pairs of unlike excitations. A brief discussion of the magnon exciton bound state in MnF_2 appears in §3.

The brief and schematic theoretical discussion presented in this section is intended only to establish language and provide a framework for interpretation of experimental results. A more complete treatment of the subject of bound states among collective excitations has recently been given by Pitaevsky (1976). The works of Abrikosov et al. (1963) and Halley (1978) are also recommended for reference.

2.4. Hybridization and interferences

When more than one type of excitation is present two types of complication to the observed spectra must be considered. First, the coupling of mode "j" to mode "i" will modify the "i" Green"s function and hence its scattering spectrum even if mode "j" has no scattering cross section of its own. Second, if both modes "i" and "j" have finite scattering cross sections (in the absence of their interaction with each other) additional structure in the scattering spectrum can result for interference effects when the i–j coupling is turned on. The first example is a special case of the second when $a_n(j) = 0$. If one is to extract the parameters of the excitations' Greens's functions from such hybridized spectra, it is essential to have a quantitative description of these mode coupling and interference effects.

Let us describe the amplitude of the ith excitation by U_i and consider the more general problem of pairwise interactions among N such excitations (Fleury 1976). The Hamiltonian will contain terms of the form $g_{ij} U_i U_j$. The coupled susceptibilities for this system can be simply derived from the equations of motion $\ddot{U}_i = \Sigma_j g_{ij} U_j = F_i$, where F_i represents the external force or field which couples linearly to the amplitude U_i. The coupling coefficients g_{ij} are in general complex, may depend on q, and seldom have been calculated a priori. The diagonal elements of the g_{ij} matrix are just the uncoupled mode susceptibilities (or "free excitation" Green functions) $g_{ii} = (G_i^0)^{-1}$. The scattered spectrum is still described by eq. (2.4), but with $\chi(\boldsymbol{q},\omega)$ replaced by χ_T, where

$$\chi_T(\boldsymbol{q},\omega)=\sum_{i,j}^{N} a(i)a(j)G_{ij}(\boldsymbol{q},\omega). \tag{2.17}$$

Here $G_{ij}=[g^{-1}]_{ij}$. Thus the formal problem simply requires the inversion of a $N\times N$ matrix, although a theoretical determination of G_{ij} presents a formidable problem.

When only two modes are involved G_{ij} has a simple form:

$$G_{ii}=\frac{G_i^0}{1-g_{12}^2G_1^0G_2^0}; \qquad G_{ij}=\frac{g_{12}G_1^0G_2^0}{1-g_{12}^2G_1^0G_2^0}. \tag{2.18}$$

This simple case has been realized quite often in the coupling between optic phonons and between optic and acoustic phonons when the optic mode frequency is strongly temperature dependent, as occurs near structural phase transitions. However, we note that the above hybridization is not restricted to the case where G_i^0 describes a single excitation. In particular, several examples have been found experimentally where G_i^0 describes a pair excitation. For example, the light scattering spectra in SiO_2 and $AlPO_4$ exhibit hybridization and interference effects (Scott 1968, 1970) describable by eqs. (2.17) and (2.18), where G_1^0 describes a single optic phonon and G_2^0 refers to a two-phonon state. Also neutron scattering (Cowley 1978) in liquid helium for $q\simeq 2.7\,\text{Å}^{-1}$ reveals unusual structure which has been interpreted by Ruvalds and Zawadowski (1970b) as a hybridization of a finite q roton-pair state with the single roton excitation at $q=2.7\,\text{Å}^{-1}$.

3. *Pair excitations in magnetic systems*

3.1. Magnon pairs at zero temperature

Since the initial observation of light scattering from spin waves (Fleury et al. 1966), where the second-order process was found to be surprisingly strong, most of the effort and interest in this field has centered upon the second-order or magnon-pair scattering, primarily for the unique information it provides about the dynamic four-spin correlation function. The rich and varied family of transition metal (Fe, Mn, Ni, Co) fluoride compounds has been the most extensively studied (Fleury 1974). Practical reasons for the favoritism of these materials center upon their transparency to visible light, their convenient magnetic transition temperatures, their relative freedom from interfering phonon scattering processes (particularly when compared to their oxide relatives) and their availability as single crystals of good optical quality. Mixed crystals are also prepared with relative ease. Fortunately from a theoretical point of view, the variety of spins and crystal structures available to this family maps well onto the set of theoretically most interesting and tractable spin Hamiltonians. All of these antiferromagnetic systems can be described by the spin Hamiltonian (Fleury 1974)

$$\mathcal{H}_s = \sum_{ij} J_{ij} \boldsymbol{S}_i \cdot \boldsymbol{S}_j + \sum_j \Big\{ D(S_j^z)^2 - E\big[(S_j^x)^2 - (S_j^y)^2\big] + \beta \boldsymbol{S}_j \cdot \mathbf{g} \cdot \boldsymbol{H} \Big\} + \sum_j \{(E \rightarrow -E)\}, \tag{3.1}$$

where J_{ij} represents the exchange interaction between magnetic ions i and j, $\mathbf{g}$ is the gyromagnetic tensor, $\boldsymbol{H}$ the applied magnetic field, D the anisotropy energy determining the direction of sublattice magnetization and E the anisotropy energy tending to cant the sublattice spins to 90° relative orientation. In cases where all of these exchange and anisotropy parameters are non-zero (e.g. NiF_2 of space group D_{4h}^{14}) the magnon dispersion curves obtained from this spin Hamiltonian are shown schematically in fig. 3. The explicit relations between the magnon frequencies and the parameters in eq. (3.1) are $\omega_0^- = 2E, \omega_0^+ = (16DJ_2 + E^2)^{1/2}$, $2\omega_M = D + 16(J_2 - J_3), 2\omega_R = D + 16J_2 - 8(J_1 + J_3)$, where J_i denotes ith nearest-neighbor exchange. Observations of both one- and two-magnon scattering have been made in the perovskite structure $NaNiF_3$ (Pisarev et al. 1973) and the rutile structure NiF_2 (Fleury 1969), from which the spin Hamiltonian parameters have been obtained.

Of primary interest here are magnon-pair spectra such as illustrated in fig. 4, which reveal that even at zero temperature, magnon–magnon interactions are not negligible. That is, the four-spin correlation function is not trivially related to the two-spin correlation function. To appreciate this, consider the lowest-order expression for the two-magnon Green's function G_{II}^0, in which there is no interaction between the members of the magnon pair (Elliott and Thorpe 1969):

$$G_{\text{II}}^0(\boldsymbol{q}=0,\omega) = \frac{\text{const}}{N} \sum_k \frac{\Phi_{\alpha\beta}^2(k)}{\omega^2 - 4\omega_k^2} \qquad (T=0). \tag{3.2}$$

This is essentially the two-magnon density of states, weighted by the trigonometric factors $\Phi_{\alpha\beta}^2$ which result from crystal symmetry requirements. The resultant lineshape is indicated in fig. 4 by the dashed curve and fails to agree quantitatively with the observed low-temperature

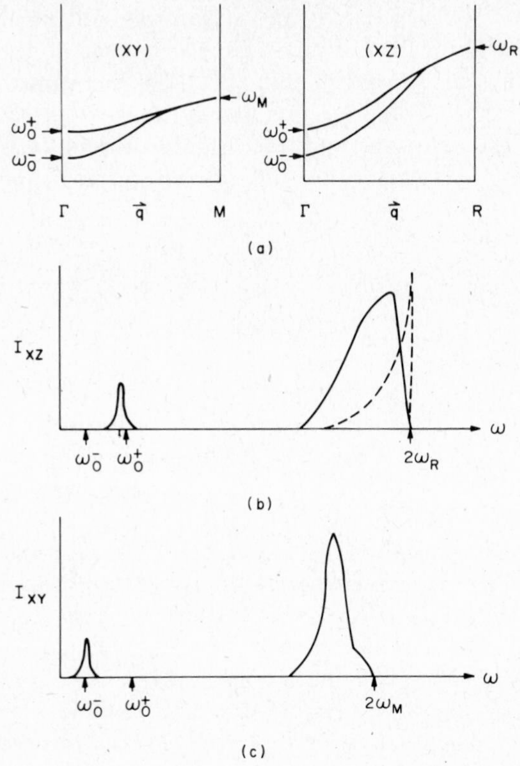

Fig. 3. Schematic of dispersion relations (a) and magnetic light scattering spectra for system with spin Hamiltonian in eq. (3.1). (b) and (c) illustrate the effect of polarization selection rules on magnon features observed; Subscripts on I indicate polarization directions of incident and scattered light.

spectrum (Fleury 1968). It was first pointed out by Elliott and Thorpe (1969) that magnon interaction effects upon G_{II} can be included, and (with no adjustable parameters) produce essentially perfect agreement with experiment at low temperature. The resulting expression is

$$G_{\text{II}} = \frac{G_{\text{II}}^0}{1 - g_4 G_{\text{II}}^0}, \tag{3.3}$$

where $g_4 = -4J_2(D + 2\langle S\rangle(4J_2 - J_1))$ and $\langle S\rangle$ is the sublattice magnetization. The quantitative expression for g_4 in the magnetic case is given in terms of the parameters in the spin Hamiltonian eq. (3.1). It results from a straightforward recovery of the leading terms normally neglected in the low-temperature approximation relating spin and magnon operators

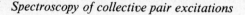

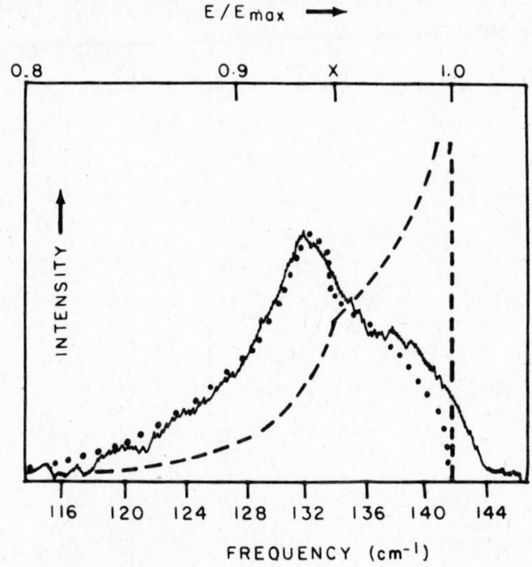

Fig. 4. Magnon-pair light scattering in $RbMnF_3$ at ~10 K. Solid line is experimental observation (Fleury 1968). Dashed curve is theory without magnon–magnon interactions. Dotted curve is theory including interactions.

through the Holstein–Primakoff transformation. The resultant predicted magnon-pair lineshape is illustrated by the dotted curve in fig. 4. This advance permits magnon-pair scattering to be used for quantitative determination of spin Hamiltonian parameters. Thorpe (1970) has computed magnon interaction corrections for MnF_2 where the magnon-pair lineshape dependence on polarization is appreciable.

The crystal structure for $KNiF_3$ (cubic perovskite) is identical to that for $RbMnF_3$ and the low-temperature effects of magnon interaction are formally the same in the two materials (Chinn et al. 1972). Interestingly, a closely related set of crystals typified by K_2NiF_4 exhibit two-dimensional rather than three-dimensional spin excitation spectra. In the two-dimensional case the Green's functions analogous to eqs. (3.2) and (3.3) can be expressed in closed analytical form. Parkinson's (1969) calculated magnon-pair lineshapes including magnon interactions are essentially identical to the experimental observations (Fleury and Guggenheim 1970). Again, no adjustable parameters are required.

The present situation regarding ($q=0$) magnon pairs in pure systems at zero temperature may therefore be regarded as quantitatively satisfactory. In crystals of lower symmetry, having more complicated spin Hamiltonians, the magnon dispersion curves (and consequently the one- and two-magnon spectra) will be correspondingly more complex. However, from an experimental point of view it is possible to sort out much of this complexity. Specifically, through the terms in eq. (3.2), magnon pairs from different regions of the Brillouin zone are emphasized in spectra with different polarization selection rules (Fleury and Loudon 1968).

As can be seen from fig. 4, the magnon-pair state appears as a resonance rather than a split off bound state. As discussed in §2, this is expected because g_4 is *negative* and the Brillouin zone boundary represents a *maximum* in the one-magnon dispersion curves. Before considering the effects of finite temperature on magnon pairs, let us consider an example of a magnetic bound pair state: the magnon–exciton pair in MnF_2.

3.2. Magnon–exciton pairs at zero temperature

Below the Neél temperature, $T_N = 67\,K$, MnF_2 is a tetragonal (D_{4h}^{14}) antiferromagnet, whose magnon dispersion curves are only slightly more complicated than those for $RbMnF_3$. The magnon-pair spectra of MnF_2 have been extensively studied both by light scattering (Fleury and Loudon 1968) and absorption (Allen et al. 1966) and the effects of magnon–magnon interactions found to be quite similar to those discussed above for $RbMnF_3$: e.g. no bound states.

A bound state has been found, however, to exist for the pair formed by one zone-boundary magnon and one zone-boundary exciton. In fig. 5 is

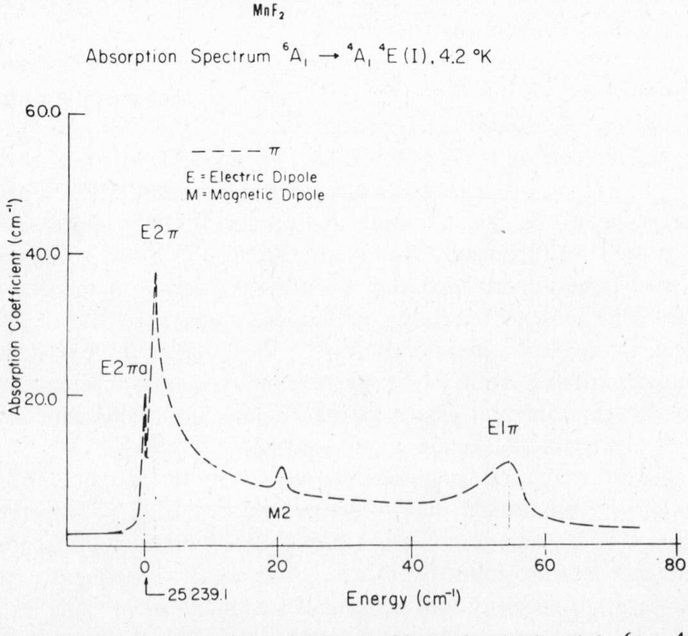

Fig. 5. Optical absorption spectrum in MnF_2 at 4.2 K, in the vicinity of the $^6A_1 \rightarrow {}^4A_1, {}^4E(I)$ transition. After Meltzer et al. (1966).

shown the absorption spectrum of MnF_2 by Meltzer et al. (1966) in the frequency range near 25200 cm^{-1}; corresponding to the $^6A_1 \to {}^4A_1$ E(I) transitions. Only the spectrum for electric vector parallel to the magnetic, c-axis (dashed curve) is displayed. The small peak labeled M2 (25260.1 cm^{-1}) is the magnetic dipole absorption by a single exciton at $k=0$. The much stronger feature at 25241 cm^{-1} labeled E2π is the associated magnon sideband. It is an electric dipole transition and arises from the simultaneous creation of a zone-boundary magnon and a zone-boundary exciton. The fact that the pair peak lies some 20 cm^{-1} below the M2 peak implies that the M2 exciton experiences a $\sim$75 cm^{-1} negative dispersion between $k=0$ and $k=\pi/c$, since the z directed magnon has a 55 cm^{-1} frequency at the zone boundary. The pair band continuum extends from 25268.8 $cm^{-1} = \Omega^m(0) + \Omega^e(0)$ down to 25241 $cm^{-1} = \Omega^m(\pi/c) + \Omega^e(\pi/c)$, with the joint magnon–exciton density of states peaking at E2π=25241 cm^{-1}. Thus an ideal situation exists for an attractive exciton–magnon interaction to split off a bound state below this continuum. This is indeed the interpretation which has been given to the very sharp feature labeled E2πa at 25239.1 cm^{-1}. The bound state stands out clearly from the pair continuum by an amount equal to the magnon–exciton pair binding energy: 1.6 cm^{-1}. An extensive and detailed theory for exciton–magnon interactions in magnetic insulators has been given by Freeman (1968); but the qualitative features are easily understood in terms of our discussion in §2.

3.3. Magnetic pair excitations at finite temperatures

At finite temperatures even the non-interacting pair propagator cannot be accurately represented by a simple convolution of single excitation propagators of the form eq. (2.8). In other words, even if g_4 is zero the pair spectrum will be broadened owing to the finite lifetime of the individual excitations. Such effects can easily destroy a weakly bound pair state like that in MnF_2, where indeed it cannot be observed above $\sim$8 K. However, for resonances such as the antiferromagnetic magnon pairs discussed in §3.1 the effects of finite temperature are less dramatic, so that it has been possible to observe these pairs even above T_N, where the long-range magnetic order is absent (Fleury 1969, 1970).

The temperature dependence of magnon-pair spectra has presented some interesting theoretical challenges. Fig. 6 shows the temperature evolution of both one- and two-magnon spectra in NiF_2. The one-magnon peak arises from scattering by a single long-wavelength ($k \approx 0$) magnon. The behavior is typical in that the magnon-pair frequency decreases at a less rapid rate with increasing T than does ω_0, the frequency of a single zone center magnon, and in that the magnon-pair mode remains well defined far into the paramagnetic regime. This behavior indicates that (1)

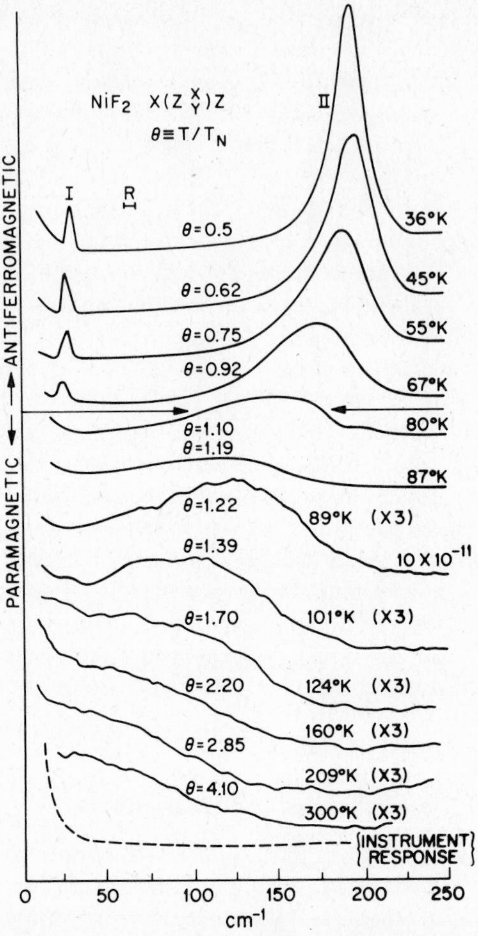

Fig. 6. Temperature dependence of single-magnon (I) and magnon-pair (II) spectra in NiF_2 (Fleury 1969). Note persistance of pair mode into paramagnetic phase, and gain change above $T = 87$ K.

the renormalization of magnon frequency with temperature is strongly dependent on wave vector and (2) the persistence of short-range magnetic order above T_N sustains collective excitations whose wavelength is small compared to the range of that order. Although it was recognized very early that the magnon-pair lineshape was influenced by the k-dependent temperature normalization of individual magnon frequencies, by temperature-dependent one-magnon lifetimes, and by magnon–magnon interactions, it has not been possible until relatively recently to extend the successful zero temperature theory to finite temperatures – much less into the paramagnetic phase. Within the last couple of years a variety of attempts at this extension were made. Those which can be regarded as replacing G^0_{II} in eq. (3.2) by the form

$$G^0_{\text{II}}(T,\omega,q\simeq 0)=\frac{\text{const}}{N}\sum_k \frac{R(T)\Phi^2_{\alpha\beta}(k)\coth\left(\frac{1}{2}\beta\hbar\omega_k\right)}{\omega^2-4\omega_k^2}, \tag{3.4}$$

where $\omega_k=\omega_k(T)$, are reviewed and compared by Natoli and Ranninger (1973). They noted there that addition of $4\mathrm{i}\omega\Gamma_k(T)$ to the denominator in eq. (2.8) would produce satisfactory agreement with experiment. Unfortunately *calculated* values of the temperature and wave-vector-dependent one-magnon damping $\Gamma_k(T)$, available to those authors were considerably smaller than required to fit the observed spectra.

More recently several groups have made significant progress on this problem. Calculations of $\Gamma_k(T)$ carried to higher order by Balucani and Tognetti have produced good agreement with experiment for both the magnon-pair frequency and lineshape for temperatures up to about $0.8\,T_{\text{N}}$. The computed lineshapes for $KNiF_3$ and corresponding experimental curves are shown in figs. 7 and 8. The theory of Balucani and Tognetti (1973) has been applied to the two-dimensional case of K_2NiF_4 by van der Pol et al. (1976), to achieve equally good agreement with experiment. Fig. 9

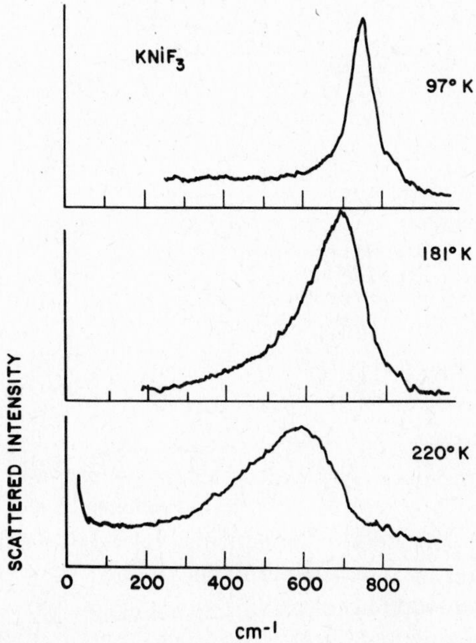

Fig. 7. Experimental magnon-pair spectra in $KNiF_3$, $T_{\text{N}}\simeq 253$ K. Gain for 97K spectrum is approximately half that for 181K and 220K spectra.

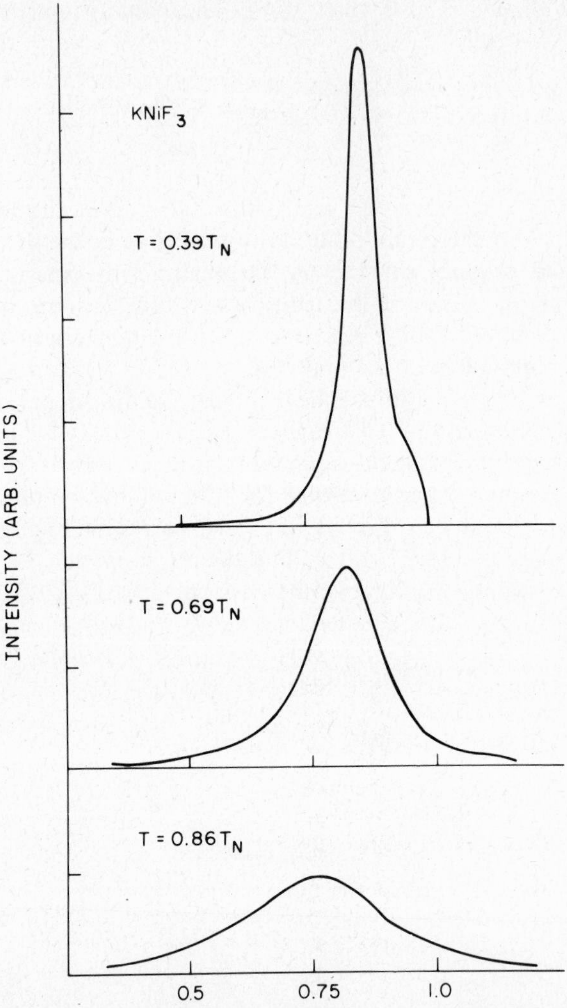

Fig. 8. Calculated magnon-pair spectra from Balucani and Tognetti (1973) for $KNiF_3$. Intensity scales are same for all displayed temperatures.

shows the measured magnon-pair frequency and linewidth as a function of T for the two-dimensional K_2NiF_4 and the closely related three-dimensional $KNiF_3$ (Fleury 1974). The van der Pol calculations for the former produce good agreement with experiment.

Still more recently Balucani and Tognetti (1977) have used a modified Markoffian approximation for the Mori continued fraction representation to obtain the four-spin correlation function lineshape. Fig. 10 compares

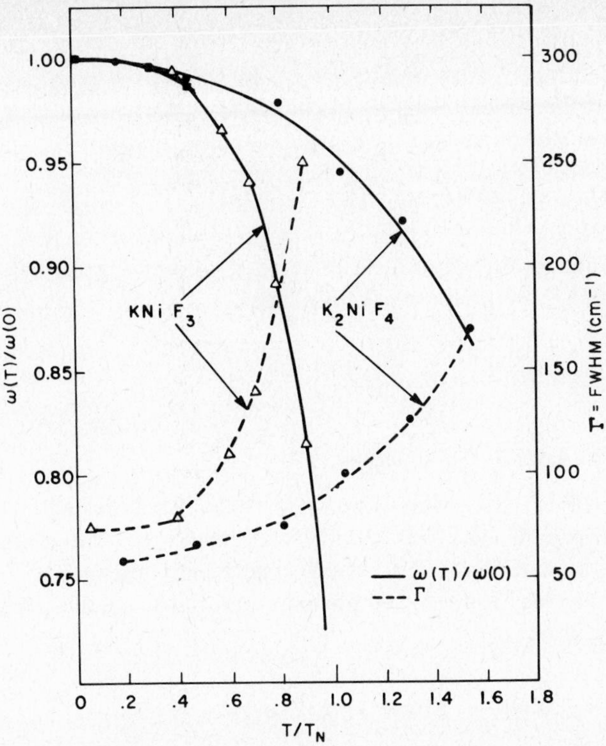

Fig. 9. Comparison of temperature dependences of magnon-pair frequency ($\omega(T)$) and linewidth (Γ) for $S=1$ antiferromagnets which are two-(K_2NiF_4) and three-($KNiF_3$) dimensional. Frequencies are normalized to the zero temperature pair mode frequency $\omega(0)$ in both cases (Fleury 1970).

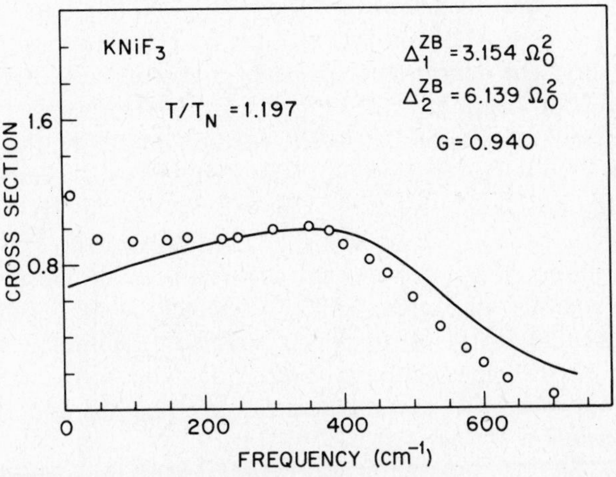

Fig. 10. Comparison of experiment (●) and theory (solid curve) for $KNiF_3$ pair mode scattering in the paramagnetic phase. From Balucani and Tognetti (1977). Overall intensity is adjusted in the calculation.

their calculations with experiments on $KNiF_3$ in the paramagnetic phase. The recent theoretical progress in calculations of the four-spin correlations at temperatures throughout the ordered phase and into the paramagnetic phase establishes light scattering as a quantitative probe of zone-boundary magnon frequencies, lifetimes and interactions. A common feature of these investigations is that no explicit temperature dependence for the magnon–magnon interaction g_4 need be invoked to produce good agreement with the observed finite temperature lineshapes.

4. *Pair excitations in liquids*

4.1. Roton pairs in liquid helium

The possibility that roton pairs in superfluid helium might be observed by light scattering was first raised by Halley (1969) and calculated independently by Stephen (1969). The formal expression for the second-order spectrum written down by Stephen is quite close to the general one we considered in §2:

$$I(\boldsymbol{q}_1-\boldsymbol{q}_2,\omega)=\sum_{q_1q_2}\int \mathrm{d}t\,\mathrm{e}^{\mathrm{i}\omega t}T(\boldsymbol{q}_1)T(\boldsymbol{q}_2)\langle\delta n_{q_1}(t)\,\delta n_{-q_1}(t)\,\delta n_{q_2}(0)\,\delta n_{-q_2}(0)\rangle, \tag{4.1}$$

where $T(\boldsymbol{q})=\tau(q)\hat{P}_s\cdot(3\hat{q}\hat{q}-1)\cdot\hat{P}_i$ expresses the dipole-induced dipole form of the pair polarizability and $\tau(q)$ is inserted to account for the fact that a helium atom does not polarize itself. The form of $\tau(q)$ assumed by Stephen introduces a cutoff at high q into the sum in eq. (4.1), and constitutes an uncontrolled approximation which must be clarified in a proper microscopic theory. The density fluctuation δn_q is expressed in terms of creation and destruction operators b_q, b_q^+.

The original calculations neglected interactions among the excitations so that the four-point correlation function in eq. (4.1) was factored into a sum of products of pair correlation functions as in eq. (2.13). The initial experiments of Greytak and Yan (1969) confirmed the existence of two-roton scattering, but noted evident departures in the observed lineshape from the density of states profile calculated from the phonon–roton dispersion curve, fig. 11. A rough comparison of the lineshapes can be seen from fig. 12. The most striking discrepancy – the virtual absence of the peak at $27\,\mathrm{K}\approx 2\omega_{\max}$ – was initially attributed (Greytak and Yan 1969) to a marked variation of $T(\boldsymbol{q})$ with q.

More refined theoretical and experimental work have since revealed that roton–roton interactions play a crucial role in the observed roton-pair lineshape. Ruvalds and Zawadowski (1970b) and Iwamoto (1970) obtained a good fit to the spectra of fig. 12 using theory sketched in §2 with an attractive roton-roton interaction of $g_4 = 0.37 \pm 0.1$ K. This value of g_4 was inferred largely from the difference in frequency between the roton–pair peak 17.0 K and twice the single-roton frequency measured by neutron scattering (Cowley and Woods 1971): 17.3 K. However, owing to finite

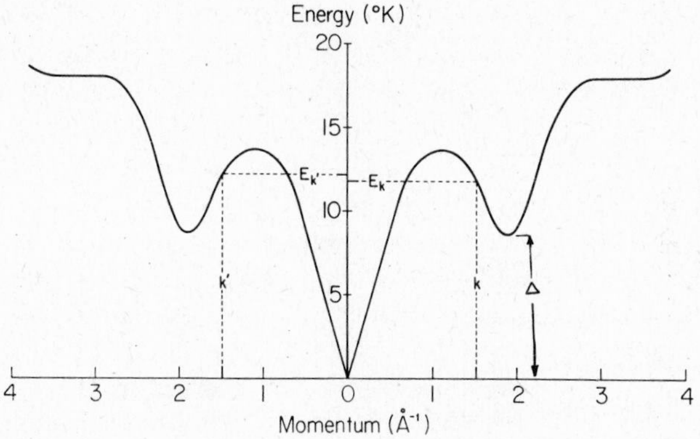

Fig. 11. Phonon–roton dispersion curve in liquid helium. Excitations near the minimum at $k \approx 2\,\text{Å}^{-1}$ are called rotons. Energy is measured in deg K. For comparison with other spectra note $1.5\,\text{K} \simeq 1\,\text{cm}^{-1} = 30$ GHz.

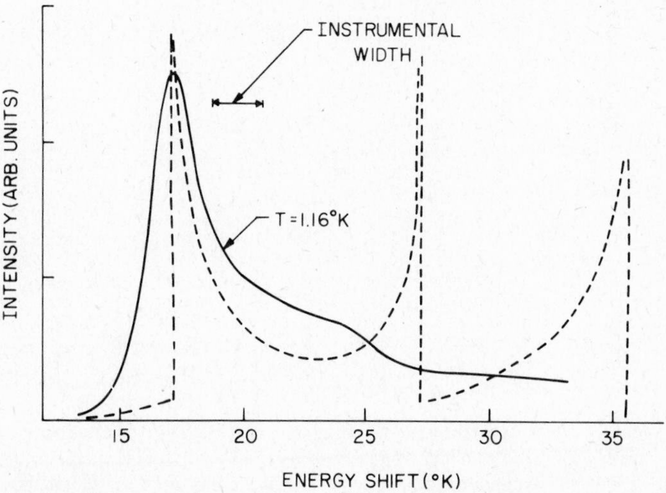

Fig. 12. Roton-pair spectrum (solid curve) as observed by Greytak and Yan (1969); compared with density of states for non-interacting rotons (dashed curve).

resolution effects in both the light and neutron scattering experiments, the difference was barely, if at all, outside experimental error. Thus the question of a possible *bound* two-roton state remained open.

The most recent experiments (Murray et al. 1975) have been extended to lower temperature (0.6 K) and higher instrumental resolution (0.13 K = 2.76 GHz FWHM) and are shown in fig. 13. The peak position is 16.95 K and the lineshape is relatively symmetrical. Both these facts can be used to extract a $g_4 \cong 0.25$ K from a theoretical analysis and from comparison with the latest neutron scattering determinations of the single-roton energy, $\Delta = 8.618 \pm 0.009$ K by Woods et al. (1977). For example by comparing the position of the roton pair peak (16.95 K) with twice the energy of the single roton (as precisely determined in the new neutron scattering measurements) Woods et al. (1977) infer a roton-pair binding energy of $g_4 = 0.27 \pm 0.04$ K. On the other hand, Murray et al. (1975) infer a value of $g_4 = 0.22 \pm 0.07$ K from an analysis of the roton-pair lineshape alone, without having to compare the peak positions observed in the neutron and light scattering experiments. The similarity in values of g_4 derived from these different analyses is gratifying. The value of $g_4 \simeq 0.25$ K would point to the existence of a *bound* roton pair in the absence of other interactions.

The experimental fact is, however, that even at 0.6 K the pair linewidth is measureable (0.11 K) and much larger than twice that of a single roton at that temperature. Thus the observed roton pair is not a true bound state.

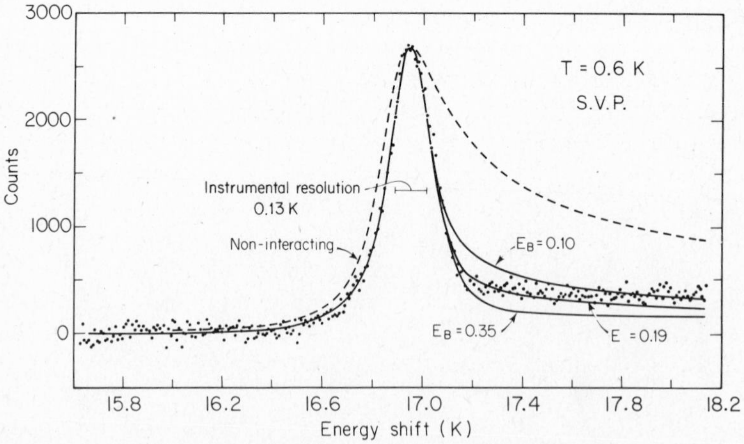

Fig. 13. Roton-pair lineshape observed by Murray et al. (1975) at 0.6 K and s.v.p., ●. All the theoretical curves have been calculated for an intrinsic full width 0.11 K of the roton pair. For the sake of lineshape comparison, each theoretical curve has been shifted so that its peak position corresponds to that of the data. Symmetric lineshape alone suggests a binding energy of ∼0.2 K.

Strictly speaking it is not expected to be, since the shape of the dispersion curve in fig. 11 shows that a $q=0$ roton pair with energy $\approx 2\Delta$ can decay into a $q=0$ phonon pair of the same energy. Although ignored in the earlier theories of interacting roton pairs, this decay channel has been shown to be experimentally significant. Latham and Kobe (1975) first included this effect phenomenologically in fitting the data of Murray et al. (1975). Woerner and Stephen (1975) have treated the coupled roton–phonon system more completely. They include phenomenologically the additional possibilities of phonon–phonon and phonon–roton interactions as well as the differences in coupling of the light to phonon- and roton-like excitations. Their description, although physically the most complete, also involves the introduction of additional adjustable parameters.

Although some slight numerical ambiguities remain, it appears that the following conclusions can safely be drawn regarding roton pairs in liquid helium:

1. The pair spectral intensity peaks about 0.3 K below twice the frequency of a single roton, implying an attractive roton–roton interaction of 0.2–0.3 K.

2. The lineshape of the observed spectrum is grossly accounted for by this roton–roton interaction, but an additional interaction describing the decay of roton pairs into phonon pairs is needed.

3. This additional interaction imparts a finite lifetime (even at zero temperature) to the $q=0$ roton-pair state so that strictly speaking it is a *resonance* rather than a bound state.

The problem can probably be regarded as closed from an experimental point of view. However, there remains the significant theoretical question of deriving a *microscopic* theory for the roton-pair light scattering. The form for $T(\boldsymbol{q})$ in eq. (4.1) is clearly an approximation, and the assumptions on $\tau(q)$ employed to date have to be improved upon. Campbell and Pinski (1978) have recently addressed the problem through a microscopic theory for the integrated scattered intensity. The final theoretical notes on the problem of light scattering from liquid helium have, however, yet to be sounded. For a comprehensive discussion of the problem of light scattering from liquid helium we refer the reader to Greytak (1978).

4.2. Second-order scattering in other simple fluids

Helium is the only liquid whose dispersion curve of elementary excitations remains well defined to such large values of q as displayed in fig. 11. Nevertheless, the process of inelastic light scattering by pairs of short-wavelength density fluctuations has been observed in simple classical liquids as well as in helium and hydrogen. Partly for historical reasons related to similarities with collision-induced light scattering (McTague and Birnbaum 1968, 1971) in gases, the scattering in simple fluids has sometimes been referred to as "intermolecular" light scattering (Fleury et al. 1971). Its relation to second-order scattering from collective excitations can be seen formally through the appropriate expression for $\alpha_{ij}(\boldsymbol{r},t)$:

$$\alpha(r,t)=\sum_{ij}\alpha_{ij}(\boldsymbol{r},t)=\sum_{ij}\int \mathrm{d}^3r'\rho_i(\boldsymbol{r}',t)\beta(\boldsymbol{r}-\boldsymbol{r}')\rho_j(\boldsymbol{r}-\boldsymbol{r}',t), \tag{4.2}$$

the dynamic polarizability associated with colliding pairs of atoms, where $\rho_i(\boldsymbol{r},t)=\rho_0\delta(\boldsymbol{r}-\boldsymbol{r}_i)\delta(t-t_i)$ and β is an intermolecular polarizability. Inserting eq. (4.2) into eq. (2.1) results in an expression identical to eq. (4.1) used for the second-order spectrum of helium. Thus the process of intermolecular scattering in dilute or dense fluids involves the same four-point correlation functions as in roton-pair scattering. The evaluation of these functions would require a full dynamic theory of the liquid state, which does not yet exist. Nevertheless, some insight into the short-time, short-wavelength dynamics of simple liquids can be gained from examining their intermolecular spectra.

Fig. 14 shows the depolarized intermolecular spectrum of liquid argon. As with all classical monatomic fluids the observed spectrum is a broad band centered at zero frequency and approximately exponential in frequency. In fact the spectra $I(\omega)$ over a remarkable range of temperature and density (Fleury et al. 1971, 1973) can be described by

$$I(\omega)=I_0\exp[-\omega/\Delta(\rho,T)]. \tag{4.3}$$

Furthermore, it has been shown that a simple corresponding states expression for $\Delta(\rho,T)$ of the form

$$\Delta(\rho,T)=\frac{3}{2\pi c}\left(\frac{\varepsilon}{m\sigma^2}\right)^{1/2}\left(\frac{kT}{\varepsilon}\right)^{1/2}\left(1+\left(\frac{2\sigma^3}{m}\right)^2\rho^2\right) \tag{4.4}$$

adequately describes all spectra in the rare gas fluids for $0.7<kT/\varepsilon<4$ and $0.01<\rho\sigma^3/m<1$, where m, σ and ε are the atomic mass, Lennard–Jones distance and energy parameters, respectively; c is the velocity of light so that Δ is expressed in cm^{-1} (Fleury et al. 1973).

McTague et al. (1969) have applied the Stephen formalism to the problem of intermolecular lineshape in liquid argon. That is, they made an approximation equivalent to eq. (2.13) and used an approximation to the neutron observed $S(q,\omega)$ for argon in an attempt to calculate $I(\omega)$. Fig. 14

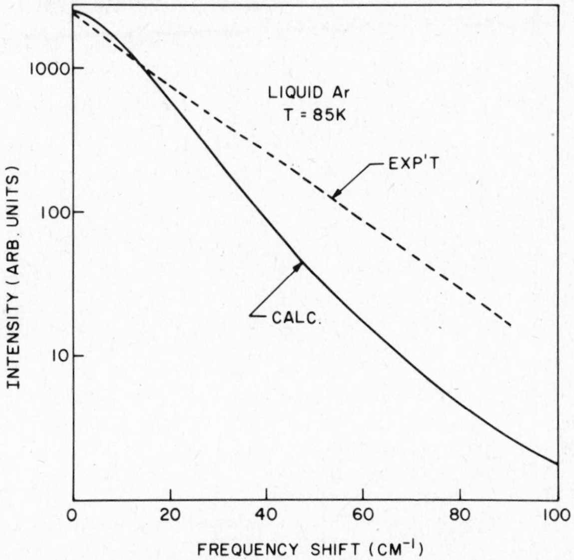

Fig. 14. Second-order Raman spectrum in liquid argon (McTague et al. 1969). Dashed line is experiment. Solid curve is approximate calculation as discussed in the text. Note the logarithmic intensity scale.

shows the result of this attempt, which suggests that the idea of pair scattering, even from overdamped pairs, has some validity and that the interaction terms left out by factoring the four-point correlation function contribute ~20% corrections.

The case of liquid hydrogen is intermediate between helium and the classical liquids. Fig. 15 shows the intermolecular spectra (Fleury and McTague 1973) for various ortho–parahydrogen liquid mixtures. The narrow, $<20\ \mathrm{cm}^{-1}$, concentration-dependent peak is due to $\Delta J=0$ rotational transitions on individual orthohydrogen molecules. The broad second-order spectrum is unchanged for different ortho-para ratios. Earlier inelastic neutron spectra by Carneiro et al. (1973) on pure parahydrogen revealed some fairly well defined high-q single collective excitations. The complete $S(q,\omega)$ has not been obtained, but the qualitative behavior indicated in fig. 16 is certainly consistent with the formation of the less-than-overdamped pair spectrum observed in fig. 15.

Finally it is instructive to compare the pair spectra observed in several of the simple liquids to the second-order Raman scattering seen in the corresponding crystalline solids. Fig. 17 shows the second-order Raman spectrum from solid Xe. The f.c.c. rare gas solids have only one atom/unit cell and therefore no first-order Raman spectrum. At low temperatures, the

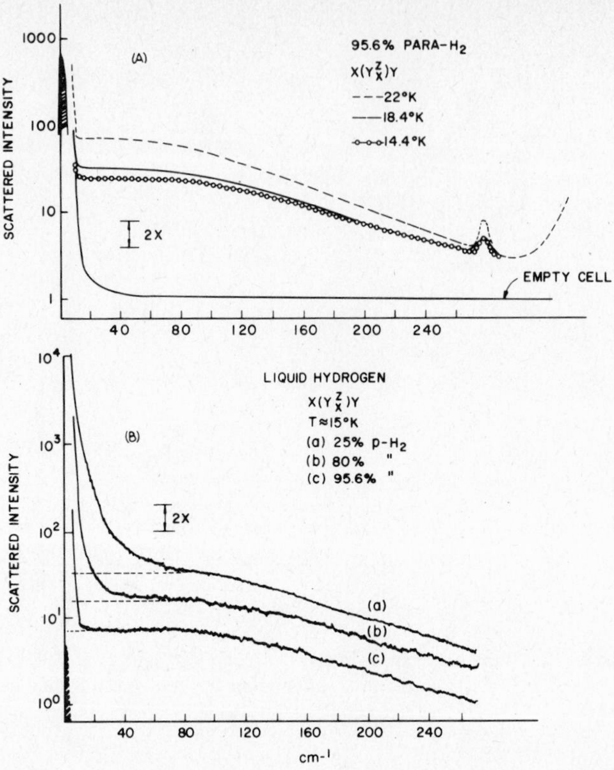

Fig. 15. Depolarized intermolecular scattering in liquid hydrogen. (A) Temperature variation in 95.6% parahydrogen. (B) Variation with ortho–para composition at constant temperature (15 K). Spectra are displaced vertically for display. After Fleury and McTague (1973).

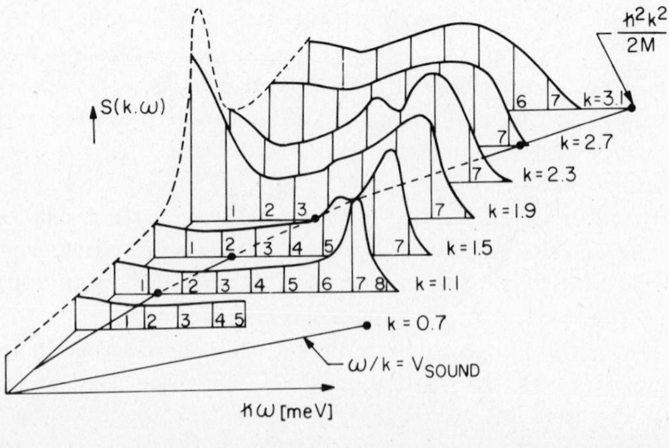

Fig. 16. Neutron scattering function $S(k,\omega)$ for liquid parahydrogen, $T = 14.7$ K. Complete spectra could not be observed below $k = 1.1\,\text{Å}^{-1}$ because of the high velocity of sound. For comparison with other spectra, note 1 meV $= 8.066\ \text{cm}^{-1}$, k is measured here in Å^{-1}. After Carneiro et al. (1973).

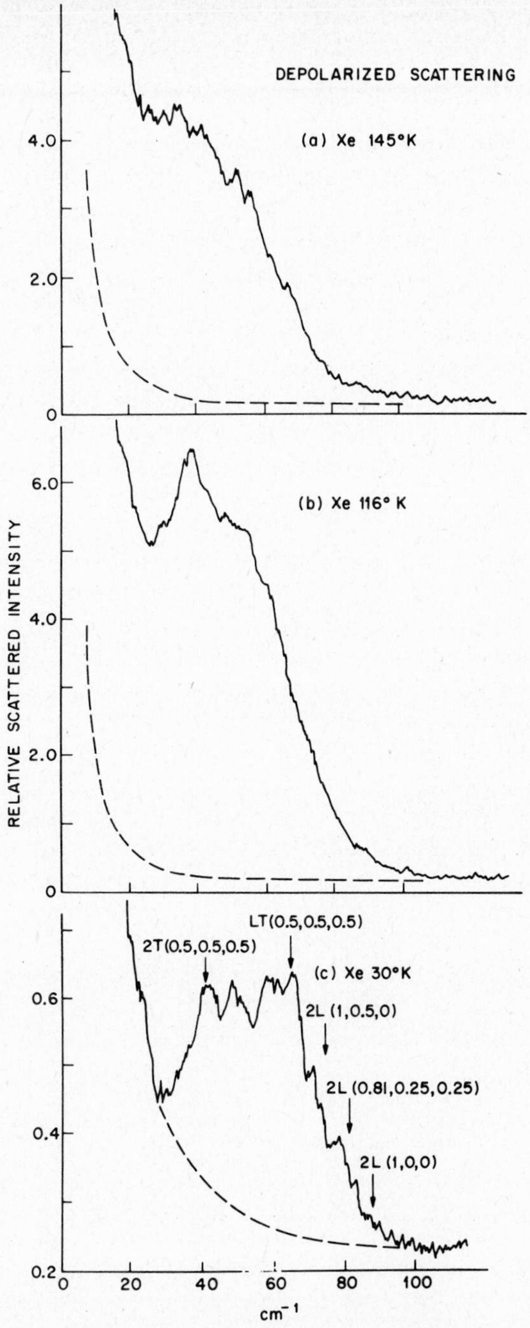

Fig. 17. Second-order Raman spectra in solid Xe. Dashed curve represents instrumental background taken in the empty scattering cell. Arrows in (c) indicate calculated frequencies of two-phonon features (Werthamer et al. 1970). After Fleury et al. (1973).

second-order spectrum is weak but clearly exhibits structure associated with phonon pairs from various points in the Brillouin zone. Spectra in solid Ar and solid Kr are similar in appearance when their frequencies are scaled according to the material's Debye frequency. With increasing temperature the second-order spectrum gradually loses its sharp structure, with the 2TA peaks persisting to higher temperatures than the 2LA or LA–TA peaks. In addition, the relatively sharp high-frequency cutoff (at $\sim 88\ \mathrm{cm}^{-1}$ for Xe) becomes smeared out and eventually assumes a nearly exponential shape.

Molecular dynamics calculations using a variety of potentials, including a Lennard–Jones potential, were also presented by Alder et al. (1973), giving a good account of the observed spectra near melting.

In fig. 18 the solid and liquid spectra are compared at the triple point (Fleury et al. 1973). Spectra are shown with no change in intensity or frequency scales between the liquid and the solid. Note that as far as the high-frequency dynamics ($\omega \gtrsim 2\omega_D$) are concerned the liquid and solid are indistinguishable. At low frequencies, however, the liquid spectra continue to increase exponentially while the solid spectra level off, so that the integrated intensities in the liquids are all much greater than in the corresponding solids. The solid integrated intensities were obtained assuming the spectrum remains constant for $\omega < 20\ \mathrm{cm}^{-1}$.

In table 1 various aspects of the second-order spectra for the condensed phases of the classical rare gases Ar, Kr and Xe are listed and experiment is compared with theory. The "calculated" characteristic frequency Δ is obtained from the corresponding states formula eq. (4.4). The calculated total scattered intensity h_T for the solid is obtained from a simplified finite temperature generalization (Fleury et al. 1973) of the theory of Werthamer et al. (1970):

where α_0 is the atomic polarizability, a the lattice parameter, Z the number of nearest neighbors, ω_i the laser frequency, n_i the optical refractive index and $f(x) = (1-e^{-x})^{-2}$. In view of the crude arguments used in this finite

$$h_T = \frac{N}{V}\left(\frac{\alpha_0}{a^3}\right)^4\left(\frac{\omega_i}{c}\right)^4 a^6 Z^2\left(\frac{\hbar}{m\omega_D a^2}\right)^2\left(\frac{n_i^2+2}{3}\right)^4 f\left(\frac{\hbar\omega_D}{kT}\right), \tag{4.5}$$

Table 1

	ω_D (cm^{-1})	$\Delta_{\mathrm{liq.}}^{\mathrm{calc}}$ (cm^{-1})	$\Delta_{\mathrm{liq.}}^{\mathrm{obs}}$ (cm^{-1})	h_T^{obs}(liq.) $(\mathrm{cm}^{-1}\mathrm{sr}^{-1})$	h_T^{obs}(sol.) $(\mathrm{cm}^{-1}\mathrm{sr}^{-1})$	h_T^{calc}(sol.) $(\mathrm{cm}^{-1}\mathrm{sr}^{-1})$
Ar	62	22.1	22	18×10^{-10}	4.6×10^{-10}	6.2×10^{-10}
Kr	46	16.7	17	83×10^{-10}	9.4×10^{-10}	10.4×10^{-10}
Xe	40	14.4	14	276×10^{-10}	33.6×10^{-10}	36.6×10^{-10}

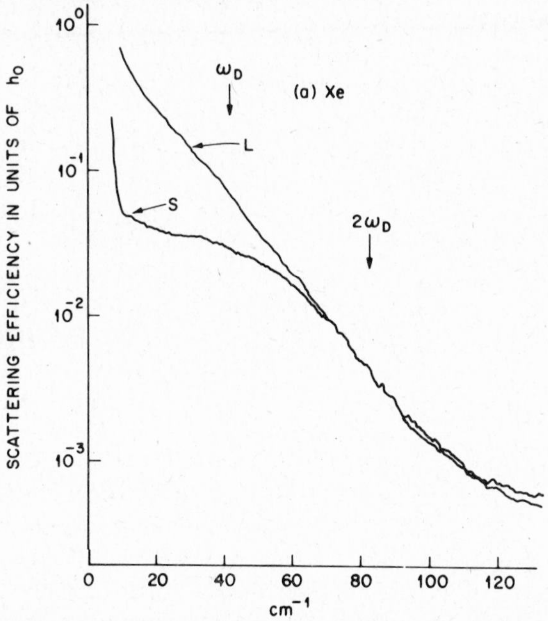

Fig. 18. Comparison of second-order Raman spectra in Xe just below (S) and just above (L) the melting temperature. Note logarithmic intensity scale. After Fleury et al. (1973).

temperature extrapolation we would not expect eq. (4.5) to be quantitatively valid to better than about a factor of 2. The considerably better agreement with experiment indicated in table 1 is probably fortuitous.

Nevertheless, the agreement between experiment (Fleury et al. 1973) and theory regarding both intensity and frequency range leaves little doubt that the observed spectra in the classical rare gas solids are essentially described by two-phonon processes. The smooth evolution of the high-frequency spectra upon melting also supports the view that in simple classical fluids the spectra can be attributed to scattering from pairs of highly damped density fluctuations. The large increase in spectral intensity at low frequencies for the liquid relative to the solid accounts for all of the rather large difference in integrated intensities between the two phases. However, this striking difference has not yet received a satisfactory explanation. Finally we mention the molecular dynamic calculations of the band shape of rare gas solids at high temperatures by Alder et al. (1976). After fitting the intensity at $\nu = 100\ \mathrm{cm}^{-1}$ to the experimental results (Fleury et al. 1973), excellent agreement is obtained with the experiments (fig. 18) over the whole frequency range using a Lennard–Jones potential. These authors show that the scattering is dominated by two-phonon processes.

More recent high resolution experiments in solid Xe have observed two phonon difference scattering (Lyons et al. 1979) in qualitative, but not quantitative, agreement with available calculations (Leese and Horton 1979).

Experimental studies of higher-order spectra have been conducted in the condensed phases of ^{3}He, ^{4}He and their mixtures by Surko and Slusher (1976). While several aspects of the spectra are similar to those just discussed for the classical materials, there are some important differences. We have already discussed the sharp structure associated with roton pairs in superfluid ^{4}He, and now concentrate on the more general features seen over wide ranges of temperature, density and isotopic composition. Typical spectra are shown in fig. 19 for solid and liquid ^{4}He. Except for a diminished spectral intensity at low frequencies these spectra appear at first glance rather similar to those for Xe in fig. 18. That is, the high-frequency lineshapes are essentially identical in the liquid and solid states and there is an appreciable increase in low-frequency intensity for the liquid, relative to the solid. As seen by the solid curve in fig. 19, the observed peak in the spectrum corresponds rather well to the two-phonon Raman scattering (shown as the solid curve) calculated by Werthamer et al. (1970). Its persistence into the liquid state is a strong argument for the

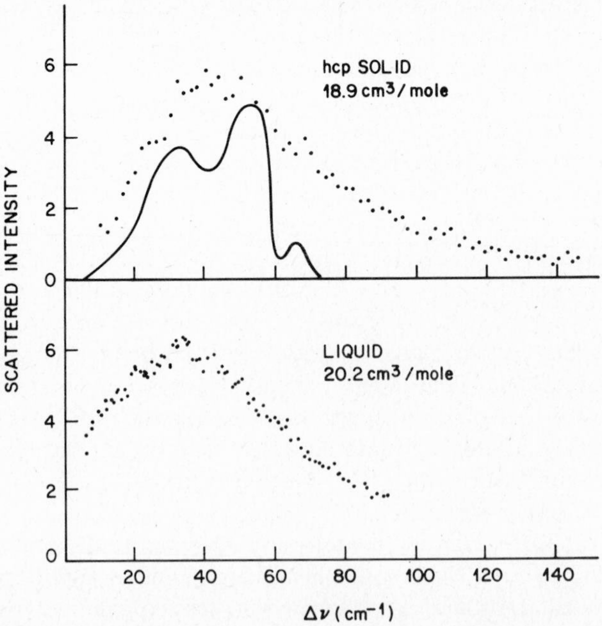

Fig. 19. Higher-order Raman spectra of solid and liquid helium (points) at 3–4 K. Solid curve is the calculated two-phonon Raman spectrum. After Surko and Slusher (1976).

existence of relatively well-defined short-wavelength phonon modes in the liquid.

A striking difference between the helium spectra and those of the classical materials is the appreciable *additional intensity* observed at frequencies well above the predicted two-phonon cutoff. Within a phonon type of description such intensity would require four-, or even six-phonon processes. Surko and Slusher (1976) have argued that this additional high-frequency scattering can successfully be viewed instead as due to creation of pairs of nearly free particle excitations. In He single free particle excitations at large energies ($>20\ \mathrm{cm}^{-1}$) and momenta ($>2.5\ \mathrm{Å}^{-1}$) have been observed by neutron scattering, and obey the dispersion relation $\varepsilon_k = \hbar^2 k^2/2m^*$ with m^* nearly equal to the mass of a bare helium atom. By taking the final single particle states at wave vector k to be distributed above $\hbar^2k^2/2m^*$ with a gaussian of width $\hbar k\Gamma$, they calculated the spectral response associated with excitation of pairs (at $\boldsymbol{k}$ and $-\boldsymbol{k}+\boldsymbol{q}$) summed over $\boldsymbol{k}$. The value of Γ was adjusted to give the best fit ($\Gamma = 8.7\ \mathrm{K}/\mathrm{Å}^{-1}$). The success of this approach indicates that pairs of both single-particle and collective (phonon) excitations contribute significantly to the light scattering spectra in the condensed phases of helium. For the heavier rare gas solids and liquids the single-particle pairs are much less important, presumably because for these materials the ratio of binding energy E_b to Debye frequency $\hbar\omega_\mathrm{D}$ is much greater than in He. That is, the single-particle pairs should contribute for frequencies $\hbar\omega > 2E_\mathrm{b}$ a spectral component of the form $I_\mathrm{SPP}(\omega) = C\mathrm{e}^{-\omega/\omega_\mathrm{D}}$. Surko and Slusher (1976) estimate that for argon, C is $\simeq 15$ times the maximum of the two-phonon intensity; but since $E_\mathrm{b} = 715\,\mathrm{K}$ and $\omega_\mathrm{D} = 93\,\mathrm{K}$, the ratio $I_\mathrm{SPP}/I_\mathrm{2Ph} \sim 15\ \mathrm{e}^{-1430/93} \approx 2\times 10^{-6}$ is negligible indeed.

5. *Phonon interactions in crystalline solids*

5.1. Diamond, graphite and the rare gas solids

The discussion of §2 suggests that one might expect bound states of phonon pairs to occur relatively frequently, given the wide variety of phonon dispersion curves and the range of anharmonicities displayed by the vast family of crystalline solids. Unfortunately, it appears that two-phonon bound states occur more frequently in theoretical calculations than they do in experimental observations.

The most thoroughly considered candidate for a two-phonon bound state is the $2770\ \mathrm{cm}^{-1}$ feature in the second-order Raman spectrum of diamond. Solin and Ramdas (1970) first emphasized the following unusual properties of this feature: (a) it lies slightly above twice the zone-center one-phonon frequency of $1333.5\ \mathrm{cm}^{-1}$; (b) it is strongly polarized; (c) its

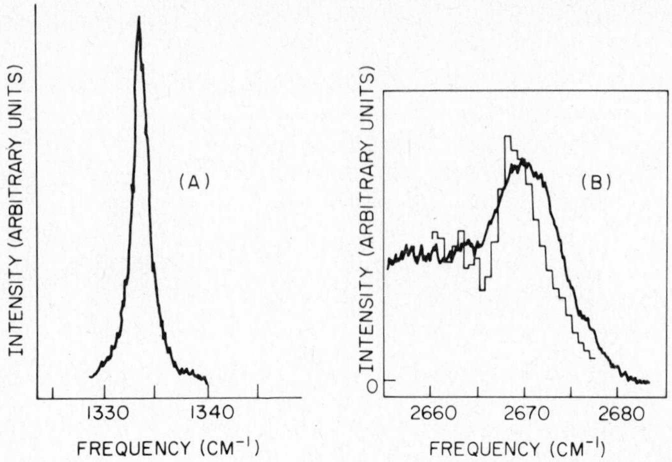

Fig. 20. (A) First-order Raman line at 1333 cm^{-1} in diamond observed by Washington and Cummins (1977) with 0.5 cm^{-1} instrumental resolution. (B) Second-order Raman feature at 2670 cm^{-1} in diamond (solid curve is experiment); compared with calculation (histogram) of Tubino and Birman (1977).

intensity scales linearly with scattering volume; (d) its width is approximately twice that of the one-phonon Raman line (see fig. 20). Cohen and Ruvalds (1969) have interpreted this feature as a two-phonon bound state split off above the two-phonon continuum by a repulsive g_4 interaction. Such an interaction could produce a bound state if the maximum in the one-phonon dispersion curves were to occur at the (Γ) Brillouin zone center.

This interpretation has, however, not been universally accepted. Opponents argue, for example, that no such feature is observed in the similar but more strongly anharmonic crystals Si and Ge; and that the diamond peak remains well defined to quite high temperatures (1200 K) relative to any phonon-pair binding energy (Go et al. 1975). Also a simple alternative explanation in terms of "overbanding" of the one-phonon dispersion curve can account for the observed spectra even within a purely harmonic description. Most recently Tubino and Birman (1977) have theoretically re-examined diamond within the harmonic framework using a bond polarizability model to calculate the phonon dispersion curves. Fig. 21 shows the results of their calculations for the dispersion curve along Δ in both diamond (a) and silicon (b). The calculated maximum in diamond occurs not at Γ but along Δ and produces a P_3 type critical point (absent in silicon) which may be responsible for the sharp peak at 2270 cm^{-1} observed in diamond. Because of the approximation involved in the model

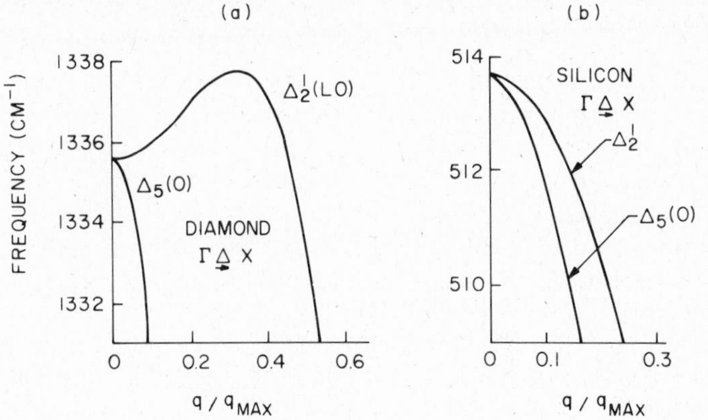

Fig. 21. Comparison of calculated dispersion curves along Γ→X for (a) diamond and (b) silicon showing presence of overbanding in diamond. (Tubino and Birman 1977).

calculations and the extremely small amount of frequency dispersion required to account for the spectroscopic observations, it cannot be said with certainty that no two-phonon binding is involved. However, the Tubino–Birman theory does a creditable job in accounting for the observed lineshape, as seen in fig. 20.

Another possible candidate for a two-phonon bound state has recently been identified by Nemanich and Solin (1977) in the spectrum of graphite. There the single phonon line at 1581 cm^{-1} is observed to have a full width at half maximum of 14 cm^{-1}, while the "two-phonon" feature at 3248 cm^{-1} is only 10 cm^{-1} wide. Thus not only is this feature shifted by more than 85 cm^{-1} above twice the one-phonon frequency, its linewidth is much less than twice the one-phonon linewidth. While the narrow width suggests a tendency toward phonon binding, Nemanich and Solin do not attribute the 3248 cm^{-1} line to a bound state of two zone-center phonons. First, the implied binding energy of 86 cm^{-1} would be fully 5% of the one-phonon energy – unusually large; and second, their own calculations of the graphite dispersion curves reveal an over-banding in the Σ direction, which would produce a peak in the two-phonon density of states at about 3230 cm^{-1}. As in the case of diamond, however, there are no direct measurements of the high-frequency phonon dispersion curves and the numerical results of such model calculations are not sufficiently precise to confirm the existence of a bound state on the grounds of frequency position alone. The narrow width and symmetric lineshape of the 3248 cm^{-1} line does argue for significant interaction effects, leaving open the possibility that it is a two-phonon bound state.

The rare gas solids present a different problem. There the complete phonon dispersion curves are well characterized both theoretically and experimentally. Further, the knowledge of the interatomic potentials is sufficient to permit more accurate estimates of the anharmonic force constants leading to g_4. Recently Jindal and Pathak (1977) have investigated theoretically the possibility of two-phonon bound states in crystals of the solid rare gases. They have solved approximately the integral equation for the two-phonon Green's function using the Fredholm method. Bound states are determined by zeroes in the Fredholm denominator which they approximate as

$$D = 1 - \frac{12\pi}{\hbar} \sum_{k_1 k_2} V_4(-k_1 j_1, -k_2 j_2 | k_1 j_1, k_2 j_2) G^0_{\mathrm{II}}(k_1 j_1, k_2 j_2; \omega), \qquad (5.1)$$

where the j's index the $j_1 j_2$ phonon branches: The explicit form of V_4 is explored in considerable detail and is obtained without having to resort to the very simplest approximation employed by Ruvalds and others for problems we have already discussed. The sums in eq. (5.1) were evaluated numerically using the anharmonic parameters derived from the well-known Lennard–Jones interatomic potential

$$\phi(r) = \varepsilon\left[(\sigma/r)^{12} - 2(\sigma/r)^6\right]. \qquad (5.2)$$

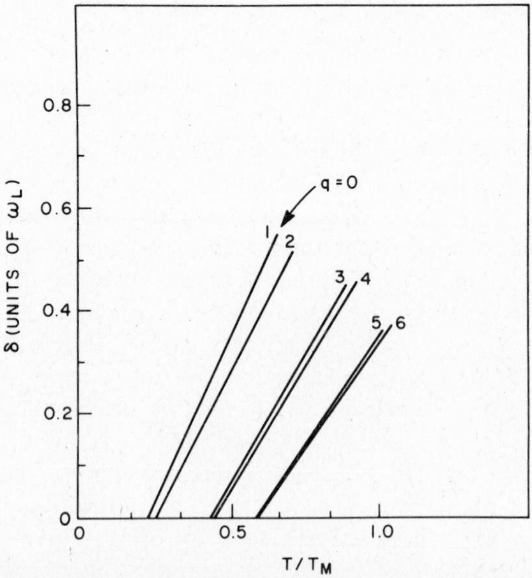

Fig. 22. Position of calculated bound states for phonon pairs of different total wave vector as a function of temperature relative to the melting temperature T_M. (Jindal and Pathak 1977). δ measures the shift above $2\omega_D$ in units of $\omega_D = \omega_L$, for $q=0$ (curve 1); (0.4,0,0) curve 2, (0.6,0,0) curve 3, (0.4,0.4,0) curve 4, (0.6,0.6,0) curve 5, and (0.4,0.4,0.4) curve 6.

Solutions to the condition $D=0$ were obtained (Jindal and Pathak 1977) for various temperatures: $\frac{1}{3}, \frac{1}{2}, \frac{2}{3}, \frac{5}{6} T_m$, where T_m is the melting temperature.

The general conclusion was reached that two-phonon bound states should exist for several q values (including $q=0$) at sufficiently high temperatures. Fig. 22 shows the calculated bound state frequency δ as a function of temperature for several values of the total wave vector q. Of particular interest to us is the $q=0$ case, which should be relevant to light scattering. According to fig. 22, the bound state should appear for $T > \frac{1}{4} T_m$ at $2\omega_L$ and move to higher frequency with increasing temperature. At $T \approx 0.6 T_m$ the frequency should be $2.6\omega_L$, or about 30% above the two-phonon cutoff. In xenon $2\omega_L \cong 88\ \text{cm}^{-1}$ so that the bound state should appear at $(2\omega_L + \delta) \cong 112\ \text{cm}^{-1}$. The observed spectrum in fig. 17, however, shows no detectable scattering in that region. Furthermore, at high temperatures $T \sim 0.99 T_m$ the logarithmic spectrum of fig. 18 provides an even more definitive negative result.

The theoretical treatment has not compared directly the spectral weight expected for the bound state with that in the two-phonon continuum. However, Jindal and Pathak (1977) have stated that "the strength is...finite and increases with increasing temperature". Our own estimates based on comparisons of the Jindal–Pathak results with the two-phonon calculations of Werthamer et al. (1970) indicate that the bound state scattering strength calculated by JP should be of the same order of magnitude as that in the two-phonon continuum. Therefore, it should have been easily detected in the data of fig. 17 and in similar experiments on solid argon under pressure by Crawford et al. (1978). The failure of a bound state to appear in these spectra calls for a re-examination of the theory.

5.2. Hybridization of single and pair excitations

Several situations involving the interactions of single excitations with pair excitations of the same momenta have been identified in both light- and neutron scattering spectra. We may cite four examples: (1) the one-optic-phonon–two-acoustic-phonon hybridization observed in the $q=0$ Raman spectrum of quartz; (2) the one-polariton–two-vibrational-phonon hybridization observed in small-angle Raman scattering from NH_4Cl; (3) the one-roton–two-roton hybridization observed in neutron scattering from liquid helium for $q \simeq 2.75\ \text{Å}^{-1}$; and (4) the hybridization between the coupled acoustic–soft optic phonon and the two-phonon difference process (phonon density fluctuations) responsible for the dynamic central peak in ferroelectric lead germanate. All of those cases can be described by special cases of eq. (2.17) involving either two or three pairwise coupled excitations.

Table 2

Material	He	NH_4Cl	SiO_2	$Pb_5Ge_3O_{11}$
Mode 1	roton	polariton	phonon	soft phonon
Mode 2	roton pair	phonon pair	TA phonon pair	PDF
Mode 3	—	—	—	LA phonon
g_{12}	g_3	g_3	g_a	δ^2
g_{13}	0	0	0	(A^2)
g_{23}	0	0	0	0
$a(1)$	$a(1)$	$a(1)$	$a(1)$	$a(1)$
$a(2)$	0	$a(2)$	0	0
$a(3)$	—	—	0	$a(3)$

Table 2 summarizes the parameters (from eqs. (2.17) and (2.18)) and identifies the excitations relevant to each example.

Except for the neutron scattering in helium all of the examples are concerned with $q=0$ excitations. Let us begin with the helium example because it has been interpreted assuming that $a(1)\neq 0$, so that any anomalous spectral structure arises from the roton-pair contributions to the single roton self-energy. Fig. 23a shows the observed roton dispersion curve displaying a peculiar double branch region for momenta between 2.7 Å^{-1} and 3 Å^{-1}. Constant k energy scans in this region reveal a broad and anomalously shaped response, as sketched on the right of fig. 23b. Ruvalds and Zawadowski (1970a) have accounted for this structure through the interaction of a single finite k roton with a roton-pair excitation at essentially the same energy ($\approx 2\Delta$) and wave vector (2.75 Å^{-1}).

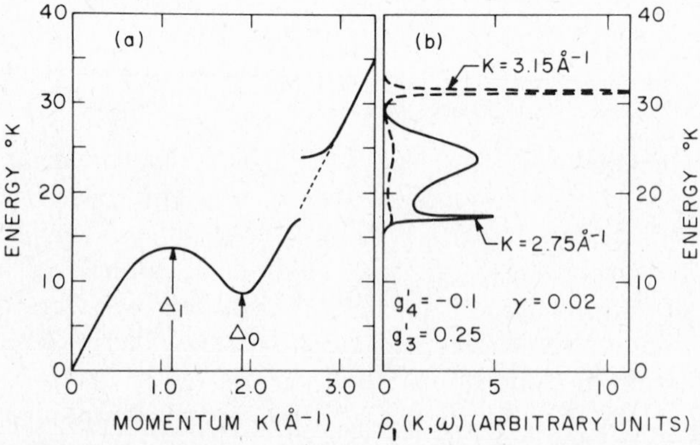

Fig. 23. (a) Phonon–roton dispersion curve in liquid helium, showing anomalous region for momenta near 2.7–3.0 Å^{-1}. (b) First-order scattering spectrum $\rho_1(k,\omega)$ shown for two different momenta (dashed and solid curves), as calculated by Ruvalds and Zawadowski (1970a) using the indicated coupling strengths for g'_4 and g'_3.

This interaction is of third order and enters into the one-roton Green's function as

$$G_{\mathrm{I}}(k,\omega)=\frac{G_{\mathrm{I}}^{0}(k,\omega)}{1-g_3^2 G_{\mathrm{II}}(k,\omega)}. \tag{5.3}$$

Approximating $G_{\mathrm{II}}(k)$ by the $G_{\mathrm{II}}^{0}(k)$ for a non-interacting roton pair ($g_4=0$) is shown by fig. 24 to be inadequate in that $G_{\mathrm{II}}^{0}(k)$ shows not a resonant peak but rather a shoulder at $\omega\simeq 2\Delta$. An attractive roton–roton interaction g_4, discussed in §4, introduces the required peak into G_{II}. That is

$$G_{\mathrm{II}}(k)=\frac{G_{\mathrm{II}}^{0}(k)}{1-g_4 G_{\mathrm{II}}^{0}(k)}, \tag{5.4}$$

so that when G_{II} of the form eq. (5.4) is introduced into eq. (5.3) the observed anomalous structure is reproduced, at least qualitatively.

Some caution is urged again because in the absence of a microscopic theory for liquid helium, the coupling between neutrons and rotons is not known. Therefore $a(1)$ and $a(2)$ must be considered phenomenological parameters and the assumption that $a(2)=0$ is an uncontrolled approximation. Further the agreement between the calculated lineshape (fig. 23) and that observed is only qualitative.

Scott (1968) has explained the previously anomalous low-frequency Raman spectrum ($\omega<300\ \mathrm{cm}^{-1}$) in crystalline quartz (SiO_2) as due to a

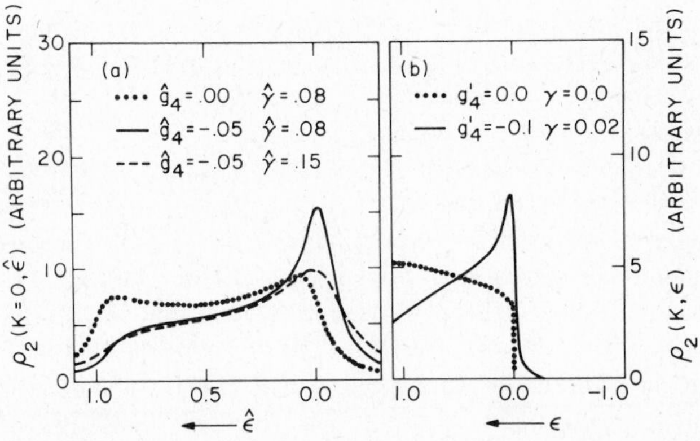

Fig. 24. (a) Calculated roton pair spectra ρ_2 for $K=0$ pairs as observable in light scattering, showing effects of increasing the attractive roton–roton interaction, g_4. (b) Effect of g_4' on *finite* K roton-pair spectra; $\rho_2(K\neq 0)$. After Ruvalds and Zawadowski (1970a).

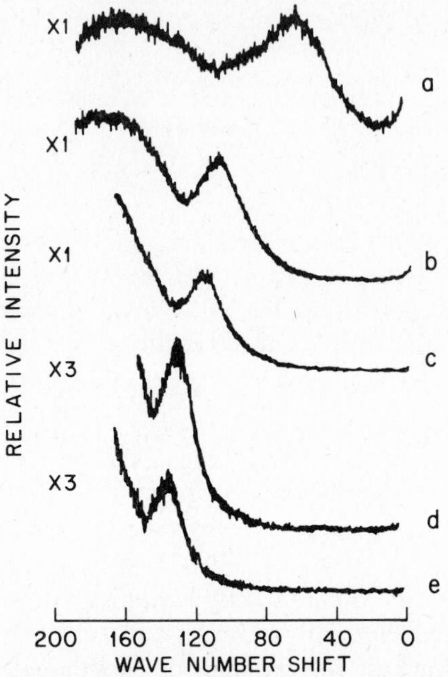

Fig. 25. Raman spectrum of SiO_2 showing effects of one-phonon–two-phonon hybridization. As temperature is increased toward the 846 K phase transition (e = 300 K, d = 400 K, c = 600 K, b = 700 K, a = 800 K) the hybridized one-phonon peak moves toward lower energy (i.e. toward the right in this figure). After Scott (1974).

hybridization of a single soft optic phonon with a phonon pair comprised of two Brillouin zone-boundary phonons on the transverse acoustic branch. The quartz α–β structural phase transition at 846 K is preceded by the softening of the 207 cm^{-1} optic phonon as the transition is approached from below. However, additional scattering was observed near 165 cm^{-1} which persisted above T_c in apparent violation of the selection rules for first-order Raman scattering in the quartz β phase. The interpretation of this behavior and the temperature-dependent frequencies as a one-phonon–two-phonon hybridization produced the agreement shown in fig. 25. Subsequently, Ruvalds and Zawadowski (1970b) produced theoretical lineshape calculations for this case using a formulation essentially identical to that discussed immediately above for the roton hybridizations in helium. Their calculated lineshapes in fig. 26 are in fairly good agreement with observations in quartz (Scott 1974). The detailed lineshapes of the hybridized spectra will again depend upon interactions between members of the pair, i.e. g_4. In particular, for $g_4 > g_4^c$ a pair bound state could be formed (see eq. (2.16)). However, even though the value of g_4 assumed in generating fig. 26 exceeds g_4^c, the comparison with the experimental lineshape is

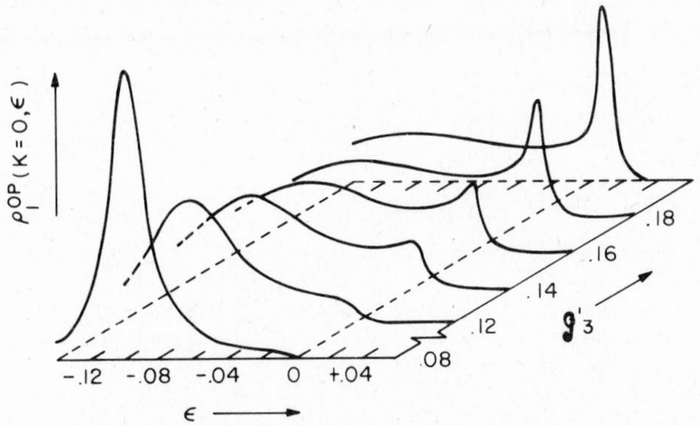

Fig. 26. Calculated hybridized lineshapes for the one-phonon–two-phonon system (Ruvalds and Zawadowski 1970b) using various values of g_3', the cubic anharmonic coupling. For this calculation $a(2)=0$ was assumed, and the quartic anharmonicity g_4 was taken to be sufficient to form a two-phonon bound state above $\varepsilon=0$.

not unique. Thus no definitive statement can be made at this point as to whether the phonon pairs in SiO_2 are bound or not.

A rather convincing case for existence of phonon-pair bound states has been made from studies of the polariton spectra in NH_4Cl (Gorelek et al. 1977). A polariton is a mixed electromagnetic–mechanical excitation. The most familiar examples involve the hybridization of the photon with a single infrared active transverse optic phonon. For small k, the optic phonon dispersion relation is $\omega_{TO}(k)=\omega_{TO}$, while that for the photon is $\omega_P(k)=\varepsilon_0^{-1/2}ck$. Thus the dispersion curves of the mixed system appear as shown in fig. 27. More generally the mechanical excitation need not be a single phonon, but could also be a pair mode (bound or free). Gorelek et al. (1977) have interpreted certain features in the angle-dependent Raman spectrum of NH_4Cl as due to polaritons arising from phonon-pair excitations – including a bound state. Fig. 28 shows a portion of the $\theta=90°$ Raman spectrum. The features labeled $\nu_{1\ell}=3122\ cm^{-1}$ and $\nu_{3\ell}=3159\ cm^{-1}$ are single phonon excitations. The broad feature at 2850 cm^{-1} is a two-phonon overtone $=2\nu_4(F_2)$. Two-phonon combination bands appear at $\nu_2(E)+\nu_{4t}(F_2)=3120\ cm^{-1}$, $\nu_2(E)+\nu_{4L}(F_2)=3140\ cm^{-1}$ and $\nu_2(E)+\nu_4(F_2)=3096\ cm^{-1}$. The sharp features at $\nu_t'=3052\ cm^{-1}$ and $\nu_\ell'=3070\ cm^{-1}$ are attributed to bound states split off from their $[\nu_2(E)+\nu_4(F_2)]$ pair continuum. The polariton dispersion of this frequency region is shown in fig. 29. The fact that appreciable dispersion is observed associated with both the two-phonon continuum at 3096 cm^{-1} and ν_t' at 3052 cm^{-1} (as shown by the solid curves) clearly demonstrates appreciable electromagnetic content in these mixed pair excitations. The bound state nature of the ν_t' and ν_ℓ' features is argued primarily from their narrow widths and the fact that they cannot arise from single excitations since the latter are all already

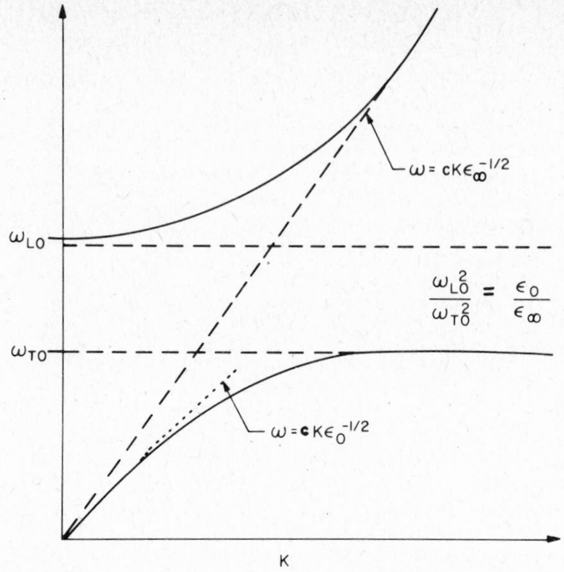

Fig. 27. Schematic dispersion curve of a coupled photon–phonon system. The uncoupled mode frequencies are shown by dashed lines. After coupling the polaritons (solid lines) result. ε_0 and ε_∞ represent the low and high frequency dielectric constants. ω_{LO} and ω_{TO} give the longitudinal and transverse optic phonon frequencies.

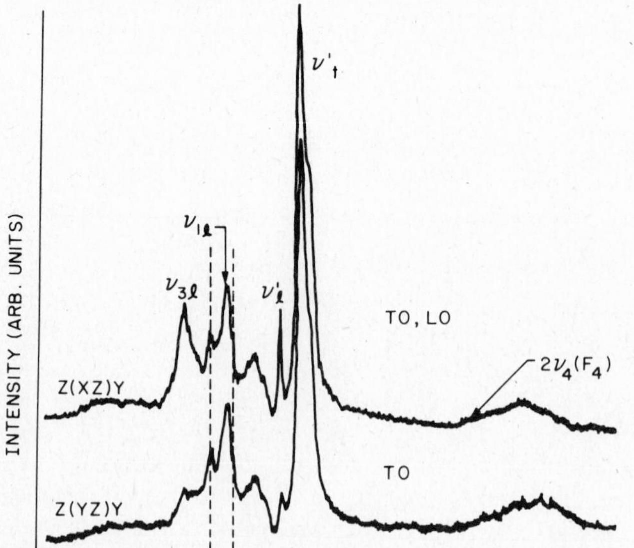

Fig. 28. Right angle Raman spectrum of NH_4Cl in the 2800–3300 cm^{-1} region for two scattering geometries. After Gorelik et al. (1977). Features are labeled according to identifications described in text.

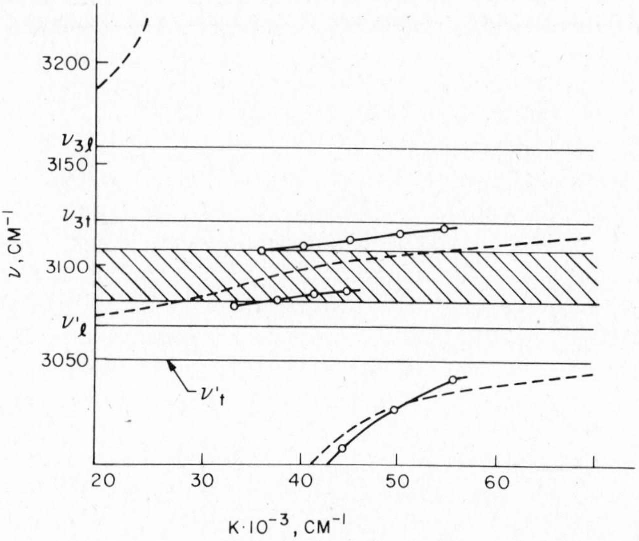

Fig. 29. Dispersion curves in NH_4Cl obtained by varying the Raman scattering angle ($K \sim \sin\frac{1}{2}\theta$). Solid lines show varying positions of features identified in fig. 28. Shaded region shows two-phonon continuum from which ν'_{ℓ} and ν'_{t} are split off as bound states. Dashed curves show calculated polariton dispersions in absence of phonon-pair hybridization. After Gorelik et al. (1977).

identified with other features in the spectra. The situation in NH_4Cl is obviously complicated, and alternative identifications of the spectral features which do not require invocation of phonon bound states *may* be possible. The interested reader is referred to the original experimental papers and to a detailed theoretical treatment by Agranovich and Lalov (1972) for complete discussions.

At this point one may say with confidence that several instances of one-phonon–two-phonon hybridization have been substantiated, but that the evidence for bound phonon pairs in these instances is less than completely convincing.

Finally we consider another class of pair excitations which are quite important in the dynamics of structural phase transitions: the so-called phonon density fluctuations (PDF). The origin of the term can perhaps most easily be appreciated from a consideration of the pair Green's function which arises from the correlation function of eq. (2.10). Expressing the U's as in eq. (2.8) clearly yields a term of the form $\langle b_k^+ b_k b_{k'}^+ b_{k'} \rangle$, where the b's are phonon creation/destruction operators. Since the phonon occupation number $n_k = \langle b_k^+ b_k \rangle$, it is clear how one may view part of the $G_{\rm II}$ as expressing the correlation function between fluctuations in the phonon number density. Generalizing this slightly to $n_k(q,t) = b_{k+q}^+(0) b_k(t)$ permits a description in terms of excitations of the phonon gas, which under certain conditions may have an interesting dynamics of their own. For example, the long-wavelength propagating component of the phonon density fluctuations gives rise to second sound, which has recently been observed directly in NaF by an elegant light scattering technique (Pohl and Irniger 1976).

More simply, light scattering from phonon density fluctuations can be viewed as two-phonon difference scattering, in which a phonon on branch i of wave vector $\boldsymbol{k}$ is created while a phonon of wave vector $-\boldsymbol{k}+\boldsymbol{q}$ on the same branch is destroyed (and vice versa). The spectral lineshape of such a scattering process depends upon (a) the shape of the one-phonon dispersion curve, (b) the k-dependent one-phonon lifetime and (c) phonon–phonon interactions. As discussed theoretically by Wehner and Klein (1972) for phonons which are under-damped and non-interacting this process will produce a spectral peak at $\omega \approx \langle \boldsymbol{v}_g(k) \cdot \boldsymbol{q} \rangle$, where $\boldsymbol{v}_g(k)$ is the phonon group velocity, $\boldsymbol{q}$ the scattering wave vector and $\langle\ \rangle$ represents a Brillouin zone average. Small $\boldsymbol{k}$ phonons on the acoustic branch will therefore contribute a peak near the Brillouin frequency $\omega_B = \boldsymbol{v}_{ph} \cdot \boldsymbol{q} (\boldsymbol{v}_{ph} =$ acoustic phonon phase velocity). For phonons on optic branches or near the Brillouin zone boundary, $v_g \approx 0$ and a peak at $\omega \approx 0$ will result. The width and shape of such a *central peak* will be determined by weighted Brillouin zone averages of individual phonon lifetimes and by possible phonon–phonon interaction effects.

Such a central peak may have the proper symmetry and lie in the proper frequency range to hybridize with low-frequency single-phonon excitations.

The phonon density fluctuations may play an important role in the dynamics of structural phase transitions. Let us consider for example the ferroelectric transition in lead germanate (Fleury and Lyons 1976). This transition is prototypical of the soft mode or displacive type of structural phase transitions, characterized by the instability of a particular normal mode of the lattice vibrations upon approach to T_c. This normal mode, whose eigenvector provides the static connection between the crystal structures above and below T_c, and whose eigenfrequencies describe the dynamics of fluctuations in the order parameter is called the "soft mode". The simple quasi-harmonic soft mode description – in which the critical dynamics are described completely by $G_s(\omega) = [\omega^2 - \omega_s^2(T) + i\omega\Gamma_s]^{-1}$ with $\omega_s^2(T) = A(T - T_c)^\gamma$ – has been shown to be inadequate for a number of structural phase transitions as scattering experiments have probed ever closer to T_c. Usually this inadequacy has been manifested by additional low-frequency structure in the scattered spectrum which exhibited singular behavior (usually diverging intensity) near T_c (Riste 1974). However, until quite recently it was not possible to demonstrate unambiguously the dynamic nature of this anomalous central peak, because the spectral resolving power was inadequate to measuring the relatively narrow linewidths associated with the central peaks. Recent use of high-contrast, high-resolution light scattering techniques has permitted detailed lineshape measurements of the soft mode–central peak spectrum in lead germanate and in strontium titanate (Lyons and Fleury 1977, 1978). The spectra – shown in fig. 30 for lead germanate – are quantitatively describable in terms of a hybridization between the acoustic phonons and the soft mode. The soft mode, however, is significantly modified by contributions

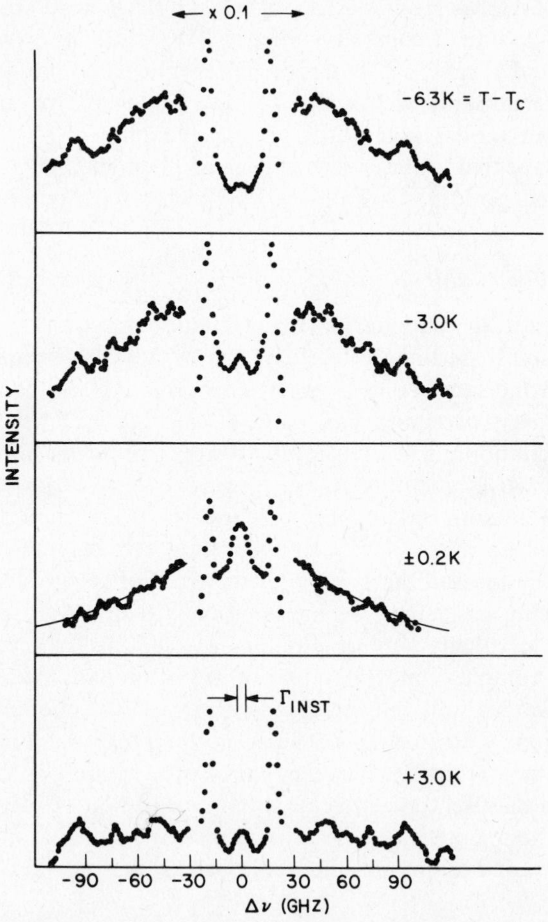

Fig. 30. High resolution Raman–Brillouin spectrum of lead germanate near the ferroelectric transition temperature, $T_c = 451$ K. Note gain reduction of 10× for small frequency shifts ($\Delta\nu < 30$ GHz). The strong peaks at ~20 GHz are due to LA phonons; the broad wing to the overdamped soft optic mode; and the singular narrow central peak due to the interaction of these with phonon density fluctuations. Special experimental techniques have been used to remove all stray light contributions near $\Delta\nu = 0$. After Lyons and Fleury (1978).

to its self-energy from phonon density fluctuations. This contribution has the effect of replacing Γ_s in G_s above by

$$\Gamma_s \to \Gamma_s + \frac{\tau\delta^2}{1 - i\omega\tau}.$$

The spectra in fig. 30, including the lineshape anomalies arising from interactions with the acoustic modes can be quantitatively reproduced using $\omega_s^2/\Gamma_s = W'(T - T_c)$, $W' = 12$ GHz/K; $\tau^{-1} = 29 \pm 2$ GHz, $\delta = 60$ GHz, $\Gamma_s = 240$ GHz; $a(1)/a(3) = 1.8$; $\omega_a = 18$ GHz; $A^2 = 30\ \mathrm{cm}^{-4}$; $\Gamma_a = 0.5$ GHz. As detailed in Lyons and Fleury (1978) all but two of these parameters, δ and τ, can be determined independently so the spectra in fig. 30 are fitted remarkably well with only two adjustable parameters. The influence of pair modes (via PDF) on the dynamics of structural phase transitions is expected to be a rather general phenomenon. A theoretical discussion of this point has been given by Cowley and Coombs (1973).

6. *Concluding remarks*

We have seen in the chapter that the excitation spectra of a remarkably diverse set of condensed matter systems can be at least semiquantitatively described from the same point of view. This point of view considers pairs of elementary excitations and their interactions with each other as well as with single excitations. Even the simplest approximations have provided adequate qualitative and sometimes quantitative (e.g. low-temperature magnon pairs) descriptions of observed spectra. They have also pointed out instances where more systematic theoretical refinements are required, as in classical liquids and the paramagnetic phase of spin systems. Phonon pairs in crystalline solids have been treated only qualitatively, expecially in that microscopic calculations of phonon interactions have not yet been carried out. A complete understanding of the roton-pair spectra in superfluid helium awaits a fully microscopic treatment of the interaction of both light and neutrons with the helium, although the progress already made in characterizing the excitations has been impressive. Hopefully the presentation made here may serve to increase cross fertilization of ideas regarding excitations in the magnetic, the liquid and the solid states of materials and to encourage needed additional theoretical and experimental work.

References

Abrikosov, A. A., L. P. Gorkov and I. E. Dzyaloshinski (1963), *Quantum field theory in statistical physics* (Prentice Hall, Englewood Cliffs, New Jersey).
Agranovich, V. M. and I. I. Lalov (1972), Sov. Phys.-JETP **34**, 350.

Alder, B. J., H. L. Strauss and J. J. Weis (1973), J. Chem. Phys. **59**, 1002.

Alder, B. J., H. L. Strauss, J. J. Weis, J. P. Hansen and M. L. Klein (1976), Physica **83B**, 249.

Allen, S. J., R. Loudon and P. L. Richards (1966), Phys. Rev. Lett. **16**, 463.

Balucani, U. and V. Tognetti (1973), Phys. Rev. B **8**, 4247.

Balucani, U. and V. Tognetti (1976), Riv. Nuovo Cimento **6**, 39.

Balucani, U. and V. Tognetti (1977), Phys. Rev. B, in press.

Bohnen, A., C. R. Natoli and J. Ranninger (1974), J. Phys. C **7**, 947.

Campbell, C. and F. Pinski (1978), in *Correlation functions and quasiparticle interactions in condensed matter*, Ed. by J. W. Halley (Plenum Press, New York).

Carneiro, K., M. Nielsen and J. P. McTague (1973), Phys. Rev. Lett. **30**, 481.

Chinn, S. R., R. W. Davies and H. J. Zeiger (1972), in *A.I.P. conference proceedings, Vol. 5 Magnetism and magnetic materials*, 1972, Ed. by C. D. Graham and J. J. Rhyne, p. 317.

Cohen, M. H. and J. Ruvalds (1969), Phys. Rev. Lett. **23**, 1378.

Cowley, R. A. (1978), in *Correlation functions and quasiparticle interactions in condensed matter*, Ed. by J. W. Halley (Plenum Press, New York).

Cowley, R. A. and G. J. Coombs (1973), J. Phys. C **6**, 121.

Cowley, R. A. and A. D. B. Woods (1971), Can. J. Phys. **49**, 177.

Crawford, K., D. G. Burns, D. A. Gallagher and M. V. Klein (1978), Phys. Rev. B **17**, 4871.

Elliot, R. J. and M. F. Thorpe (1969), J. Phys. C **2**, 1630.

Fleury, P. A. (1968), Phys. Rev. Lett. **21**, 151.

Fleury, P. A. (1969), Phys. Rev. **180**, 591.

Fleury, P. A. (1970), Int. J. Magnetism, **1**, 75.

Fleury, P. A. (1972), *Comments on solid state physics* **IV**, 149 and 167.

Fleury, P. A. (1974), *Proceedings of international conference on magnetism* (Nauka, Moscow), Vol. 1, p. 80.

Fleury, P. A. (1976), in *Light scattering in solids*, Ed. by M. Balkanski, R. C. C. Leite and S. P. S. Porto (Flammarion, Paris), p. 747.

Fleury, P. A. (1978), in *Correlation functions and quasiparticle interactions in condensed matter*, Ed. by J. W. Halley (Plenum Press, New York).

Fleury, P. A. and H. J. Guggenheim (1970), Phys. Rev. Lett. **24**, 1346.

Fleury, P. A. and R. Loudon (1968), Phys. Rev. **166**, 514.

Fleury, P. A. and K. B. Lyons (1976), Phys. Rev. Lett. **37**, 1088.

Fleury, P. A. and J. P. McTague (1969), Optics Commun. **1**, 164.

Fleury, P. A. and J. P. McTague (1973), Phys. Rev. Lett. **31**, 914.

Fleury, P. A. et al. (1966), Phys. Rev. Lett. **17**, 84.

Fleury, P. A., W. B. Daniels and J. M. Worlock (1971), Phys. Rev. Lett. **27**, 1493.

Fleury, P. A., J. M. Worlock and H. L. Carter (1973), Phys. Rev. Lett. **30**, 591.

Freeman, S. (1973), Phys. Rev. B **7**, 3960.

Go, S., H. Bilz and M. Cardona (1975), Phys. Rev. Lett. **34**, 580.

Gorelek et al. (1977), Solid State Commun. **21**, 615.

Greytak, T. J. (1978), in *Quantum liquids*, Ed. by J. Ruvalds and T. Regge (North-Holland, New York).

Greytak, T. J. and J. Yan (1969), Phys. Rev. Lett. **22**, 987.

Halley, J. W. (1969), in *Light scattering spectra of solids*, Ed. by G. B. Wright (Springer-Verlag, New York), p. 177.

Halley, J. W. (1978), in *Correlation functions and quasiparticle interactions in condensed matter*, Ed. by J. W. Halley (Plenum Press, New York).

Iwamoto, F. (1970), Prog. Theor. Phys. **44**, 1121.

Jindal, V. K. and K. N. Pathak (1977), Phys. Rev. B **15**, 1202.

Latham, W. P. and D. H. Kobe (1975), J. Phys. C **8**, L461.

Leese, J. G. and G. K. Horton (1979), J. Low Temp. Phys. **35**, 205.

Lyons, K. B. and P. A. Fleury (1977), Solid State Commun. **23**, 477.

Lyons, K. B. and P. A. Fleury (1978), Phys. Rev. B **17**, 2403.
Lyons, K. B., P. A. Fleury and H. L. Carter (1979), Phys. Rev. B (to be published).
Meltzer, R. S., M. Lowe and D. S. McClure (1966), Phys. Rev. **180**, 463.
McTague, J. P. and G. Birnbaum (1968), Phys. Rev. Lett. **21**, 661.
McTague, J. P. and G. Birnbaum (1971), Phys. Rev. A **3**, 1376.
McTague, J. P., P. A. Fleury and D. B. DuPre (1969), Phys. Rev. **188**, 303.
Murray, C. A., R. L. Woerner and T. J. Greytak (1975), J. Phys. C **8**, L90.
Natoli, C. R. and J. Ranninger (1973), J. Phys. C **6**, 345.
Nemanich, R. J. and S. A. Solin (1977), **23**, 417.
Parkinson, J. B. (1969), J. Phys. C **2**, 2012.
Pisarev, R. V., P. Moch and C. Dugautier (1973), Phys. Rev. B **7**, 4185.
Pitaevsky, L. P. (1976), in *Theory of light scattering in condensed matter*, Ed. by B. Bendow, J. L. Birman and V. M. Agranovich (Plenum Press, New York), p. 89.
Pohl, D. W. and V. Irniger (1976), Phys. Rev. Lett. **36**, 480.
Riste, T., Ed. (1974), *Anharmonic lattices, structural transitions and melting* (Noordhoff, Leiden).
Ruvalds, J. and A. Zawadowski (1970a), Phys. Rev. B **2**, 1172.
Ruvalds, J. and A. Zawadowski (1970b), Phys. Rev. Lett. **25**, 333.
Scott, J. F. (1968), Phys. Rev. Lett. **21**, 907.
Scott, J. F. (1970), Phys. Rev. Lett. **24**, 1107.
Scott, J. F. (1974), Rev. Mod. Phys. **46**, 83.
Solin, S. A. and A. K. Ramdas (1970), Phys. Rev. B **1**, 1687.
Stephen, M. J. (1969), Phys. Rev. **187**, 279.
Surko, C. M. and R. E. Slusher (1976), Phys. Rev. B **13**, 1095.
Thorpe, M. F. (1970), Phys. Rev. B **2**, 2690.
Tubino, R. and J. L. Birman (1977), Phys. Rev. B **15**, 5843.
Van der Pol, A., et al. (1976), Solid State Commun. **19**, 177.
Washington, M. A. and H. Z. Cummins (1977), Phys. Rev. B **15**, 5840.
Wehner, R. K. and R. Klein (1972), Physica **62**, 161.
Werthamer, N. R., R. L. Gray and T. R. Koehler (1970), Phys. Rev. B **2**, 4199.
Woerner, R. L. and M. J. Stephen (1975), J. Phys. C **8**, L461.
Woods, A. D. B. and R. A. Cowley (1978), *Proceedings of symposium on inelastic neutron scattering* (Copenhagen, Denmark).
Woods, A. D. B., P. A. Hilton, R. Scherm and W. G. Stirling (1977), J. Phys. C **10**, L45.
Zawadowski, A. and J. Ruvalds (1970), Phys. Rev. Lett. **24**, 1111.

CHAPTER 4

Interaction of Magnetic Ions with Phonons

B. LÜTHI

Physikalisches Institut der Universität Frankfurt
Frankfurt, F. R. Germany

Dynamical Properties of Solids, edited by
G. K. Horton and A. A. Maradudin

Contents

1. *Introduction*

Interacting ions with unfilled 3d, 4f or 5f shells show a rich variety of magnetic properties (Rado and Suhl 1963–1973, Elliott 1972, Freeman and Darley 1974). Exchange interactions in transition metal compounds, rare earth ion compounds and actinide compounds give rise to ferromagnetic, antiferromagnetic, ferrimagnetic and more complicated structures. Many of the magnetic properties can be understood with simple model Hamiltonians (e.g. Heisenberg Hamiltonian, RKKY interaction, Dzyaloshinsky mechanism, crystal electric field (CEF) potential). Although a microscopic description of the fundamental parameters, such as exchange and crystal field coefficients, is far from being complete, many physical effects, described by these Hamiltonians, have been thoroughly discussed and tested (e.g. spin waves, crystal field effects, phase transition aspects).

In crystalline substances the magnetic ion is located at a lattice site of given symmetry. The crystal field potential and the exchange interaction depend on the lattice sites of neighboring ions and on the ion–ion distance. Any displacement of these ions will therefore change the parameters in these Hamiltonians. This gives rise to an interaction between the displacement (strain coordinates or phonon coordinates) and the magnetic ion via a modulation of the exchange interaction, the crystal field potential etc. This interaction is generally called the magnetoelastic interaction. It leads to shifts in equilibrium positions of the ions for the static part of the interaction and to magnetic ion–phonon effects for the dynamic part. The various effects which result from this interaction are the subject of this chapter. We will show how this interaction leads to changes in the phonon and magnon spectra in the case of magnetically ordered materials, how it can give rise to structural transitions in the case of strong magnetoelastic coupling, how it can lead to critical effects for sound propagation near phase transitions and to various other effects.

We shall treat only the case of localized magnetic ions, e.g. rare earth or actinide compounds and insulating transition metal compounds. We omit the case of itinerant magnetism completely, because the magnetoelastic Hamiltonian is not so well characterized and the effects difficult to interpret. We also concentrate on the case of concentrated magnetic compounds, and we omit the dilute case where one has single-ion effects. There exist a number of reviews (Sturge 1967, Tucker 1967) on this latter topic, and a thorough discussion of single-ion effects would be outside the scope of this chapter.

In the next section we discuss the basic interaction Hamiltonian, which will be used in the quantitative explanation of the various effects. In §3 the various experimental techniques, which are used in the study of these effects, are discussed. The presentation of the magnetoelastic experiments we divide into 2 sections: §4 deals with effects in the paramagnetic region. For example, we describe the influence of magnetic ions on various thermodynamic properties, such as thermal expansion, elastic constants and phonon spectra. Some consequences of these magnetoelastic interactions are discussed (cooperative Jahn–Teller effect, critical elastic phenomena, magnetoelastic experiments in the presence of a magnetic field). In §5 we discuss the interaction of phonons with other collective excitations in the magnetically ordered region, especially the magnon–phonon interaction.

2. *Basic Hamiltonian*

For our purposes we consider only localized magnetic ions with Russell–Saunders coupling whose ground state is given by Hund's rule. The ground state term can be characterized by ${}^{2S+1}L_J$, where the total spin angular momentum S has the maximum value, the total orbital angular momentum L takes its maximum value in accordance with the exclusion principle, and where the total angular momentum J takes the value $|L-S|$ for less than half-filled shells and $L+S$ for more than half-filled shells. For example for 3d electrons one can have only the following ground state terms: ^{2}D, ^{3}F, ^{4}F, ^{5}D, ^{6}S. For 4f electrons the ground state J can take the following values: 0, $\frac{5}{2}$, $\frac{7}{2}$, 4, $\frac{9}{2}$, 6, $\frac{15}{2}$, 8. Because of the different strengths of spin–orbit and crystal fields in 3d- and 4f-compounds, L and J characterize the ground state level in these two cases (see below and fig. 1).

2.1. Magnetic Hamiltonian

The basic Hamiltonian which describes the properties of magnetic ions can be separated as

$$\mathcal{H} = \mathcal{H}_{\text{s.o.}} + \mathcal{H}_{\text{CEF}} + \mathcal{H}_{\text{ex}} + \mathcal{H}_{\text{dip}}. \tag{2.1}$$

For the spin–orbit interaction one can write $\mathcal{H}_{\text{s.o.}} = \lambda L \cdot S$ as long as the spin–orbit splitting is small compared to the term splitting (Tinkham 1964,

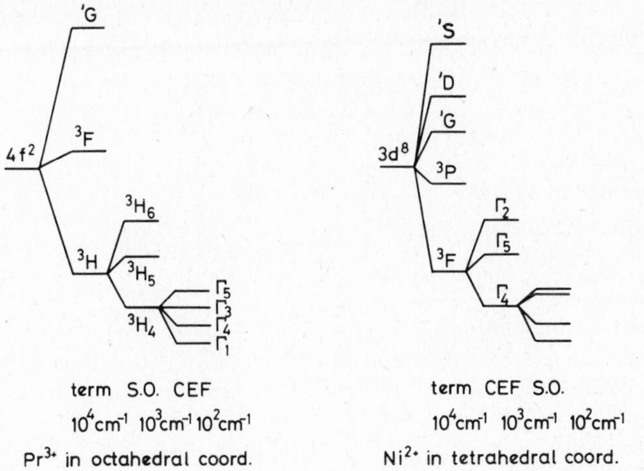

Fig. 1. Energy splittings for Pr^{3+} in octahedral symmetry and for Ni^{2+} in tetrahedral symmetry. Typical figures for overall splittings are also given. Note the different strength of spin orbit (s.o.) and crystal field (CEF) splitting for the two ions.

Abragam and Bleaney 1970). As pointed out above, this interaction is stronger for the heavier 4f and 5f ions than for the 3d ions. It gives rise to the J-multiplets in the case of 4f ions. For 3d ions it is often less important because of quenching effects due to the crystal field. In the following we consider for 4f electrons the lowest J-level and in the case of 3d ions we especially mention spin–orbit effects whenever necessary.

$\mathcal{H}_{CEF}$ is due to the crystalline Stark effect from surrounding ions. Owing to the more extended 3d wavefunctions this effect is more important for transition metal compounds than for 4f compounds. As typical examples of term, spin–orbit and crystal field splitting we show the case of Ni^{2+} and Pr^{3+} for cubic point symmetry in fig. 1. For a full exposition of CEF theory we refer to existing review articles (Hutchings 1964, Fulde 1978). If one expands the CEF potential in powers of the coordinates (x,y,z) of the magnetic ion, one gets the following expression for cubic point symmetry:

$$\begin{aligned}\mathcal{H}_{CEF} = {} & b_4\left[x^4+y^4+z^4-\tfrac{3}{5}r^4\right]\\ & +b_6\left[x^6+y^6+z^6-\tfrac{15}{14}r^6\right.\\ & \left.+\tfrac{15}{4}(x^2y^4+x^2z^4+y^2x^4+y^2z^4+z^2x^4+z^2y^4)\right].\end{aligned} \tag{2.2}$$

For 3d compounds $b_6=0$, but for 4f compounds one has in general to consider both terms. As one descends in symmetry more terms involving x^2 etc. occur and one needs more parameters b_i to characterize $\mathcal{H}_{CEF}$.

Confining the attention to the ground state J or L level, it is customary to introduce an equivalent operator formalism (Stevens 1952), $(x_i \to J_i)$ and to express $\mathcal{H}_{\mathrm{CEF}}$ in terms of new operators

$$\mathcal{H}_{\mathrm{CEF}} = B_4(O_4^0 + 5O_4^4) + B_6(O_6^0 - 21O_6^4), \tag{2.2a}$$

where $O_4^0 = 35J_z^4 - 30J(J+1)J_z^2 + 25J_z^2 + 6J(J+1) + 3J^2(J+1)^2$, $O_4^4 = \frac{1}{2}(J_+^4 + J_-^4)$, and the others can be found in the literature cited above. We shall use this formalism also for the magnetoelastic Hamiltonian discussed below. We stressed the case of cubic point symmetry because only a few parameters are needed and because there exist many classes of materials, with a cubic (octahedral, tetrahedral, etc.) point symmetry for the magnetic ions. For S-state ions ($L=0$) the CEF parameters are very small (Taylor 1975).

For the exchange interaction the most widely used form is the Heisenberg exchange Hamiltonian

$$\mathcal{H}_{\mathrm{ex}} = \sum_{ij}{}' J_{ij} S_i S_j, \tag{2.3}$$

where J_{ij} is the exchange integral, originating from direct overlap of wavefunctions, or from superexchange or from indirect exchange via conduction electrons. In many cases one can have additional terms (biquadratic exchange, Dzyaloshinsky mechanism). These terms will be discussed in §§4.2 and 5.2.

The dipolar interaction $\mathcal{H}_{\mathrm{dip}}$ is usually small and is important only indirectly in the form of demagnetizing fields, etc.

2.2. Strains and elastic constants

Before treating the magnetoelastic interaction, we first discuss elements of the theory of elasticity (Ludwig 1970, Wallace 1972). In an elastic deformation a volume at site $\boldsymbol{R}$ is shifted to $\boldsymbol{R}' = \boldsymbol{R} + \boldsymbol{u}(\boldsymbol{R}, t)$. The displacement vector $\boldsymbol{u}(\boldsymbol{R}, t)$ can be expanded: $\boldsymbol{u} = \partial \boldsymbol{u}/\partial \boldsymbol{R} \cdot \boldsymbol{R} = \mathbf{v}\boldsymbol{R}$. $v_{ij} = \partial u_i/\partial R_j$ is a component of the deformation tensor. Here we have chosen the Lagrange or material description, with R_i denoting the material coordinate. The finite strain tensor can be obtained by calculating the scalar product of the vectors

$$\boldsymbol{R}'_{mn} \cdot \boldsymbol{R}'_{mn} - \boldsymbol{R}_{mn} \cdot \boldsymbol{R}_{mn} = \sum_{\alpha\beta} 2\eta_{\alpha\beta} R^{\alpha}_{mn} R^{\beta}_{mn} \quad \text{with} \quad \boldsymbol{R}_{mn} = \boldsymbol{R}_m - \boldsymbol{R}_n.$$

This gives

$$\eta_{ij} = \tfrac{1}{2}\Big(v_{ij} + v_{ji} + \sum_{\alpha} v_{\alpha i} v_{\alpha j} \Big). \tag{2.4}$$

$\eta \neq 0$ means a pure strain deformation. A general deformation in a solid can be thought of as a pure strain deformation, followed by a rotation, e.g.

$$(I+\mathbf{v})=\mathbf{R}(1+2\eta)^{1/2}, \tag{2.5}$$

or for the rotation tensor

$$\mathbf{R}=(\mathbf{I}+\mathbf{v})(1+2\eta)^{-1/2}. \tag{2.5a}$$

For small deformations η reduces to the infinitesimal strain tensor

$$\varepsilon_{ij}=\tfrac{1}{2}(v_{ij}+v_{ji}), \tag{2.5b}$$

and $\mathbf{R}$ reduces to $R_{ij}=\delta_{ij}+\omega_{ij}$, where

$$\omega_{ij}=\tfrac{1}{2}(v_{ij}-v_{ji})=-\omega_{ji} \tag{2.5c}$$

is the antisymmetric part of the deformation tensor. If the deformation tensor does not depend on spatial coordinates one speaks of homogeneous deformations. The equation of motion for non-homogeneous deformations (phonons) in an elastic continuum can be obtained by considering the stresses acting on a volume element. Applications of Newton's law gives

$$\rho\frac{\partial^2 u_i}{\partial t^2}=\frac{\partial T_{ji}}{\partial R_j}. \tag{2.6}$$

With the linearized stress–strain relation (Hooke law)

$$T_{ij}=c_{ijkl}v_{kl} \tag{2.7}$$

and for plane waves

$$u_i Ue_i(k)e^{i(kx-\omega t)}$$

we get the eigenfrequencies and normal modes

$$\left[\rho\omega_k^2\delta_{il}-c_{ijlm}k_mk_j\right]e_l(k)=0. \tag{2.8}$$

Here $\boldsymbol{e}(k)$ is the polarization vector of the plane wave and U the amplitude. T_{ij} is a component of the stress tensor and the c_{ijkl} are the elastic constants, which can be written in the contracted notation as c_{mn} ($11\to 1, 22\to 2, 33\to 3, 23\to 4, 13\to 5, 12\to 6$). In this way Hooke's law (2.7) reads $T_n=c_{nm}\varepsilon_m, c_{nm}=c_{mn}$.

2.3. Magnetic ion–lattice interaction

All the parameters in eqs. (2.2), (2.2a), (2.3) are functions of lattice positions. For example, the exchange interaction $J_{ij}=J(|R_i-R_j|)$, where R_i is the lattice vector for magnetic ion i. As another example, the coefficient B_4 in eq. (2.2a) has the following form in the point charge model (PCM) for cubic NaCl and CsCl structures

$$\begin{aligned} &\text{coordination 6 (NaCl)} \quad B_4=\frac{7}{16}\frac{Ze^2}{R^5}\langle r^4\rangle\beta, \\ &\text{coordination 8 (CsCl)} \quad B_4=-\frac{7}{18}\frac{Ze^2}{R^5}\langle r^4\rangle\beta. \end{aligned} \tag{2.9}$$

Here β is the Stevens factor, which arises from the transformation $(x\to J_x)$ to equivalent operators (Hutchings 1964, Fulde 1978), $\langle r^4\rangle$ is the r^4 integral over 4f wavefunctions, Z is the effective ligand charge and R the magnetic ion–ligand distance. A modulation due to a phonon of $\mathcal{H}_{\mathrm{CEF}}$ or $\mathcal{H}_{\mathrm{ex}}$ can therefore give rise to a magnetic ion–lattice coupling. Depending on whether one has effectively a one-ion Hamiltonian, ($\mathcal{H}_{\mathrm{CEF}}$), or whether one has a two-ion Hamiltonian ($\mathcal{H}_{\mathrm{ex}}, \mathcal{H}_{\mathrm{dip}}$), the magnetoelastic coupling is also of one-ion or two-ion nature. These two types of coupling are often called the Van Vleck mechanism and the Waller mechanisms (Van Vleck 1939, Waller 1932).

The magnetoelastic interaction can now be obtained by expanding $\mathcal{H}_{\mathrm{CEF}}$ and $\mathcal{H}_{\mathrm{ex}}$ in terms of the deformation tensor v_{ij}. One gets in lowest order

$$\mathcal{H}_{\mathrm{me}}=\sum_{\alpha,\beta} G_{\alpha\beta}\eta_{\alpha\beta}. \tag{2.10}$$

Strictly speaking one has from eq. (2.5) also a rotational contribution to the magnetoelastic interaction which in lowest order is

$$\mathcal{H}_{\mathrm{me}}^{\mathrm{rot}}=\sum_{\alpha,\beta} V_{\alpha\beta}\omega_{\alpha\beta}. \tag{2.11}$$

We postpone discussion and evaluation of eq. (2.11) till §4.4 where we discuss rotational invariant magnetoelastic phenomena. The actual form of eq. (2.10), i.e. the determination of the tensor components G_{ij} which are non-zero, is given by symmetry considerations. This has been given for the important crystal systems (cubic, hexagonal, tetragonal, orthorhombic, etc.) in the literature (Callen and Callen 1963, 1965, Mason 1954). We give here the derivation for cubic symmetry. From the 3 elastic constants c_{ij} (c_{11}, c_{12}, c_{44}) and the 6 strains η_{ij} or ε_{ij} we can form 3 symmetry elastic constants and 6 symmetry strains, which belong to the 3 irreducible representations Γ_1, Γ_3 and Γ_5. A singlet Γ_1 associated with the bulk modulus $c_B = (c_{11} + 2c_{12})/3$ has the fully symmetric volume strain $\varepsilon_v = \varepsilon_{xx} + \varepsilon_{yy} + \varepsilon_{zz}$. The doublet Γ_3 with $c_{11} - c_{12}$ has as basis functions $\varepsilon_2 = (\varepsilon_{xx} - \varepsilon_{yy})/\sqrt{2}$ and $\varepsilon_3 = (2\varepsilon_{zz} - \varepsilon_{xx} - \varepsilon_{yy})/\sqrt{6}$. The triplet Γ_5 for c_{44} has 3 strains $\varepsilon_{xy}, \varepsilon_{xz}, \varepsilon_{yz}$. The antisymmetric strains ω_{ij} (eq. (2.5c)) belong to the Γ_4 representation and Γ_2 appears only if third-order terms in ε_{ij} are considered (Thalmeier and Fulde 1975). In fig. 2 the various symmetry strains are shown schematically for cubic symmetry. With these symmetry strains one can construct the magnetoelastic Hamiltonian, which must transform as the fully symmetric irreducible representation Γ_1, by forming direct products of the strain representations with corresponding combinations of the equivalent operators J_m. To lowest order this gives for the one-ion case ($\mathcal{H}_{\text{CEF}}$)

$$\begin{aligned}
\mathcal{H}_{\text{me}}(c_B) &= -g_1 \varepsilon_v N, \\
\mathcal{H}_{\text{me}}(c_{11} - c_{12}) &= -g_2 \sum_i \left(\sqrt{3}\, \varepsilon_2 O_2^2 + \varepsilon_3 O_2^0\right)_i, \qquad (2.12)\\
\mathcal{H}_{\text{me}}(c_{44}) &= -g_3 \sum_i \varepsilon_{xy}(J_x J_y + J_y J_x)_i + \ldots,
\end{aligned}$$

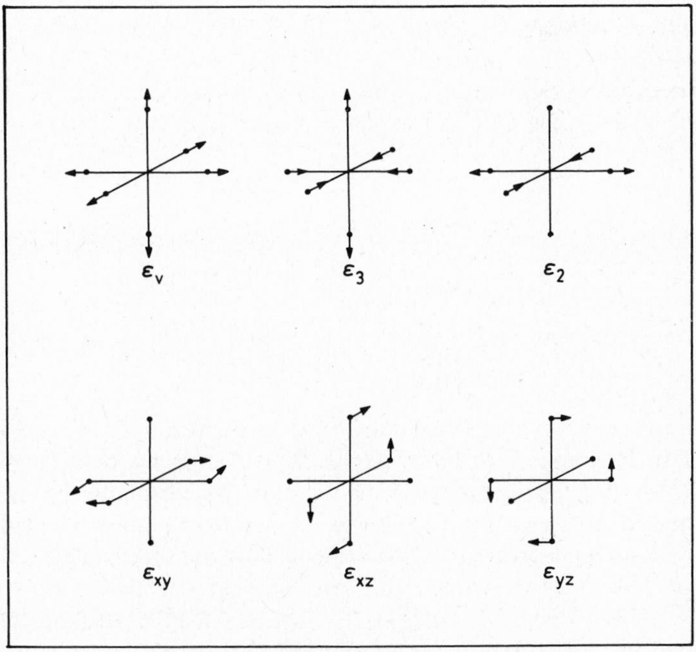

Fig. 2. Symmetry strains for cubic symmetry.

where $O_2^2 = J_x^2 - J_y^2$, $O_2^0 = 2J_z^2 - J_x^2 - J_y^2$. Because of time reversal symmetry, the strain components couple to quadrupole operators and in higher order to octupole operators. For ε_v in cubic symmetry there is no coupling to quadrupole operators.

For the two-ion case one could write down similar expressions as in eqs. (2.12), the only difference being the replacement of $(J_iJ_k)_n$ by $J_i(n)J_k(m)$, where n, m, are lattice sites. The most interesting case is the coupling to ε_v:

$$\mathcal{H}_{me} = \sum_{i,j} D_{ij}\varepsilon_v(\boldsymbol{S}_i \cdot \boldsymbol{S}_j), \tag{2.12a}$$

which results from an expansion of $\mathcal{H}_{ex}$, eq. (2.3).

In §§4 and 5 we shall study various applications of these magnetoelastic Hamiltonians. The single-ion $\mathcal{H}_{me}$ will influence the elasticity and thermal expansion in the paramagnetic phase and will give rise to magnon–phonon interactions in the magnetically ordered phase. The two-ion $\mathcal{H}_{me}$ leads to volume magnetostriction, thermal expansion anomalies near magnetic phase transitions and critical ultrasonic effects.

The elastic energy density in terms of the symmetry strains can be written for cubic symmetry

$$E_{el} = \tfrac{1}{2}c_B\varepsilon_v^2 + \tfrac{1}{2}(c_{11} - c_{12})(\varepsilon_2^2 + \varepsilon_3^2) \\ + 2c_{44}(\varepsilon_{xy}^2 + \varepsilon_{xz}^2 + \varepsilon_{yz}^2). \tag{2.13}$$

In the magnetoelastic Hamiltonian the symmetry strains denote macroscopic strains (long-wavelength acoustic modes). But for substances with more than one ion per unit cell ε_Γ can also describe internal deformations of proper symmetry.

A coupling between internal distortions and the macroscopic deformations is possible. We shall mention some cases of magnetoelastic coupling to internal modes (see §§4.1.2 and 4.2).

3. *Experimental techniques*

Many different experimental techniques are available for studying the various manifestations of the interaction of magnetic ions with lattice modes. For static phenomena such as thermal expansion and magnetostriction one has dilatometric and X-ray techniques at one's disposal. For dynamic experiments such as elastic constant measurements, ultrasonic propagation and phonon dispersion one can use sensitive sound velocity measurements, inelastic neutron scattering and light scattering techniques. In this section we would like to review these experimental techniques. We leave out the neutron scattering techniques, because these were discussed in a previous volume (Dolling 1974).

3.1. Dilatometer

For measuring relative length changes, one can use a variety of different techniques: strain gauges, interferometric methods, double grating with photocell arrangements, and capacitive methods (White 1961, Brändli and Griessen 1973, Ott and Lüthi 1977). All these methods have their particular advantages. The most widely used method is now the capacitance method. It has the advantage of high accuracy even with small samples and it can be used without corrections in external magnetic fields. Therefore it is suitable for the study of both thermal expansion and magnetostriction, especially at low temperatures. Capacitance bridges of high sensitivity are commercially available (e.g. General Radio, Type 1615-A). In fig. 3 we give a schematic drawing of a sample holder, used for capacitive length measurements (Brändli and Griessen 1973, Ott and Lüthi 1977). Such a capacitance dilatometer has a sensitivity of the order of 10^{-9} for relative length changes. It is therefore ideally suited for many applications, such as length change corrections for elastic constant measurements, thermal expansion anomalies near phase transition temperatures, Schottky anomalies in thermal expansion, etc.

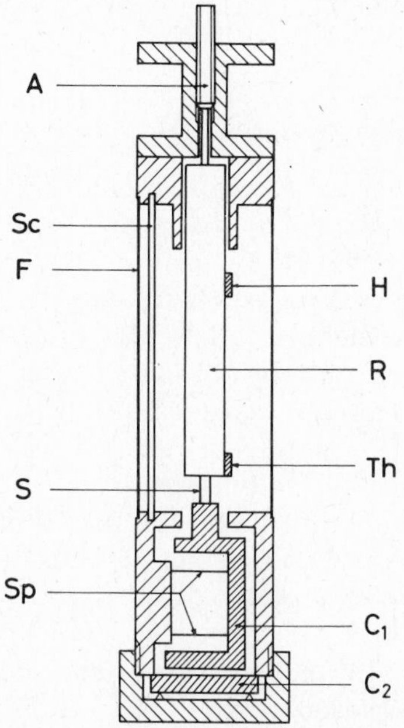

Fig. 3. Capacitance dilatometer: Schematic drawing of sample holder. A = screw to adjust capacitor gap; C_1, C_2 = capacitor plates; F = copper frame; H = heater; R = copper rod; S = sample; Sc = screw fixing copper frame; Sp = springs; Th = thermometer (from Ott and Lüthi 1977).

3.2. X-ray techniques

In order to study crystal structure changes at structural transitions, or small lattice constant changes due to magnetic ordering, or any internal lattice deformation, one has to use diffraction methods using neutrons or X-rays. Accurate lattice constant determinations are usually made with a Guinier camera using powder samples (Azároff 1968). For measuring structural changes one employs either a 4-circle- single-crystal diffractometer or Debye–Scherrer powder diffractometry using monochromatic X-rays. Domain formation at structural phase transitions excludes the former (single-crystal) method unless special attention is paid to this problem. The resolution in lattice constant determination of these methods is typically one part in 10^{-5}. Such experiments with a variety of external parameters (temperature, magnetic field, pressure) have been widely used in recent years in the study of structural transitions or magnetostrictive deformations. We shall discuss representative cases in §4.2.

Nuclear Bragg reflections using thermal neutrons can give similar information. Special mention should be made of a high-energy resolution method utilizing the Doppler effect with two crystals for measuring small lattice constant changes (Alefeld 1969). Another new experiment, especially for measuring crystal structure changes of off-center ions is particle channeling (Kollewe and Gibson 1978).

3.3. Sound velocity and attenuation

In the last few decades considerable progress has been made in the field of high-resolution sound velocity and sound attenuation measurements. There is a wide variety of different methods used in this field. Numerous books and review articles describe this topic. In the low-frequency regime, where the sound wavelength is of the dimension of the specimen, the elastic moduli (Young's modulus Y and shear modulus G) can be determined by a c.w. resonance method or by measuring flexural and torsional oscillations. These techniques are described by Read et al. (1974). In the ultrasonic regime one can determine the elastic constants c_{ij} using pulse or c.w. techniques (Truell et al. 1969, Bolef and Miller 1971, Fuller et al. 1974). The most widely used method to measure sound velocities and attenuation is the pulse superposition technique or variations of it (Truell et al. 1969, Fuller et al. 1974).

Here we describe briefly another system which has been widely used for velocity measurements (Eastman 1966, Moran and Lüthi 1969). It uses also a pulse method, but it can be easily adapted to c.w. measurements. It has the advantage of having a relatively high resolution even in the presence of strong attenuation, and it can be used on a wide frequency range. The principle, based on a phase comparison method, is shown in fig. 4. The output of a stable c.w. signal generator is split into two channels, one of which is gated to form pulses, which are amplified and fed into a quartz transducer. The received echo train and the c.w. reference are beat down to 30 MHz in separate mixers with the same local oscillator. Both channels are then amplified and phase detected. By nulling a single echo's phase by adjustment of the source frequency, one can determine the relative velocity change by $\Delta v/v = \Delta f/f + \Delta l/l$, where the last term represents the relative length change, which one has to correct for by separate thermal expansion measurements. Except for very small sound velocity effects, this correction is usually negligible. Typical resolutions, depending on the transit time of the pulse, range from one part in 10^5 to one part in 10^7.

3.4. Light scattering

In recent years, with the advent of the laser, light scattering has become an important tool for studying collective excitations. We shall subsequently discuss experiments in which Raman and Brillouin scattering techniques

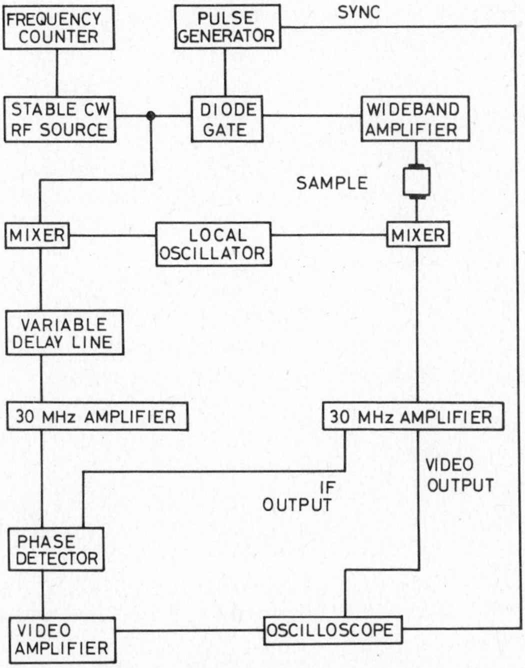

Fig. 4. Block diagram of apparatus for sound velocity measurements using a phase comparison method (from Moran and Lüthi 1969).

have been used to measure optical phonon modes and high-frequency acoustic phonons. The Raman scattering technique is for many cases advantageous over conventional non-coherent optical measurements, because of the good resolution and the easy access to excitations in the infrared and far infrared region of the spectrum. The Brillouin scattering technique is practically the only tool (except for inelastic neutron scattering) for studying high-frequency acoustic phonons, where conventional ultrasonic and hypersonic experiments are very difficult to perform because of high attenuation and stringent requirements for acoustical quality of specimens and specimen surfaces.

In light scattering (Cardona 1975), one can distinguish 3 regimes depending on the frequency shift $\Delta\omega$ of the scattered light. The elastic Rayleigh scattering ($\Delta\omega = 0$) can be measured using light beating spectroscopy or Fabry–Perot interferometry. To our knowledge there are no measurements of Rayleigh linewidths in solids containing magnetic ions. Brillouin scattering techniques ($\Delta\omega \sim 10^9 \uparrow 10^{11}$ Hz) use Fabry–Perot interferometric spectroscopy. With the recent development of multipass spectrometers (Sandercock 1975), even imperfect or opaque crystals which show low contrast for Brillouin scattering can now be studied. For the Raman scattering ($\Delta\omega \sim 10^{11}-10^{15}$ Hz) one usually uses monochromators with various modifications.

4. *Effects in the paramagnetic region*

We start our survey of experimental results with effects in the paramagnetic region. In this region one can observe effects under rather ideal conditions, without disturbing influences due to domains. Furthermore one can often interpret the results neglecting many additional parameters (exchange constant, etc.).

4.1. Crystal electric field effects

In this section we shall discuss the influence of crystal electric fields (CEF) on various physical properties. We shall mainly consider rare earth compounds, because for 3d ions the CEF-splitting is usually too large to have a noticeable effect on thermal and lattice properties (fig. 1), and for actinide compounds no CEF effects have been observed so far with one exception (Shamir et al. 1978). For S-state ions the CEF-splitting is so small (Taylor 1975), that for temperatures $T > 1$ K the levels are saturated.

One can divide CEF effects conveniently into three categories (Lüthi 1976): resonance or spectroscopic effects, thermal properties and transport properties. Here we discuss some thermal effects which describe the magnetic ion–lattice coupling, such as thermal expansion, magnetostriction and elastic constants. At the end of this section we shall also mention magnetoelastic effects involving optical phonons.

In order to stress the close analogy of these effects with other thermodynamic quantities, we discuss them together with the specific heat and the magnetic susceptibility. We start with the case of non-interacting ions. The Helmholtz free energy density of such a system can be written

$$F = -kTN \ln \sum_i \exp(-E_i/kT), \tag{4.1}$$

where N is the number of ions per unit volume and $E_i = E(\Gamma_i)$ denotes the CEF split energy level with representation index Γ_i. For the different thermodynamic effects, one needs to know the strain or magnetic field dependence of these energy levels. Knowing the undisturbed $E_0(\Gamma_i)$ from either eq. (2.2a) or its equivalents, or from experiment, one can obtain $E(\Gamma_i, \varepsilon_{ij}, H)$ either by perturbation theory with respect to the perturbation $\mathcal{H}'$ or by solving the secular equation $|E_0(\Gamma_i) - E - \langle\Gamma_j|\mathcal{H}'|\Gamma_i\rangle| = 0$. Here $\mathcal{H}'$ is either the Zeeman term $\mathcal{H}'_z = g\mu_B J^z H^z$, for the magnetic case, or the magnetoelastic Hamiltonian. (eq. (2.12)), for the strain dependence.

4.1.1 *Specific heat and thermal expansion*

The specific heat contribution due to CEF split ions is given by

$$C = -T\frac{\partial^2 F}{\partial T^2} = \frac{N}{kT^2}(\langle E^2\rangle - \langle E\rangle^2), \tag{4.2}$$

where the statistical average $\langle X\rangle$ is defined as

$$\langle X\rangle = \frac{\sum_i X_i \exp(-E_i/kT)}{\sum_i \exp(-E_i/kT)}.$$

Likewise the volume thermal expansion β due to magnetic ions is (Ott and Lüthi 1977)

$$\beta = \frac{1}{V}\left(\frac{\partial V}{\partial T}\right)_P = -\kappa V\frac{\partial^2 F}{\partial V \partial T} = \frac{\kappa N}{kT^2}(\langle E^2\gamma\rangle - \langle E\rangle\langle E\gamma\rangle). \tag{4.3}$$

Here $\kappa = 1/c_B$ is the compressibility and $\gamma_i = -\partial \ln E_i/\partial \ln V$ is the Grüneisen parameter for the CEF level E_i. Eqs. (4.2) and (4.3) have a similar

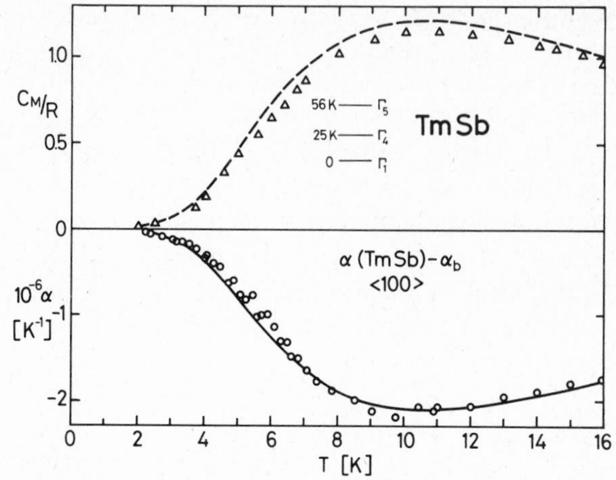

Fig. 5. Magnetic part of specific heat C_M and linear thermal expansion coefficient $\alpha - \alpha_b$ as a function of temperature, for TmSb single crystal. α_b is background expansion determined from LaSb and scaled appropriately. Full lines are calculations based on eqs. (4.2), (4.3) (from Ott and Lüthi 1977).

structure. β and C are proportional to each other as a function of temperature if all the γ_i which contribute to β are the same. This is, for example, the case for $J=\frac{5}{2}$ in cubic symmetry (e.g. SmSb, CeTe), where the ground state multiplet is split into two levels only.

In fig. 5 we give experimental results for the specific heat and thermal expansion for the cubic compound TmSb. Tm^{3+} has a $J=6$ ground state, which for octahedral symmetry splits into a singlet Γ_1, a triplet Γ_4 at 25 K and a triplet Γ_5 at 56 K, as determined by inelastic neutron scattering (Birgeneau et al. 1971). All other levels lie higher than 100 K and do not contribute to β or C in this temperature range. In this compound, exchange interactions are negligible, as indicated by the absence of magnetic or structural transitions and by other properties. TmSb is therefore an ideal paramagnet. The magnetic part of the specific heat C_M exhibits a well-defined Schottky anomaly. The dashed line is calculated using eq. (4.2) and experimentally determined CEF levels.

In the same fig. 5 we show experimental results for the linear thermal expansion coefficient $\alpha=\beta/3$. These results are a clear manifestation of CEF effects. The solid line is calculated using eq. (4.2) with $\gamma_{\Gamma_4}=\gamma_{\Gamma_5}=-1.2$. The agreement with experiment is very good, as is the proportionality between α and C. Note that the γ's are different from the point charge value $\gamma=+\frac{5}{3}$. From the Grüneisen parameters one can deduce corresponding magnetoelastic coupling constants. Including higher-order contributions in eq. (2.12) (Ott and Lüthi 1977):

$$\mathcal{H}_{me}(C_B) = -g_1 N\varepsilon_v - g_{11}\sum_i \varepsilon_v(O_4^0 + 5O_4^4)_i - g_{12}\sum_i \varepsilon_v(O_6^0 - 21O_6^4)_i, \tag{4.4}$$

we get for TmSb $g_{11} = -16\,\mathrm{mK}$.

The importance of CEF effects on thermal expansion lies in the determination of the Grüneisen parameters γ or of the corresponding magnetoelastic coupling constants g_{11}, g_{12}, which are difficult to determine otherwise and which are important for the pressure and volume dependence of crystal field parameters and magnetic susceptibility. Since the original measurements on TmSb, additional rare earth compounds have been investigated with similar results (Ott and Lüthi 1977): PrSb, SmSb, ErSb, CeTe, TmTe. A related effect has also been observed for Fe^{2+} ions in ZnS (Sheard et al. 1977).

4.1.2. *Crystal field effects on elastic constants*

Here we first show the close analogy between magnetic and strain susceptibility. The magnetic susceptibility measures the response of a system of magnetic ions to an applied magnetic field. With the Zeeman term for $\mathcal{H}_z'$ one can determine $E(\Gamma_i, H)$ as described above. For the magnetic susceptibility we get

$$\chi_m = \frac{N\partial\langle J^z\rangle}{\partial H} = -\frac{\partial^2 F}{\partial H^2}$$
$$= -N\left\{\left\langle\frac{\partial^2 E}{\partial H^2}\right\rangle - \frac{1}{kT}\left\langle\left(\frac{\partial E}{\partial H}\right)^2\right\rangle + \frac{1}{kT}\left\langle\frac{\partial E}{\partial H}\right\rangle^2\right\}. \tag{4.5}$$

In this expression the first term is the famous Van Vleck term, which probes the off-diagonal magnetic dipole matrix elements. The next two terms are the so-called Curie terms. They are due to the diagonal matrix elements and are particularly important for Kramers ions. The effect of exchange can be easily incorporated into eq. (4.5) in the molecular field approximation. Many examples of the CEF effect on the magnetic susceptibility are known (Van Vleck 1930, Taylor 1972), and we do not pursue this any further here.

In an analogous way the strain susceptibility is the response of the structural order parameter $\langle O_\Gamma\rangle$ to an applied strain ε_Γ. We have seen that for cubic symmetry, for example, there are 3 symmetry modes (fig. 2).

Using the magnetoelastic Hamiltonian eq. (2.12) one can determine $E(\Gamma_i, \varepsilon_i)$, and one gets for the CEF contributions to the elastic constants the corresponding strain susceptibility (Lüthi 1976)

$$c_\Gamma = \frac{\partial \langle Q_\Gamma \rangle}{\partial \varepsilon_\Gamma} = \frac{\partial^2 F}{\partial \varepsilon_\Gamma^2} = -c_0 g_\Gamma^2 \chi_s$$

$$= N \left\{ \left\langle \frac{\partial^2 E}{\partial \varepsilon_\Gamma^2} \right\rangle - \frac{1}{kT} \left\langle \left(\frac{\partial E}{\partial \varepsilon_\Gamma} \right)^2 \right\rangle + \frac{1}{kT} \left\langle \frac{\partial E}{\partial \varepsilon_\Gamma} \right\rangle^2 \right\}. \qquad (4.6)$$

Eq. (4.6) can be interpreted in the same way as eq. (4.5). c_0 is the background elastic constant. CEF effects on elastic constants can be observed best in the paramagnetic phase for magnetic ions whose CEF states have strong quadrupole matrix elements.

A typical example of a CEF effect on the elastic constants $c_{11} - c_{12}$ and c_{44} is shown in fig. 6 for SmSb. The $J = \frac{5}{2}$ state of Sm^{3+} splits in a field of octahedral symmetry into a Γ_7 and a Γ_8 level. From the results of fig. 6 we

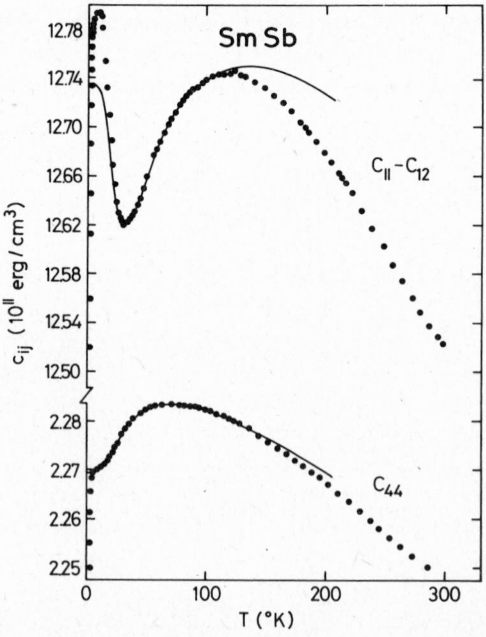

Fig. 6. Temperature dependence of elastic constants $c_{11} - c_{12}$ and c_{44} for SmSb. Full lines are calculations based on eq. (4.6) for $T > T_N = 2.1$ K (from Mullen et al. 1974).

determine a level splitting of 65 K (Mullen et al. 1974). Both the minimum of the $c_{11}-c_{12}$ mode at 30 K and the shoulder of the c_{44} mode are nicely reproduced by the calculated strain susceptibility. The deviations at higher temperatures are due to the difficulty in choosing an appropriate background c_0, which at higher temperatures also shows a temperature dependence. SmSb has a magnetic phase transition at $T_N=2.1$ K. In the ordered region the elastic constants show a strong temperature dependence, which cannot be described by the strain susceptibilities χ_s but is probably due to domain wall-stress effects. From such elastic constant measurements one can determine the square of the magnetoelastic coupling constants g_Γ^2.

The point charge model expressions for g_2 for cubic symmetry, in analogy with the corresponding B_4 (eq. (2.9)) are

$$\text{coordination 6 (NaCl)}\ g_2=3\frac{\sqrt{6}}{2}\,\alpha_J\langle r^2\rangle\frac{Ze^2}{R^3},$$

$$\text{coordination 8 (CsCl)}\ g_2=\frac{8}{\sqrt{6}}\,\alpha_J\langle r^2\rangle\frac{Ze^2}{R^3}. \tag{4.7}$$

The symbols in eq. (4.7) have the same meaning as those in eq. (2.9). α_J is the Stevens factor for quadrupole operators. The experimentally determined values of g_2, g_3 show a strong correlation with α_J.

Such CEF effects on elastic constants have been observed for many cubic compounds (rare earth pnictides and rare earth chalcogenides (Mullen et al. 1974, Lüthi 1976), and also have been found in non-cubic metals (d.h.c.p. Pr), (Greiner et al. 1973). A similar effect has been observed in $TmVO_4$ for the c_{66} mode (Melcher et al. 1973), and a related behavior has been observed even for a transition metal compound $FeCl_2$ (Gorodetsky et al. 1976). In this case the cubic part of the crystal electric field splits the $Fe^{2+}(3d^6)^5D$ level into an upper doublet and a lower triplet, with a large separation. The triplet $^5T_{2g}$ is, however, further split by an axial crystal field and spin–orbit coupling, giving rise to 3 singlets and 6 doublets between 0 and 500 K. The non-vanishing quadrupole matrix elements between these levels induce the CEF effects on the elastic constants. These effects could be quantitatively interpreted using eq. (4.6) for $T>T_N=23.5$ K.

A related effect has been recently found for optical phonons in paramagnetic rare earth compounds in the presence of magnetic fields. In $CeCl_3$, CeF_3 and NdF_3 a Raman scattering experiment showed a splitting of some $\boldsymbol{k}=0$, E_g type optical phonons upon application of a strong magnetic field (Schaack 1977). No splitting was observed for $EuCl_3$ and $GdCl_3$ (both S-state ions). This indicates that a similar magnetoelastic interaction as in the case of elastic modes is also present for these homogeneous internal modes. One can construct a magnetoelastic Hamiltonian, coupling the electric quadrupole moments of the rare earths with the optical phonon coordinates of the same symmetry (Thalmeier and Fulde 1977). Theory accounts for the field-dependent splitting (linear splitting with subsequent saturation at high fields for $H \parallel$ crystal symmetry axis, quadratic field dependent splitting for $H \perp$ symmetry axis).

We shall discuss some other applications of the CEF effect on the elasticity of a crystal in §§4.2 and 4.5.

4.1.3. *Magnetostriction*

Another manifestation of the magnetic ion–lattice interaction is the change in length of a specimen for externally applied fields. Under the influence of a field, the quadrupole moments have a non-zero expectation value, i.e. they show ordering, and via the magnetoelastic coupling eqs. (2.12) give rise to a macroscopic strain. One should emphasize that such an effect arises also for cubic crystals. This effect is well known for ferromagnetic or ferrimagnetic materials (Callen 1968, Comstock 1965), but it is also present for paramagnets. For cubic paramagnets the field-induced deformation reduces the symmetry to tetragonal for applied fields in (100) directions and to trigonal for H in (111) directions. Using eqs. (2.12) and (2.13) we get for the magnetostrictive strain $\varepsilon(H)=(\partial l/l)_H-(\partial l/l)_0$ for H applied along (100) (Lüthi et al. 1977, Morin et al. 1978)

$$\varepsilon(H)=\sqrt{\frac{2}{3c_\Gamma^0}}\left[\left(\frac{\partial F}{\partial \varepsilon_\Gamma}\right)_H-\left(\frac{\partial F}{\partial \varepsilon_\Gamma}\right)_0\right]$$

$$=-\sqrt{\frac{2}{3c_\Gamma^0}}\left\langle\frac{\partial F}{\partial \varepsilon_3}\right\rangle_H=\sqrt{\frac{2}{3}}\,g_2\frac{N}{c_\Gamma^0}\langle O_2^0\rangle_H. \tag{4.8}$$

Note that $\langle O_2^0\rangle_0=0$ for cubic materials in the paramagnetic state. For fields small compared with the crystal field splitting eq. (4.8) predicts a quadratic field dependence.

Apart from magnetostriction experiments in the paramagnetic region of ferromagnets, ferrimagnets and rare earth metals, such experiments have recently been performed for cubic rare earth compounds with NaCl and CsCl structure. With such experiments one can determine the magnetoelastic coupling constants g_2, g_3 with sign. The results of these measurements show that the exchange striction contribution, such as described in eq. (2.12a), as well as pseudodipolar two-ion contributions are also important in interpreting these magnetostriction experiments. A systematic study of magnetostriction due to rare earth impurities in a metal has been recently carried out (Nickolson et al. 1978).

Another interesting approach to determine magnetoelastic coupling constants is by measuring the strain dependence of the electron paramagnetic resonance for dilute concentrations of magnetic ions in a diamagnetic matrix. Such experiments have been performed for transition metal ions Ni^{2+}, Mn^{2+}, Fe^{2+}, Co^{2+} in MgO and for rare earth ions Gd^{3+}, Yb^{3+} and Er^{3+} in the garnets YAlG and YGaG (Phillips and White 1967a, b). In these experiments one determines not only the strain modulation of the CEF terms but also of the g-tensor. The system can be described by a spin Hamiltonian, derived from spin–orbit and Zeeman terms. The different dependence of the g factor and the CEF parameters on the effective spin enables one to determine these contributions separately. This being a single-ion effect, one can extrapolate these results to the concentrated compounds (e.g. transition metal oxides, rare earth garnets). One usually finds the same order of magnitude as from magnetostriction experiments in the concentrated compounds.

A somewhat related method is the measurement of antiferromagnetic resonance in single-crystal antiferromagnets which are subjected to uniaxial pressure (Smith and Jones 1963, Eastman 1967). In antiferromagnets, the length change method is not suitable.

Before leaving the subject of magnetostriction, one should mention a new class of materials, the rare earth transition metal compounds of the cubic Laves or C15 phase, which exhibit gigantic magnetostrictive strains, even at room temperature in the ordered phase. For example $TbFe_2$ gives a strain of the order of 10^{-3}. It is believed that a coupling of internal to external strains is responsible for this effect (Clark 1974, Cullen and Clark 1977).

4.2. Structural transitions

The concepts developed in §4.1 can be applied to the investigation of a structural transition, the so-called cooperative Jahn–Teller transition. This structural transition was originally discussed for spinel compounds, containing transition metal ions (Dunitz and Orgel 1957). Since then many more compounds, especially rare earth compounds, have been found which exhibit such a phase transition. Progress in this field has been very rapid and there exist now several review articles covering this topic (Gehring and Gehring 1975, Melcher 1976, Thomas 1977). We give a brief exposition of this effect and refer for many details to these reviews.

4.2.1. Theory of the cooperative Jahn–Teller effect

A simple description of the elastic constants and equilibrium strains for compounds exhibiting a cooperative Jahn–Teller effect can be obtained, using the results discussed in §4.1.2. Considering the strain susceptibility (eq. (4.6)), one notices, that the Curie term $(1/T)\langle(\partial E^2/\partial\varepsilon_\Gamma)\rangle$ can become very large at low temperatures for ions with a non-Kramers degeneracy ($\Gamma_3, \Gamma_4, \Gamma_5, \Gamma_8$ for cubic point symmetry). For such systems, a temperature T_a exists, where $c_\Gamma = c_0 - c_0 g_\Gamma^2 \chi_s = 0$. Physically this means that the ions no longer have a restoring force under the strain ε_Γ. They shift to new equilibrium positions, i.e. a structural transition occurs at T_a. In this case the coupling between magnetic ions is mediated through the strain field, leading to a softening of the symmetry elastic constant c_Γ, to an ordering of the quadrupole moments $\langle O_\Gamma\rangle$, as well as to a spontaneous strain ε_{Γ_3} for $T \leqslant T_a$. This effect may occur not only for orbitally degenerate ground state ions but also for others, as long as $c_\Gamma \to 0$. In this latter case one can speak of an induced cooperative Jahn–Teller effect. Examples are, e.g. $TbVO_4$, $PrAlO_3$, DySb, $PrCu_2$.

So far we have considered only a coupling of the magnetic ion to the homogeneous elastic strain. One can have additional coupling to internal strains (optical or acoustical phonon modes) or conduction electrons in the case of intermetallic compounds. In this latter case the coupling is called the aspherical Coulomb charge scattering (Fulde 1978). One can have in principle also a direct biquadratic exchange. For calculating elastic constants, one replaces these additional interactions by a quadrupolar coupling between magnetic ions:

$$\mathcal{H} = -g_\Gamma \varepsilon_\Gamma \sum_\Gamma O_{\Gamma_i} - \sum_i{}' G_{ij} O_{\Gamma_i} O_{\Gamma_j}. \tag{4.9}$$

However, this simple form can be derived rigorously only for special cases (Gehring and Gehring 1975). In general the interaction between magnetic ions takes on more complicated forms. From eq. (4.9) one can calculate the symmetry elastic constants (see reviews cited above and Allen (1968), Kataoka and Kanamori (1972), Levy (1973) and Mullen et al. (1974)):

$$c_\Gamma / c_0 = \frac{1 - (g_\Gamma^2 + g')\chi_s}{1 - g'\chi_s}. \tag{4.10}$$

Eq. (4.10) was derived with the usual assumption that the order parameter $\langle O_\Gamma\rangle$ can follow the applied strain. In eq. (4.10) χ_s is the one-ion strain susceptibility (eq. (4.6)), $g' = \Sigma'_{ij} G_{ij}$ is essentially the $\boldsymbol{q} = 0$ component of the quadrupolar coupling constant G_{ij}, with the self-energy term subtracted.

Eq. (4.10) predicts again a softening of c_Γ for $g_\Gamma \neq 0$. The paraelastic Curie temperature is given again by $c_\Gamma = 0$ or

$$1 = (g_\Gamma^2 + g')\chi_s(T_a). \tag{4.11}$$

Eq. (4.11) is a generalization of the condition discussed above. T_a is the structural transition temperature in the case of a second-order phase transition. $\partial F/\partial\langle O_\Gamma\rangle = 0$ and $\partial F/\partial\varepsilon_\Gamma = 0$ determine the temperature dependence of the order parameter $\langle O_\Gamma\rangle$ and of the spontaneous strain ε_Γ^s. The two quantities are related by $\varepsilon_\Gamma = q_\Gamma N\langle O_\Gamma\rangle / c_\Gamma$.

In cases where the symmetry elastic constant and the symmetry strains belong to an irreducible representation, for which higher-order (cubic) invariants can be constructed, the phase transition will in general be first order. As an example consider the cubic–tetragonal phase transition, for which $c_{11} - c_{12}$ is the symmetry elastic constant with the tetragonal and orthorhombic symmetry strains ε_3 and ε_2 (see §2.3 and fig. 2). The third-order invariants in this case are $\partial\varepsilon_3(\varepsilon_3^2 - 3\varepsilon_2^2) + \alpha[(\varepsilon_3^2 - \varepsilon_2^2)O_2^0 - 2\varepsilon_2\varepsilon_3 O_2^2]$ (Kataoka and Kanamori 1972), which have to be added to eq. (4.9). These terms affect the order of the structural transition and the elastic constants c_Γ for $T \leqslant T_a$ in the ordered region.

4.2.2. *Survey of experimental results*

We discuss some results of the cooperative Jahn–Teller effect in transition metal-, rare earth- and actinide compounds. A more complete survey can be found in the references cited above.

In the case of transition metal ions, one finds such structural transitions for compounds having the spinel, perovskite, MnF_3 and $CsNiF_3$-structure. References to older work can be found in Goodenough (1966). The best studied systems are $NiCr_2O_4$ and $CsCuCl_3$. In the first case, the Ni^{2+}, located at a tetrahedral site of a normal spinel lattice, has a triplet Γ_4 ground state (see fig. 1). $NiCr_2O_4$ undergoes a first-order cubic–tetragonal transition at $T_a = 299$ K, which can be well described by the theory outlined above. In fig. 7 we show the temperature dependence of the spontaneous strain $c/a - 1$ and the symmetry elastic constant $c_{11} - c_{12}$ as a function of temperature, together with a theoretical fit (Kataoka and Kanamori 1972, Kino and Miyahara 1966, Kino et al. 1972, 1973, Terauchi et al. 1972). Both the order parameter and the soft mode are well reproduced by the theory. The same theory describes the structural phase diagram of $Cu_{1-x}Ni_xCr_2O_4$, $Fe_{1-x}Ni_xCr_2O_4$ and $Zn_{1-x}Ni_xCr_2O_4$ very well. Elastic constant measurements, diffuse X-ray scattering and inelastic neutron scattering indicate that the coupling responsible for the structural

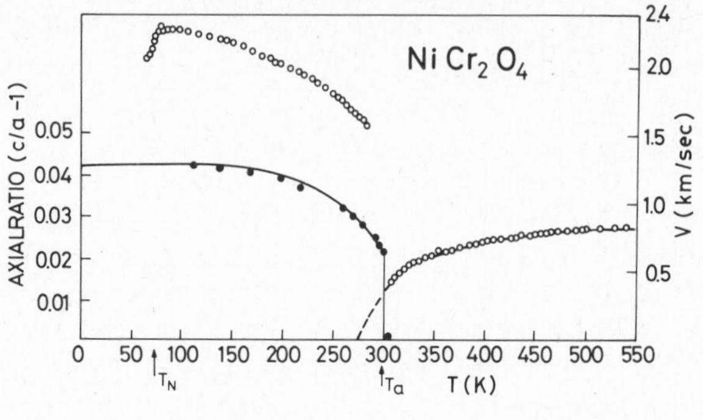

Fig. 7. $NiCr_2O_4$: Sound velocity $v=(c_{11}-c_{12})^{1/2}/2\rho$ (open circles) and $c/a-1$ (full circles) versus temperature. Full lines are calculations for v based on eq. (4.10) for $T>T_a=299\,K$ and for axial ratio $c/a-1$ from theory by Kataoka and Kanamori (1972) (adapted from Kino and Miyahara 1966, and Kino et al. 1973).

transition is the magnetic ion–strain coupling with $g'\ll g_1$. Another system which has been recently investigated by various techniques is $CsCuCl_3$, where Cu^{2+} in octahedral symmetry is the Jahn–Teller ion (Kroese and Maaskant 1974, Hirotsu 1977). At $T_a=423\,K$ the hexagonal structure distorts into another hexagonal helically distorted structure, with a new wave vector $\boldsymbol{q}_0=(0,0,\frac{2}{3}\pi c)$. Although there exists a strong coupling of the magnetic ion to a c_{44} shear strain, the main coupling, responsible for the transition, involves $\boldsymbol{q}_0$-phonons.

For compounds containing rare earth ions, one has found structural transitions in vanadates and arsenates ($LnVO_4$, $LnAsO_4$), $PrAlO_3$, $PrCu_2$ and cubic compounds with NaCl- and especially CsCl-structure. The best characterized systems are the vanadates and $PrAlO_3$, which have been investigated with neutron scattering, optical, EPR, and thermal techniques and elastic constant measurements (see reviews cited above and Birgeneau et al. 1974). In all these cases it was found that the strain coupling to the magnetic ion is a dominant mechanism for these structural transitions. One interesting feature of these systems is the absence of cubic invariants of the order parameter. This is in contrast to the cubic case mentioned above. It allows the phase transition to be of second order. A famous example of a complete elastic softening is shown in fig. 8. Here the temperature dependence of the $c_{11}-c_{12}$ mode is shown for $DyVO_4$ as measured by Brillouin scattering and ultrasonic methods (Sandercock et al. 1972, Melcher and Scott 1972). $DyVO_4$ undergoes a tetragonal–orthorhombic transition at $T_a=9.5\,K$. The full line is a fit using eq. (4.10).

Another interesting series of compounds are the CsCl-structures with Tm^{3+}-ions: TmZn, TmCd (Lüthi et al. 1973, Morin et al. 1978). TmCd has

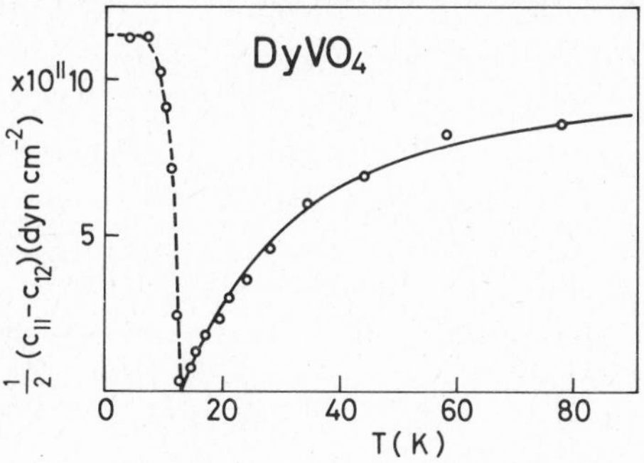

Fig. 8. Soft elastic mode $(c_{11}-c_{12})/2$ as a function of temperature, determined with Brillouin scattering (from Sandercock et al. 1972) for $DyVO_4$.

no magnetic ordering temperature but a cubic–tetragonal transition at $T_a=3.16\,K$. Again the $c_{11}-c_{12}$ mode shows strong softening for $T>T_a$. TmZn shows a similar structural transition at $T_a=8.55\,K$, followed by ferromagnetic transition at $T_c=8.2\,K$. TmAg has an AF transition at $T_N=9.5\,K$. In these intermetallic compounds the conduction electrons contribute strongly to the exchange and quadrupolar coupling.

There are many more cubic rare earth systems (pnictides and chalcogenides) where the structural and magnetic transitions coincide, i.e. $T_N=T_a$ (Lüthi 1976). One again observes elastic softening for $T>T_a$ and a spontaneous macroscopic strain for $T=T_a$: The best studied systems are the rare earth antimonides.

One has simple arguments (Stevens and Pytte 1973) to predict the easy axis of magnetization and the type of distortion to expect for cubic materials. If one considers the fourth-order CEF term (eq. (2.2a)), $-B_4(J_x^2J_y^2+J_y^2J_z^2+J_z^2J_x^2)$ predicts for $B_4<0$ a (100) easy axis, a tetragonal distortion and a $c_{11}-c_{12}$ soft mode. For $B_4>0$ one has a (111) easy axis, a trigonal distortion and a c_{44} soft mode. The sign of B_4 depends on β, Ze and on the coordination (eq. (2.9)). This argument explains easy axis, main distortion and soft elastic modes for many cubic compounds with NaCl-, CsCl- or Cu_3Au-structures, with only few exceptions.

Investigations of actinide compounds are just now being started (Freeman and Darley 1974). The most detailed study is the one of UO_2 (Brandt and Walker 1967, Allen 1968, Faber et al. 1975). The observed elastic softening of c_{44} for $T>T_N=60\,K$ can again be interpreted with eq. (4.10). For $T<T_N$ internal displacements described by $\boldsymbol{k}\neq 0$ optical phonon coordinates have been observed. The existence of discrete CEF energy levels has been observed now for UPd_3 (Shamir et al. 1978).

We have seen that elastic constant measurements, either by Brillouin scattering or by ultrasonic methods, are a powerful tool for investigating Jahn–Teller systems. Unlike neutrons, the strains couple directly to the quadrupole moments. In the high-symmetry phase, eq. (4.10) describes the temperature dependence adequately. In the low-symmetry phase, elastic constant measurements do not yield intrinsic c_{ij} ($c_{11}-c_{12}, c_{33}-c_{13}$ for a tetragonal phase), unless special precautions are made to make the sample monodomain. In multidomain samples, the elastic constants are affected by domain averaging and domain wall-stress effects. There are no elastic measurements yet for single domain structure Jahn–Teller compounds.

An interesting modern topic is the question of marginal dimensionality at these structural transitions (Als-Nielsen and Birgeneau 1977). Using the Ginzburg criterion, the marginal dimensionality d^* is defined such that one has molecular field behavior for systems whose spatial dimension d obeys $d>d^*$, logarithmic corrections in thermodynamic quantities for $d=d^*$ and strong deviations from Landau behavior for $d<d^*$. For systems where the macroscopic strain is the order parameter and a $g_\Gamma=0$ acoustic phonon the soft mode, e.g. cooperative Jahn–Teller transitions, one can show (Cowley 1976, Folk et al. 1976), that the marginal dimensionality $d^* \leqslant 3$. Therefore the structural transitions, discussed above, should exhibit molecular field behavior. Especially eq. (4.10) should be applicable even close to the transition temperature. This is indeed what has been found in the cases where one can test this conjecture, i.e. where one does not have a strong first-order transition. Notable examples are the rare earth vanadates and $PrAlO_3$ (see review articles quoted above and Birgeneau et al. 1974).

4.2.3. *Valence instabilities*

The cooperative Jahn–Teller effect, discussed above, leads to a softening of certain symmetry elastic constants and to a structural instability, which involves a lattice symmetry change but which is normally volume conserving. With the occurrence of a spontaneous strain ε_Γ^s, the volume strain $\varepsilon_v=0$. There can be secondary effects, however, which can result in $\varepsilon_v \neq 0$.

On the other hand, in valence transitions there are large volume changes observed, with no change in lattice symmetry (see Conference Proceedings on Valence Instabilities, 1977). The appropriate order parameter for such a transition is the volume strain, the reduced volume $(V-V_c)/V_c$, with V_c = critical volume. In this case it is the bulk modulus c_B which becomes soft at the transition. Two systems have been investigated so far (Penney et al. 1975, Croft and Parks 1977): $Sm_{1-x}Y_xS$ and $Ce_{1-x}Th_x$.

In $Sm_{1-x}Y_xS$ room-temperature elastic constant measurements show a strong softening of c_B as a function of x for $x>x_c=0.15$. With increasing

x, the system changes at x_c from a semiconducting phase with a Sm $4f^6$ configuration to a metallic phase with mixed valence $4f^6$ and $4f^5 5d$. For $x<x_c$, c_B is roughly constant. In $Ce_{1-x}Th_x$ for $x=0.272$ the bulk modulus c_B softens on approaching the valence instability transition temperature $T_c = 148$ K. The resulting critical exponents for c_B or the inverse compressibility are again classical, in agreement with the concept of marginal dimensionality.

4.3. Magnetic transitions: critical effects

In §4.2 we have discussed systems where structural transitions are induced through the softening of elastic modes. In this case the magnetic ion–lattice coupling is one of the important interactions. Another case occurs for systems exhibiting a magnetic phase transition where the exchange interaction is the dominant interaction and the magnetoelastic interaction plays only a secondary role. This will be the case, for example, for S-state ions (Gd^{3+}, Eu^{2+} for rare earths, Fe^{3+}, Mn^{2+} for transition metal compounds), where the single-ion coupling constants g_i (eq. (2.12)), are much smaller than the exchange striction constant $\partial J/\partial \varepsilon$ (eq. (2.12a)). There are many intermediate cases, as discussed in §4.2, where one can explain the elastic behavior in the paramagnetic phase with the strain–quadrupolar coupling, eqs. (2.12), (4.9), for the soft mode, but where exchange striction coupling can become important for the other modes.

There are two important problems involving the secondary magnetoelastic interaction for magnetic phase transitions. First one can ask, what are the effects of such interactions on static thermodynamic quantities, such as specific heat, thermal expansion, magnetostriction and on the order of the transition. Considerable attention has been given especially to the last point in recent years. For a detailed discussion of this problem with references to previous work, see Bergman and Halperin (1976). It is difficult to apply the results of these theories to real magnetic systems. The general consensus seems to be that for ideal magnetic systems, e.g. Ising systems, the predicted effects on the order of the transition are too close to T_c to be accessible for experiments. This need not be the case for non-magnetic Ising systems. The second problem is the effect this magnetoelastic, exchange striction coupling has on ultrasonic properties (attenuation, dispersion) near the critical point. It is this second aspect we want to discuss in this section. We first discuss in more detail the coupling mechanism and then we give a survey of theoretical and experimental results.

4.3.1. Coupling mechanism

As already pointed out above, the dominant mechanism for critical ultrasonic effects near magnetic phase transitions is, with few exceptions, the so-called volume magnetostrictive or exchange striction effect, which is governed by $\partial J/\partial \varepsilon$. One can test this directly experimentally (see below) or one can infer this from other experiments (pressure dependence of T_c or T_N, forced magnetostriction experiments, anomalous thermal expansion, etc.) or one can give the following estimate: for many superexchange magnetic compounds one has for the magnetic Grüneisen parameter $d\ln J/d\ln V \simeq -3$ (Bloch 1966). This gives for $D_{ij} = \partial J/\partial \varepsilon_v \sim 50\,\mathrm{K}$ for many materials with $T_c, T_N \geqslant 10\,\mathrm{K}$. This exchange striction coupling constant is at least an order of magnitude larger than the g_Γ for S-state ions. For detailed comparisons for various materials see Lüthi et al. (1970). For non-S-state ions this is no longer the case (see §4.2).

One can obtain a more detailed expression than eq. (2.12a) by expanding the exchange Hamiltonian, eq. (2.3), for a sound wave with wavevector q:

$$\mathcal{H}_{\text{s-ph}} = A \sum_{ij} \left(\frac{\partial J}{\partial \boldsymbol{R}_{ij}} \cdot \boldsymbol{e}_q \right) (\boldsymbol{R}_{ij} \cdot \boldsymbol{q}) \mathrm{e}^{\mathrm{i}(\boldsymbol{q}\cdot\boldsymbol{R}_i)} (\boldsymbol{S}_i \cdot \boldsymbol{S}_j). \tag{4.12}$$

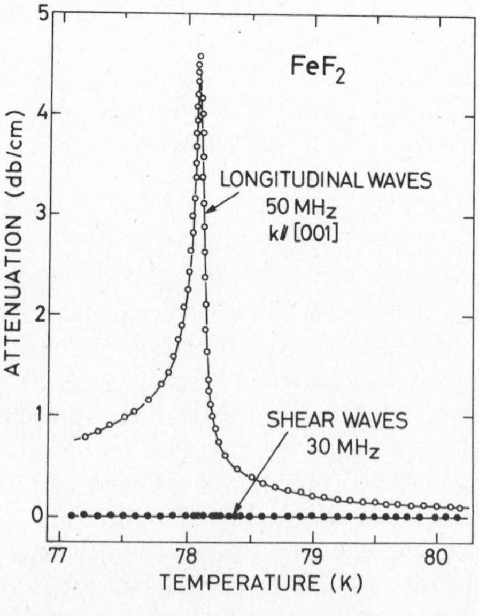

Fig. 9. Ultrasonic attenuation of longitudinal and shear waves propagating along the c-axis in FeF_2 in the vicinity of $T_N = 78.1\,\mathrm{K}$ (from Ikushima and Feigelson 1971).

Here $\boldsymbol{R}_{ij}$ is the vector connecting sites i,j; A is the sound wave amplitude and $\boldsymbol{e}_q$ the polarization vector. The occurrence of the products $[(\partial J/\partial \boldsymbol{R}_{ij})\cdot\boldsymbol{e}_q](\boldsymbol{R}_{ij}\cdot\boldsymbol{q})$ in eq. (4.12) eliminates shear waves for high symmetry propagation directions. This is what one observes normally. In fig. 9 we show experimental results for longitudinal and shear wave propagation along the tetragonal axis for FeF_2 in the vicinity of T_N (Ikushima and Feigelson 1971). These results clearly demonstrate the effect of the coupling mechanism, a large effect on longitudinal sound attenuation near T_N and a very small effect for shear waves, indicating the $D_{ij} \gg g_3$ in this case. Similar experiments have been performed for many other substances. There are a few notable exceptions (Lüthi et al. 1970, Tachiki et al. 1975): Cr, CoO, a-axis propagation in rare earth metals.

4.3.2. *Critical sound attenuation and dispersion*

In a typical experiment one measures the attenuation coefficient and velocity change for a sound wave as a function of temperature and frequency. There exist several reviews on this topic (Garland 1970, Lüthi et al. 1970, Kawasaki 1976). Critical anomalies are expected; because of the spin phonon interaction (eq. (4.12)), energy is transferred from the sound wave to the spin system, which has relaxation channels, whose relaxation times can diverge at the critical temperature. Hence this is reflected in a divergence of the attenuation α for $\omega\tau \ll 1$. From eq. (4.12) one finds the attenuation coefficient α to be proportional to the space–time Fourier transform of a four-spin correlation function. This can be written in two parts (Kawasaki 1976): for materials with short-range exchange interactions (predominantly insulators), $\partial J/\partial \varepsilon$ will also be short range and α_1 will be governed by the spin–lattice relaxation of the spin energy density $\tau_s^{-1} = \gamma/C_M$, with γ a constant and C_M the magnetic specific heat. On the other hand, for long-range exchange interactions the spin relaxation will be dominated by the critical spin relaxation time τ_c which diverges as $(T-T_c)^{-z\nu}$, where $z\nu$ ranges between 1 and $\frac{5}{3}$, depending on whether one deals with isotropic or anisotropic antiferromagnets or ferromagnets. Detailed calculations give for the paramagnetic phase for the two contributions:

$$\alpha_1 \sim \frac{\omega^2 C_M \tau_s}{1+\omega^2\tau_s^2}, \qquad \alpha_2 \sim \begin{cases} B_0\omega^2\tau_c, & \omega\tau_c \ll 1, \\ B_\infty \omega, & \omega\tau_c \gg 1. \end{cases} \tag{4.13}$$

α_1 gives a weak singularity like C_M^2 for $\omega\tau_s \ll 1$ and α_2 gives strong singularities with exponents $z\nu$ for $\omega\tau_c \ll 1$ obtained from mode–mode coupling theories. This is what one generally observes experimentally: for insulators like $RbMnF_3$, MnF_2, FeF_2, EuO, $Y_3Fe_5O_{12}$, $Cu_{0.4}Cd_{0.6}Fe_2O_4$, one observes exponents much smaller than $z\nu$ (Lüthi et al. 1970, Ikushima and Feigelson 1971, Moran and Lüthi 1971, Kamilov and Alier 1972, Kamilov et al. 1973). For some insulators the measured spin–lattice relaxation time, from a combination of critical attenuation and velocity change, $1/\tau = \omega^2 \Delta v / v_\infty^2 \alpha$, agrees rather closely with a calculated one (Huber 1971, Itoh 1975). For metallic substances on the other hand (Ni, Gd, Dy, Tb, Ho, Er), one usually observes large exponents, which are close to the appropriate value of $z\nu$. An interesting case is Cr_2O_3, where despite its being an insulator large exponents $z\nu \sim 1.3$ were found (Bachellerie and Frenois 1974). This can again be related to the rather long-range exchange interactions found in this insulating material.

For the low-frequency velocity change ($\omega\tau \ll 1$) one gets from thermodynamic arguments $\Delta v / v_\infty \sim -C_M$. Such weak singularities are found for sound velocity anomalies near magnetic phase transitions. For the higher frequencies one obtains for the dispersive part in the case of spin–lattice relaxation for $[v(\omega) - v(0)]/v(0)$ a relaxation type function.

In the ordered region a number of new effects are observed: apart from fluctuation contributions, which should be analogous to the ones in the paramagnetic region discussed above, strong domain wall-stress effects can be observed. In some favorable cases (MnF_2, MnP) the Landau–Khalatnikov order parameter relaxation can be isolated (Bachellerie et al. 1973, Golding 1975). This is being done by measuring the frequency-dependent maxima of the attenuation for $T < T_N, T_C$. The order parameter relaxation

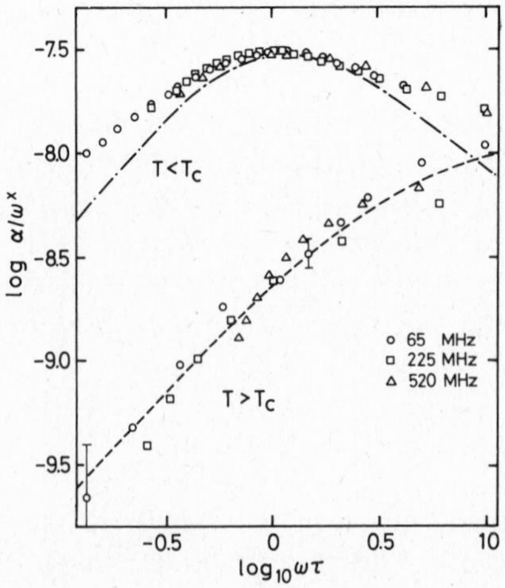

Fig. 10. Ultrasonic attenuation in MnP for $T < T_c$ (relaxation contribution) and for $T > T_c$ (fluctuation contribution) plotted in the scaling form $\alpha/\omega^{0.82}$ versus $\log \omega\tau$ (Golding 1975).

rates were found to be consistent with those from neutron scattering experiments. In fig. 10 a scaling test of ultrasonic attenuation in MnP is shown (Golding 1975). The curves $T>T_c$ and $T<T_c$ correspond to attenuation laws α_2 (eq. (4.13)) and order parameter relaxation, respectively. In addition a similar order parameter relaxation was also found in the paramagnetic phase of Ho and Tb in the presence of external magnetic fields (Maekawa et al. 1976).

Finally, recent acoustic measurements on lower-dimensional magnetic systems should be mentioned. In $CsNiCl_3$ ($T_N=4.3$ K) a broad attenuation maximum, peaked around 30 K, is interpreted, because of a similar specific heat anomaly, as due to a phonon–spin energy density coupling, leading to an attenuation α_1 discussed above (Almond and Rayne 1975). In $CsNiF_3$ an enormous softening of the Young's modulus E parallel to the linear chain was observed from room temperature down to 30 K (Barmatz et al. 1977). This is not yet explained.

4.4. Rotational invariance and magnetoelastic interaction

Another interesting application of CEF effects, discussed in §4.1, is the rotationally invariant magnetoelastic interaction. This type of interaction has now been tested for a number of magnetic systems, antiferromagnets and paramagnets. A quantitative comparison between experiment and theory, without adjustable parameters, has been recently made for the ideal paramagnet TmSb (Wang and Lüthi 1977), a substance we discussed in §4.1.

Rotational invariance in a magnetomechanical system was investigated for the first time in the famous Einstein–de Haas experiment (Einstein and de Haas 1915). In this experiment one measures the resulting rotation of a freely suspended magnetic wire, such as iron, due to the reversal of the direction of magnetization. Angular momentum conservation forces the rod to rotate if one changes the magnetization. Actually the authors took the angular momentum conservation for granted and considered this effect as an indication of the existence of Ampère molecular currents.

Total angular momentum conservation leads also to additional terms in the magnetoelastic Hamiltonian (Tiersten 1964, Brown 1965), involving the antisymmetric strain tensor ω_{ij}, which is not present in the non-invariant $\mathcal{H}_{me}$ of eq. (2.12). This becomes evident, if one derives eq. (2.12) from a point charge calculation, by expanding the CEF potential not only in terms of the symmetric strain tensor ε_{ij}, but rather in terms of the deformation tensor (§2.2). For a shear deformation (c_{44}) in octahedral symmetry one obtains in lowest order in $\varepsilon_{xz}, \omega_{xz}$, compared to eq. (2.12)

$$\mathcal{H}_{me} = -g_3\varepsilon_{xz}(J_xJ_z+J_zJ_x) - \tfrac{21}{6}g_3\varepsilon_{xz}\omega_{xz}(J_x^2-J_z^2).$$

One has a coupling, involving ω_{xz}, to new quadrupole operators. One also obtains this new magnetoelastic Hamiltonian by requiring rotational invariance for the total Hamiltonian (Dohm and Fulde 1975, Dohm 1976)

$$\mathcal{H}(\boldsymbol{J}_n,(1+\mathrm{v}_n)\boldsymbol{R}_n)=\mathcal{H}\big(\boldsymbol{J}_n,\mathrm{R}(1+2\eta)^{1/2}\boldsymbol{R}_n\big)$$

should be equivalent to

$$\mathbf{R}\mathcal{H}=\mathcal{H}\big(\mathbf{R}^{-1}\boldsymbol{J}_n,(1+2\eta)^{1/2}\boldsymbol{R}_n\big). \tag{4.14}$$

This means that the interaction of a magnetic ion with angular momentum $\boldsymbol{J}_n$, with the strained and rotated lattice, should be equal to the interaction of the reversely rotated spin $\boldsymbol{RJ}_n$ with the purely strained lattice. Here use has been made of eqs. (2.5), (2.5a). For the magnetoelastic Hamiltonian (eq. (2.12)), this leads to the rotation of J_x, J_z operators and results in

$$\mathcal{H}_{\mathrm{me}}=-g_3\varepsilon_{xz}(J_xJ_z+J_zJ_x)-2g_3\varepsilon_{xz}\omega_{xz}(J_x^2-J_z^2). \tag{4.15}$$

Eq. (4.15) is of the same form as the one resulting from the PCM calculation; however, it has a different numerical factor. This arises because, from the finite strain tensor (eq. (2.4)), one has additional contributions involving terms in g_2, which in a PCM expansion cannot be isolated from the g_3 terms.

One striking effect this rotationally invariant magnetoelastic interaction has on sound wave propagation is the lifting of the degeneracy of some elastic modes in the presence of a magnetic field, for propagation parallel and perpendicular to the applied field. This was first shown by Melcher (1970c, 1972) for MnF_2 at low temperatures in the antiferromagnetic state. In fig. 11 we show for the case of a cubic crystal and for c_{44} shear propagation along the fourfold axis [100] and [001], respectively, why this

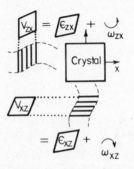

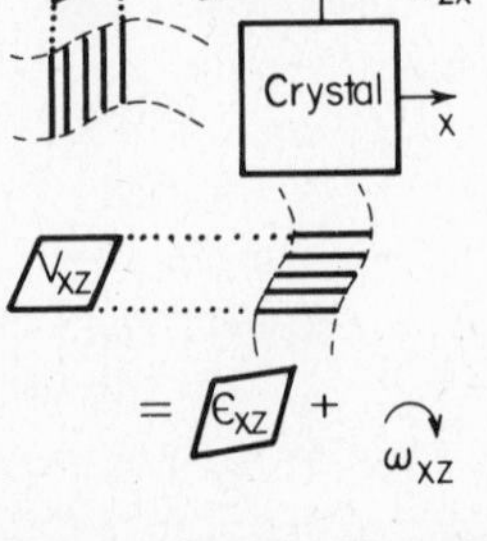

Fig. 11. c_{44}-mode propagating along x and z axis. Figure indicates that (k_z,u_x) has opposite rotational deformation to (k_x,u_z)-mode.

leads to a mearurable effect. It is seen that, for a given strain, the two plane waves have opposite rotational deformations. Therefore in a magnetic field, where the quadrupole operators $(J_x^2 - J_z^2)$, etc., have non-vanishing expectation values, or contribute to the strain susceptibility, the two terms add differently for the two modes. For non-invariant magnetoelastic interactions (eq. (2.12)), the two modes have the same magnetic field dependence, i.e. the degeneracy for this c_{44} mode is not lifted.

As an example we show experimental results for TmSb in fig. 12. Shown are relative sound velocity changes for the two c_{44} modes (k_z, e_x) and (k_x, e_z) together with theoretical calculations (k_j = component of $\boldsymbol{k}$-vector, e_i = component of polarization vector). The parameters which enter the theory (g_2, g_3, B_4) have been determined from other experiments as discussed in §4.1. Therefore there is no adjustable parameter left. The agreement, both in the field-dependent splitting of the modes as well as the field dependence of the modes themselves, is quite good. Other higher-order contributions (second-order magnetoelastic contributions, magnetoelastic coupling to octupole operators, etc.) are negligible in this case. Other modes such as $c_{11} - c_{12}$ show similar effects.

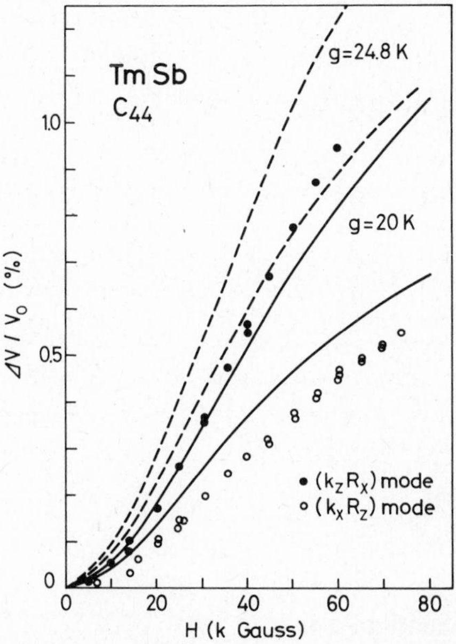

Fig. 12. Relative sound velocity changes for c_{44}-mode in TmSb at $T = 2\,\mathrm{K}$ in high magnetic fields. Full circles are mode $\boldsymbol{k} \| \boldsymbol{H}$, $\boldsymbol{u} \perp \boldsymbol{H}$, open circles $\boldsymbol{k} \perp \boldsymbol{H}$, $\boldsymbol{u} \| \boldsymbol{H}$. Full and dashed lines are calculations for different coupling constants g_3. Note that non-invariant theory would give no splitting of the two modes, but a field dependence which would be the average of the two modes (from Wang and Lüthi 1977).

Other systems have been investigated recently: $LnVO_4$ with Ln= Tm, Nd, Tb, Dy (Bonsall and Melcher 1976) and Pr_3Te_4 (Wang and Lüthi 1977). All these experiments show that rotational deformations can give elastic effects in the presence of a magnetic field, which cannot be explained by small-strain theory as given by eq. (2.12).

5. *Effects in the magnetically ordered region*

In magnetically ordered substances the interaction between phonons and spin waves leads to a number of interesting effects. These will be discussed in the subsequent sections. For the case of long-wavelength phonons and magnons it is convenient to use the continuum approximation, as we have done already for the case of sound waves before (§2.2, eqs. (2.6), (2.8)). Instead of the Hamiltonians, eqs. (2.2a), (2.3), (2.12), we use expressions for the free energy density $F = F_{el} + F_a + F_{me} + F_m$. Here F_{el} is the elastic energy, which for cubic symmetry is given by eq. (2.13). The anisotropy energy F_a, the magnetic energy F_m and the magnetoelastic energy F_{me} can again be deduced using symmetry arguments for a given crystal symmetry. For example, for cubic symmetry one gets for F_{me} in lowest order, in analogy to eq. (2.12) for a ferromagnetic or ferrimagnetic substance:

$$F_{me} = \frac{B_1}{M_0^2}\left[\left(M_x^2 - \frac{M_0^2}{3}\right)\varepsilon_{xx} + \left(M_y^2 - \frac{M_0^2}{3}\right)\varepsilon_{yy} + \left(M_z^2 - \frac{M_0^2}{3}\right)\varepsilon_{zz}\right]$$
$$+ \frac{B_2}{M_0^2}\left[M_x M_y \varepsilon_{xy} + M_x M_z \varepsilon_{xz} + M_y M_z \varepsilon_{yz}\right]. \quad (5.1)$$

Whereas the g_i of eq. (2.12) are temperature-independent coupling constants, the B_i of eq. (5.1) are temperature dependent. Eq. (5.1) can be obtained from eq. (2.12) by expanding the free energy in terms of the magnetoelastic coupling constants and factorizing the quadrupolar terms $\langle O_\Gamma \rangle \rightarrow \langle J_i \rangle \langle J_j \rangle$ (Callen and Callen 1963). For the exchange energy in the continuum approximation one gets

$$F_m = C_{ij} \frac{\partial \boldsymbol{M}}{\partial x_i} \cdot \frac{\partial \boldsymbol{M}}{\partial x_j}.$$

5.1. Magnon–phonon interaction

The coupled spin wave–phonon modes were first derived by Akhiezer et al. (1968) and by Kittel (1958). The equations of motion for the magnetization and the strain wave read

$$\frac{\partial \boldsymbol{M}}{\partial t} = \gamma[\boldsymbol{M} \times \boldsymbol{H}_{eff}], \qquad \frac{\rho \partial^2 \boldsymbol{u}}{\partial t} = \boldsymbol{f}. \quad (5.2)$$

Here H_{eff} is the field acting on the magnetization, $\boldsymbol{H}_{\text{eff}} = \boldsymbol{H}_{\text{ext}} - \partial F/\partial \boldsymbol{M}$, γ is the gyromagnetic ratio and $\boldsymbol{f}$ is given by

$$f_i = \frac{\partial T_{ij}}{\partial x_j} + \frac{\boldsymbol{M}\,\partial \boldsymbol{H}_{\text{eff}}}{\partial x_i}, \qquad T_{ij} = \frac{\partial F}{\partial \varepsilon_{ij}}.$$

The coupled sound wave–spin wave modes can be obtained by linearizing the equation of motion (5.2) and by neglecting complications due to rotational invariance. This has been done for the general case (propagation direction $\boldsymbol{k}$ making an angle θ with applied magnetic H^z_{ext}) by Schlömann (1960). We shall quote the final results below but first we try to discuss the physics using simple arguments: taking $H_{\text{ext}} = H^z_{\text{ext}}$ and setting $M_0 = M_z$ we notice from eq (5.1) that only the terms $F'_{\text{me}} = (B_2/M_0)(M_x\varepsilon_{xz} + M_y\varepsilon_{yz})$ are linear in the spin wave and strain wave amplitudes, therefore giving rise to a resonant spin wave phonon interaction. All the other terms are bilinear in the spin wave amplitudes and give therefore rise to higher-order phonon–spin wave interactions. One can distinguish various geometries with respect to sound wave propagation direction:

(a) Ferroacoustic resonance geometry: for the angle $\theta = \sphericalangle(\boldsymbol{k}, H_z) = 90°$ and taking $\boldsymbol{k} = (k_x, 0, 0)$ only ε_{xz} couples to spin waves but not ε_{yz}. From the equation of motion eq. (5.2) one obtains the coupled spin wave phonon dispersion equation

$$(\omega^2 - \omega_s^2)(\omega^2 - v^2k^2) = \frac{\gamma^2 B_2^2 k^2 H_k}{\rho M_0} \tag{5.3}$$

In this case the spin wave frequency $\omega_s = \gamma H_k(\gamma H_k + 4\pi\gamma M_0)$, $H_k = H_{\text{ext}} + H_{\text{ex}} a^2 k^2$ and v is the sound velocity. This leads to coupled spin wave phonon modes as shown in fig. 13. Near the crossover, the eigenmodes of the system have mixed phonon and magnon character, whereas far away from it the new eigenfrequencies approach again the values from the uncoupled modes. From eq. (5.3) one expects strong dispersive effects for sound wave or spin wave propagation in the region of the crossover. In addition, strong damping should be observed at the crossover. In order to observe this effect unambiguously, one has to make sure that the sample is in a single domain state for the resonance field. One has to use either high-frequency sound waves or materials with small M_0 and small anisotropy fields. Such effects have been observed in a number of substances: $Y_xGa_{3-x}Fe_5O_{12}$ (Lüthi and Oertle 1964), YIG (Lemanov et al. 1971). An example of this effect is shown in fig. 14a. The crossover frequency or ω_s can be changed with H_{ext}. The velocity dispersion has been observed for Fe_3O_4 and rare earth metals (Moran and Lüthi 1969, 1970) and $Y_3Fe_5O_{12}$ (Eastman 1966).

(b) Voigt–Cotton–Mouton geometry: closely related to the case discussed above is the linear birefringence geometry, which is the magneto-acoustic analogue of the magnetooptic Voigt–Cotton–Mouton effect.

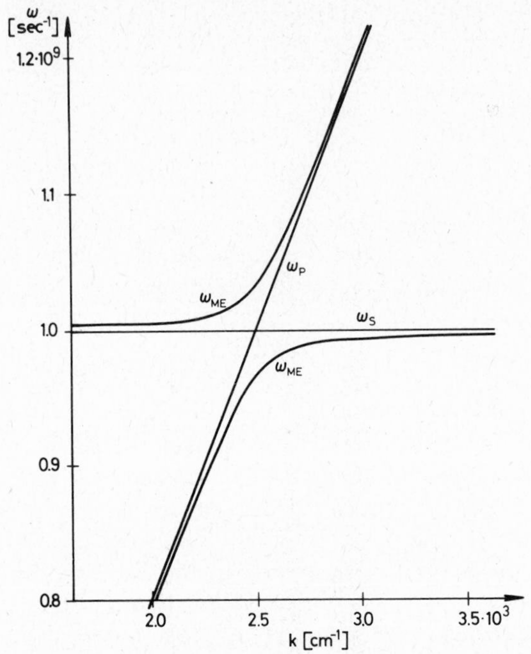

Fig. 13. Dispersion equations of spin waves ω_s, phonons ω_p and coupled magnetoelastic waves ω_{ME}, eq. (5.3). Parameters used are for YIG.

Again with $\boldsymbol{k}_x = (k_x, 0, 0)$, where the x-axis is a fourfold cubic axis, a shear wave (ε_{xz}) with polarization u_z couples to the spin waves whereas the one with u_y does not. Therefore the spin wave–phonon coupling induces birefringence. Depending on the phase between u_z and u_y (which can be changed with H_{ext}) the propagating sound wave is linearly, circularly or elliptically polarized. One can observe this effect as an amplitude modulation. It was observed in a number of systems: $Gd_3Fe_5O_{12}$, $Y_3Ga_{5-x}Fe_xO_{12}$, Fe_3O_4 (Lüthi 1965, 1966, Moran and Lüthi 1969), $RbNiF_3$ (Grishmanovskii et al. 1973). An example is shown in fig. 14b.

(c) Faraday geometry: for $\theta = 0$ one has the magnetoacoustic analogue of the magnetooptic Faraday effect (Kittel 1958). It follows from eqs. (5.1), (5.2) that the circularly polarized modes $m_s + \mathrm{i}m_y$ and $u_x + \mathrm{i}u_y$ couple. This leads to a coupled dispersion equation, similar to eq. (5.3),

$$(\omega - \omega_s)(\omega - v^2k^2) = \frac{\gamma B_2^2 k^2}{\rho M_0}, \tag{5.4}$$

where now $\omega_s = \gamma H_k$. For linearly polarized sound waves this leads to a rotation of the plane of polarization. This has been observed for a number

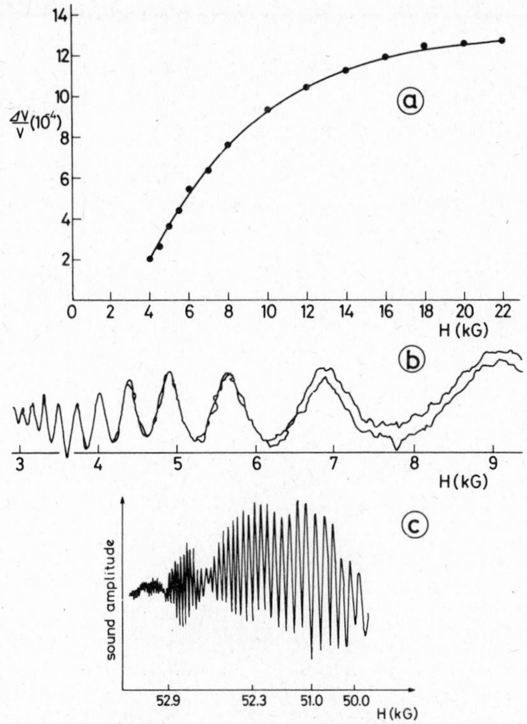

Fig. 14. Magnetoacoustic effects. (a) Sound velocity dispersion in Fe_3O_4 at $T=140$ K, $f=50$ MHz, $\boldsymbol{k}(100)$, $\boldsymbol{u}_t(001)$, $\boldsymbol{H}(001)$. Circles experimental, full line theory eq. (5.3) (adapted from Moran and Lüthi 1969). (b) Linear magnetoacoustic birefringence (Voigt geometry) for GdIG, $T=297$ K, $f=70$ MHz, echo no. 10, $\boldsymbol{k}(100)$, $\boldsymbol{u}(001)$, $\boldsymbol{H}(011)$. Phase difference between maxima is π (from Lüthi 1965). (c) Circular magnetoacoustic birefringence (Faraday geometry) for antiferromagnet Cr_2O_3, $T=4.2$ K, $f=8890$ MHz, $H \parallel c$-axis. Phase difference between maxima is $\pi/2$ (from Boiteux et al. 1971).

of ferrimagnetic substances: $Y_3Fe_5O_{12}$ (Matthews and Le Craw 1962, Lemanov et al. 1971), $Gd_3Fe_5O_{12}$ (Lüthi 1966). Fig. 14c shows a typical result for this geometry.

(d) Longitudinal sound waves couple only for $\theta \neq 0°$, $\neq 90°$ to spin waves, because only in these geometries do they have transverse components coupling to M_i. This has been experimentally verified for yttrium iron garnet, $Y_3Fe_5O_{12}$ (Lemanov et al. 1971).

From such experiments one can determine material parameters, such as the magnetoelastic coupling constants B_i and internal fields (anisotropy parameter). The agreement with values obtained from other experiments is close.

Effects similar to those discussed above for ferrimagnets can also be observed for antiferromagnetic substances, as long as the antiferromagnet has an easy axis or an easy plane of sublattice magnetization. These magnetoacoustic effects can be observed both below and above the spin–flop transition. Acoustic Faraday rotation has been observed in the uniaxial antiferromagnet Cr_2O_3 in both phases (Boiteux et al. 1971), and the magnetoacoustic Voigt effect has been observed in $MnCo_3$, which is an easy plane antiferromagnet (Gakel 1969). The magnetoacoustic Faraday effect has also been observed for a paramagnet: Ni^{2+} in MgO (Guermeur et al. 1968).

In antiferromagnets, such as $RbMnF_3$ and MnF_2, one observes strongly temperature- and field-dependent elastic constants (Melcher and Bolef 1969a, b, Melcher 1970b). These can be explained also with a magneto-elastic interaction similar to eq. (5.1), but also with coupling terms to the sublattice magnetization, which for cubic symmetry are

$$F_{\mathrm{me}}=\frac{B_1}{M_0}\sum_{i,j} M_{ij}^2\varepsilon_{ij}+\frac{B_3}{M_0^2}\sum_{j} M_{ij}M_{2j}\varepsilon_{jj}+\frac{B_2}{M_0^2}\sum_{i}(M_{ix}M_{iy}\varepsilon_{xy}+\text{c.p.})$$
$$+\frac{B_4}{M_0^2}(M_{ix}M_{2y}+M_{1y}M_{2x})\varepsilon_{xy}+\text{c.p.}, \tag{5.5}$$

where i denotes the sublattice index and j the Cartesian component. From the linearized coupled equations of motion for M_{ij} and u_i, similar to eq. (5.2), one can again obtain effective elastic constants. As an example we show the temperature dependence of the elastic constants $c_{11}-c_{12}$, c_{44} and c_{33} for tetragonal MnF_2 in fig. 15. Strongly temperature-dependent modes are shown for c_{44} in the ordered region and for $c_{11}-c_{12}$ in the paramagnetic region. This last result shows a precursor effect of a cubic–tetragonal lattice instability, similar to ZnF_2 and NiF_2 (Rimai 1977). The longitudinal mode c_{33} shows an anomaly at the phase transition T_N, an effect we discussed in §4.3.2.

A different method of studying the spin wave–phonon interaction in ordered substances opened up with the realization of pulse propagation of spin waves and magnetoelastic waves. In magnetic substances like YIG, with low acoustic and microwave losses, it was possible to launch and detect spin wave pulses. This was achieved utilizing a r.f.-dipole coupling to the r.f.-magnetization in inhomogeneous internal fields (Schlömann 1961, Eshbach 1963). With varying internal fields along the propagation direction, spin waves convert to magnetoelastic waves and vice versa. These experiments opened up many possibilities for potential applications (delay lines, parametric amplification, etc.). Almost all these experiments were performed on YIG. Reviews covering these aspects can be found in Damon and Van de Vaart (1965) and Strauss (1968).

There are many more experimental manifestations of the spin wave–phonon interaction, such as parallel pumping of magnetoelastic waves,

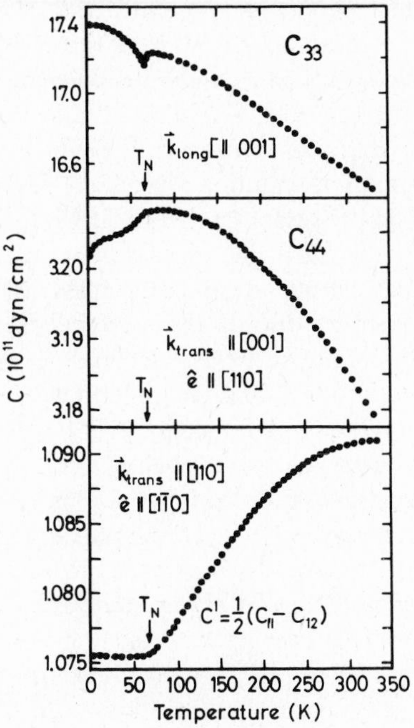

Fig. 15. Temperature dependence of elastic constants in MnF_2 (from Melcher and Bolef 1969a, b).

magnetostrictive generation of phonons, standing magnetoelastic wave resonances, optical diffraction by magnetoelastic waves, etc. (Bömmel and Dransfeld 1959, Turner 1960, Le Craw and Comstock 1965, Smith 1968, M. T. Elliott et al. 1974).

So far we have discussed long-wavelength magnon–phonon interactions. These were cases where the spin wave energy gap was relatively small, and the crossover was at such small k that the continuum approximation is still valid. There are cases where the spin wave energy gap is very large, so that the crossover is at large k-values. This is the case, for example, for magnetic compounds with unquenched orbital angular momentum (FeF_2, $FeCO_3$, etc.). In this case the resonant magnon–phonon interaction can be measured with inelastic neutron scattering. Instead of the coupled equations of eq. (5.2), one describes the resonant interaction with a spin wave–phonon interaction

$$\mathcal{H}_{\text{int}} = c_q(a_q b_q^+ + a_q^+ b_q) \tag{5.6}$$

(with a_q, b_q magnon and phonon operators) which can be diagonalized readily, giving again coupled dispersion equations similar to eqs. (5.3),

(5.4). The agreement between theory and experiment is good. These results were reviewed recently (Lovesey 1976). There should also be a crossover for spin waves with negligible energy gap, as long as the Brillouin zone boundary magnon is energetically higher than the phonon of the corresponding symmetry. Such a case seems not to have been observed yet.

Apart from the resonant interaction discussed so far, where one has one-magnon–one-phonon processes, non-resonant processes are also possible. From eqs. (5.1) and (2.12a) one can have one-phonon–two-magnon processes, which can affect both the phonon attenuation and the dispersion. Many theoretical papers discuss these effects (Kaganov and Chilkvashvili 1961, Pytte 1965, Ghatak 1972). However, it is difficult to check these predictions for various reasons (small effects, smooth temperature dependence, disturbing domain wall-stress effects, etc.). It is easier to discuss elastic constant changes using phenomenological models. If one assumes a single domain sample (by applying a magnetic field for ferro- and ferrimagnets), one can have from eq. (2.12a) exchange striction contributions to the elastic constants which have the same form as the pair contributions in the Fuchs theory of elastic constants. The c_{ij} are proportional to M^2 and the two parameters entering the expressions are $\partial J/\partial\varepsilon$, $\partial^2 J/\partial\varepsilon^2$, which can have either sign. Such contributions have been observed for a number of substances (Fe alloys, $GdAl_2$). Another effect which can occur is due to the stress-induced change of the magnitude of the spontaneous magnetization, which always leads to a lowering of the elastic constants (Alers et al. 1960, Hausch 1973).

5.2. Spin reorientation phenomena

Because of temperature-dependent anisotropy and magnetoelastic energies, the equilibrium spin configuration can also change as a function of temperature. Well-known examples of such spin reorientation phenomena are the spiral ferromagnetic spin structure changes in the heavy rare earth metals (Dy, Tb, Ho, Er) (Elliott 1972), mentioned above. Other examples are cubic magnetic crystals, where the easy axis of magnetization can change from, say, the (100) to (111) direction. Examples of this case are, e.g. magnetite Fe_3O_4 at 130 K, hematite Fe_2O_3 at 253 K. In all these cases the spin reorientation occurs discontinuously at a given temperature, involving typically a first-order orientational phase transition.

A particularly interesting spin reorientation phenomenon occurs in some rare earth orthoferrites ($LnFeO_3$, Ln = Er, Tm, Sm). In these orthorhombic compounds the competing Heisenberg and Dzyaloshinsky exchange mechanisms produce a slight canting of the antiferromagnetically coupled sublattices. The resulting weak magnetic moment $\boldsymbol{m}$ can rotate from one symmetry direction to another under the influence of temperature-dependent anisotropy energies. For example, in $ErFeO_3$, $\boldsymbol{m}$ points along the orthorhombic a-axis for $T<T_l=87\,\mathrm{K}$ and rotates continuously between $87\,\mathrm{K}<T<96\,\mathrm{K}$ in the ac-plane. For $T>T_u=96\,\mathrm{K}$, $\boldsymbol{m}$ points along the c-axis. This spin reorientation phenomenon can be described as a Landau-type second-order phase transition (Horner and Varma 1968, Levinson et al. 1969, Hornreich and Shtrikman 1976), the marginal dimensionality being $d^*=2.5$. The soft mode is the $\boldsymbol{k}=0$ acoustic spin wave mode, which, because of its coupling to elastic modes, hybridizes to a $\boldsymbol{k}=0$ magneto-elastic wave. The elastic energy for modes propagating along the c-axis can be written

$$F_{\mathrm{me}}=B\varepsilon_{zz}\sin^2\theta+b\varepsilon_{xz}\sin 2\theta. \tag{5.7}$$

Here θ is the angle $\boldsymbol{m}$ makes with the c-axis. The order parameter is θ for $T\sim T_u$ and $\frac{1}{2}\pi-\theta$ for $T\sim T_l$, while B,b are effective magnetoelastic coupling constants. It follows from eq. (5.7) that a longitudinal wave propagating along the c-axis (ε_{zz}) couples quadratically to the order parameter, whereas a shear wave ε_{xz} couples linearly. A ε_{yz} mode couples to the optical spin wave branch. The different types of coupling lead to different elastic behavior: the quadratic coupling to the order parameter leads to steplike discontinuities for the longitudinal c_{33} mode, the linear coupling for the ε_{xz} mode leads to a resonant interaction and therefore to a softening of the c_{55} mode, whereas the coupling to optical spin waves gives only small effects for the c_{44} mode. Calculations of these modes have been made (Gorodetsky and Lüthi 1970, Gorodetsky et al. 1971, 1976c), either using coupled equations of motion for spin waves and sound waves, or using thermodynamic arguments $c_{ij}=\partial^2F/\partial\varepsilon_i\,\partial\varepsilon_j$ with the order parameter equation $\partial F/\partial\theta=0$. Here $F=F_{\mathrm{el}}+F_{\mathrm{a}}+F_{\mathrm{me}}$, with F_{el} the elastic energy for an orthorhombic symmetry, $F_{\mathrm{a}}=k_u\cos 2\theta+k_b\cos 4\theta$, with k_u,k_b the two-fold and fourfold anisotropy constants. With $k_b>0$ and T-independent, and k_u approximately linearly T dependent between $T_l\leqslant T\leqslant T_u$, one obtains readily the Landau-type phase transitions sketched above and the temperature dependence of the various elastic modes. An example of such behavior is shown (Gorodetsky et al. 1976c) in fig. 16 for $TmFeO_3$, which shows very similar behavior to $ErFeO_3$. The different effects for the various elastic modes are clearly shown and the agreement with theory is very good. With applied magnetic fields one can shift the transition temperatures or even suppress the transition (Gorodetsky et al. 1971). For further details see the recent survey on spin reorientations by Belov et al. (1976).

5.3. Spin flop transitions, antiferromagnetic phase boundaries

Another application where magnetoelastic techniques prove fruitful is the investigation of antiferromagnetic phase boundaries. If one applies a

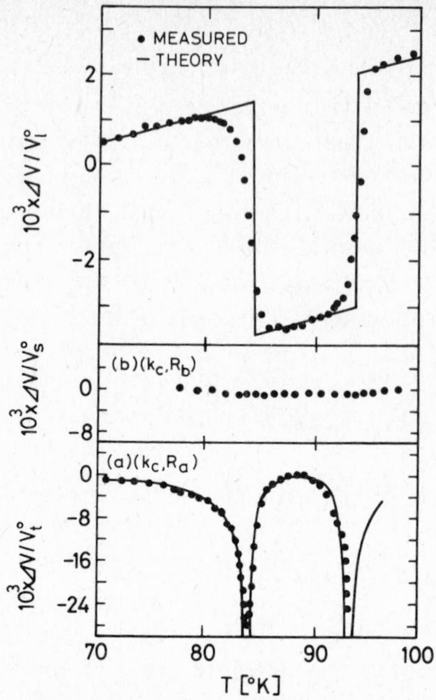

Fig. 16. Relative sound velocity changes in the spin reorientation region of $TmFeO_2$ for 3 modes, exhibiting the different couplings to the order parameter. Full lines are theoretical curves based on magnetoelastic energy eq. (5.7) (from Gorodetsky et al. 1976c).

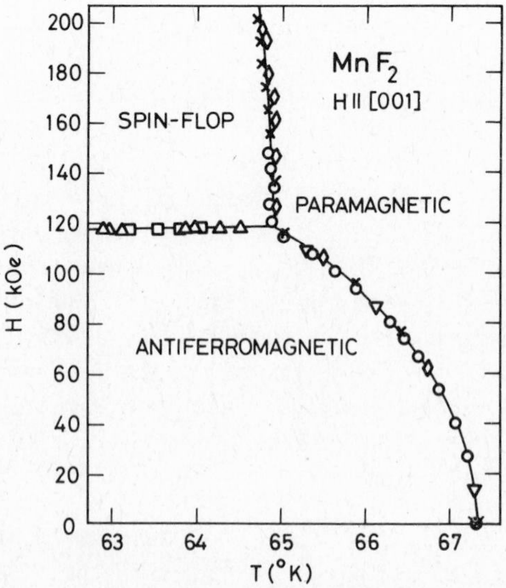

Fig. 17. Magnetic phase diagram of MnF_2 for ***H***(001). Data points from ultrasonic attenuation measurements (from Shapira 1971).

magnetic field along the easy axis of an uniaxial AF one drives the system for low enough temperatures into a canted spin flop state, because the gain in Zeeman energy outweighs the loss in exchange and anisotropy energy. As a function of temperature one obtains a phase diagram as shown in fig. 17. In this figure the magnetic phase diagram of MnF_2 with H along the easy axis (001) is shown. This phase diagram was obtained using ultrasonic attenuation. It demonstrates the usefulness of this method in this field of phase transitions, which receives considerable attention these days (Fisher 1974).

At the antiferromagnet–spin flop phase boundary one usually observes a sharp attenuation peak or an attenuation step function, accompanied by a sound velocity anomaly. Two processes have been proposed as possible mechanisms for these effects (see the review by Shapira 1971): coupling of ultrasonic waves to low-frequency spin waves and domain wall-stress effects. Which of the mechanisms is operative for a given material and a given geometry has not been established with certainty in any case, although detailed calculations have been carried out for the spin wave mechanism. Such ultrasonic anomalies at antiferromagnetic phase boundaries have been observed for MnF_2, FeF_2, CoF_2, Cr_2O_3, α-Fe_2O_3, Cr, EuTe and $GdAlO_4$ (Shapira 1971).

Such experiments, and related thermal expansion and susceptibility experiments, can be used to investigate multicritical points (bi-, tri- and tetracritical) (Fisher 1974) and the field dependence of $T_N(H)$. Such experiments have been performed on MnF_2, Cr_2O_3, $RbMnF_3$. See, for example, the recent paper by Shapira et al. (1978).

5.4. Nuclear acoustic resonance in magnetic solids

The technique of nuclear acoustic resonance (NAR) in non-magnetic solids has been reviewed by Bolef (1966). In antiferromagnetic crystals, NAR has been observed for the F^{19} nuclei in $KMnF_3$ (Denison et al. 1964), and $RbMnF_3$ (Melcher 1970a). In such experiments the absorption and dispersion of elastic waves by the F^{19} nuclear spin system can be observed only with very sensitive techniques. On the other hand, the Mn^{55} nuclear system couples strongly to the elastic system, giving rise to large NAR effects in compounds such as MnTe, $CsMnF_3$ (Walther 1971, 1973) and $RbMnF_3$ (Merry and Bolef 1969, Jimbo and Elbaum 1974). The interaction responsible for the coupling of elastic waves to the nuclear spin system in these compounds is the Silverstein mechanism (Silverstein 1963): an acoustic phonon interacts with the electronic spin system via the magnetoelastic interaction, creating a virtual magnon. The magnon is then coupled to the nuclear spin system and produces a nuclear spin transition via the hyperfine interaction.

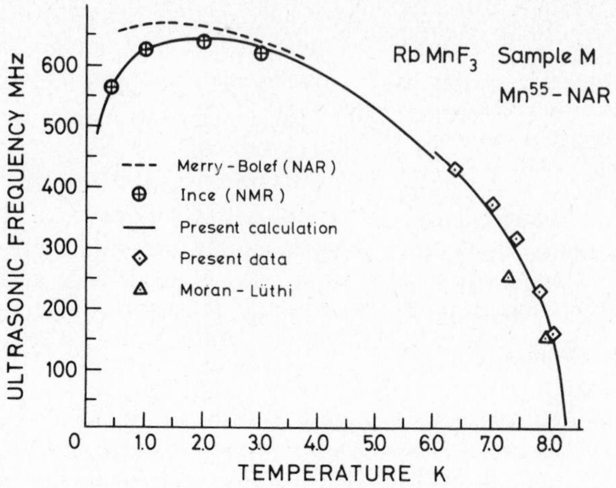

Fig. 18. Ultrasonic NAR frequency versus temperature for $RbMnF_3$ (from Jimbo and Elbaum 1974).

In fig. 18 the frequency–temperature relation for NAR in the absence of an applied field is shown for $RbMnF_3$. The strong temperature dependence can be explained as follows (de Gennes et al. 1963):

$$\nu = \nu_0\left[1 - 2\gamma_e^2 H_e H_N/\omega_{1,2}^2\right]^{1/2}. \tag{5.8}$$

Here ν_0 is the nuclear resonant frequency due to the hyperfine field, which in turn is proportional to the sublattice magnetization. γ_e is the electronic gyromagnetic ratio, H_e is the exchange field and H_N is the nuclear field due to nuclear polarization. $\omega_{1,2}$ are the two antiferromagnetic resonance modes. The temperature dependence arises through the temperature dependence of the hyperfine field, ν_0 in eq. (5.8) and from the strong coupling to the AF resonance modes $\omega_{1,2}$ (frequency pulling). Such NAR effects have not been observed yet in rare earth compounds.

Acknowledgement

Comments on the manuscript by J. Maetz, P. Morin, H. R. Ott and P. Thalmeier are gratefully acknowledged.

References

Abragam, A. and B. Bleaney (1970), *Electron paramagnetic resonance of transition ions* (Oxford University Press, Oxford).

Akhiezer, A. I., V. G. Baryakhtar and S. V. Peletminskii (1968), *Spin-waves* (North-Holland, Amsterdam).
Alefeld, B. (1969), Z. Phys. **222**, 155.
Alers, G. A., J. R. Neighbours and H. Sato (1960), J. Phys. Chem. Solids **13**, 40.
Allen, S. J. (1968), Phys. Rev. **167**, 492.
Almond, D. P. and J. A. Rayne (1975), Phys. Lett. **54A**, 295.
Als-Nielsen, J. and R. J. Birgeneau (1977), Am. J. Phys. **45**, 554.
Azaroff, L. V. (1968), *Elements of x-ray crystallography* (McGraw-Hill, New York).
Bachellerie, A. and Ch. Frenois (1974), J. de Physique **35**, 437.
Bachellerie, A., J. Joffrin and A. Levelut (1973), Phys. Rev. Lett. **30**, 617.
Barmatz, M., L. R. Testardi M. Eibschütz and H. J. Guggenheim (1977), Phys. Rev. B **15**, 4370.
Belov, K. P., A. K. Zvedzin, A. M. Kadomtseva and R. Z. Levitin (1976), Sov. Phys.-Uspekhi **19**, 574.
Bergman, D. J. and B. I. Halperin (1976), Phys. Rev. B **13**, 2145.
Birgeneau, R. J., E. Bucher, C. Passell and K. C. Turberfield (1971), Phys. Rev. B **4**, 718.
Birgeneau, R. J., J. K. Kjems, G. Shirane and L. G. van Uitert (1974), Phys. Rev. B **10**, 2512.
Bloch, D. (1966), J. Phys. Chem. Solids **27**, 881.
Boiteux, M., P. Doussineau, B. Ferry, J. Joffrin and A. Levelut (1971), Phys. Rev. B **4**, 3077.
Bolef, D. I. (1966), in *Physical acoustics*, Ed. by W. P. Mason and R. N. Thurston (Academic Press, New York), Vol. IVA.
Bolef, D. I. and J. G. Miller (1971), in *Physical acoustics*, Ed. by W. P. Mason and R. N. Thurston (Academic Press, New York), Vol. VIII.
Bömmel, H. and K. Dransfeld (1959), Phys. Rev. Lett. **3**, 83.
Bonsall, L. and R. L. Melcher (1976), Phys. Rev. B **14**, 1128.
Brändli, G. and R. Griessen (1973), Cryogenics **13**, 299.
Brandt, O. G. and C. T. Walker (1967), Phys. Rev. Lett. **18**, 11.
Brown, W. F., Jr. (1965), J. Appl. Phys. **36**, 994.
Callen, E. (1968), J. Appl. Phys. **39**, 519.
Callen, E. and H. B. Callen (1963), Phys. Rev. **129**, 578.
Callen, E. and H. B. Callen (1965), Phys. Rev. **139A**, 455.
Cardona, M., editor (1975), in *Light scattering in solids* (Springer Verlag, Berlin).
Clark, A. E. (1974), AIP Conf. Proc. **18**, 1015.
Comstock, R. L. (1965), Proc. IEEE **53**, 1508.
Cowley, R. A. (1976), Phys. Rev. B **13**, 4877.
Croft, M. C. and R. D. Parks (1977), in *Proceedings on valence instabilities and related narrow band phenomena*, Ed. by R. D. Parks (Plenum Press, New York).
Cullen, J. R. and A. E. Clark (1977), Phys. Rev. B **15**, 4510.
Damon, R. W. and H. Van de Vaart (1965), Proc. IEEE **53**, 348.
de Gennes, P. G., P. A. Pincus, F. Hartmann-Boutron and J. M. Winter (1963), Phys. Rev. **129**, 1105.
Denison, A. B., L. W. James, J. D. Currin, W. H. Tantilla and R. J. Mahler (1964), Phys. Rev. Lett. **12**, 244.
Dohm, V. (1976), Z. Phys. B **23**, 153.
Dohm, V. and P. Fulde (1975), Z. Phys. B **21**, 369.
Dolling, G. (1974), in *Dynamical properties of solids*, Ed. by G. K. Horton and A. A. Maradudin (North-Holland, Amsterdam), Vol. 1.
Dunitz, J. D. and L. E. Orgel (1957), J. Phys. Chem Solids **3**, 20.
Eastman, D. E. (1966), Phys. Rev. **148**, 530.
Eastman, D. E. (1967), Phys. Rev. **156**, 645.
Einstein, A. and W. J. de Haas (1915), Verh. d. D. Phys. Ges. **17**, 152.

Elliott, M. T., M. O'Donnell and H. A. Blackstead (1974), Phys. Rev. Lett. **32**, 734.
Elliott, R. J., editor (1972), *Magnetic properties of rare earth metals* (Plenum Press, New York).
Eshbach, J. R. (1963), J. Appl. Phys. **34**, 1298.
Faber, J., Jr., G. H. Lander and B. R. Cooper (1975), Phys. Rev Lett. **35**, 1770.
Fisher, M. E. (1974), AIP Conf. Proc. **24**, 273.
Folk, R., H. Iro and F. Schwab (1976), Z. Phys. B **25**, 69.
Freeman, A. R. and J. B. Darley, Jr., editors (1974), *The actinides*, (Academic Press, New York), Vol. I, II.
Fulde, P. (1978), in *Handbook on the physics and chemistry of rare earths*, Ed. by K. A. Gschneidner Jr. and LeRoy Eyring (North Holland, Amsterdam), ch. 17.
Fuller, E. R., Jr., A. V. Granato, J. Holder and E. R. Naimon (1974), *Methods of experimental physics*, Vol. 11.
Gakel, V. R. (1969), JETP Lett. **9**, 360.
Garland, C. W. (1970), in *Physical acoustics*, Ed. by W. P. Mason and R. N. Thurston (Academic Press, New York), Vol. VII.
Gehring, G. A. and K. A. Gehring (1975), Rep. Prog. Phys. **38**, 1.
Ghatak, S. K. (1972), Phys. Rev. B **5**, 3705.
Golding, B. (1975), Phys. Rev. Lett. **34**, 1102.
Goodenough, J. B. (1966), *Magnetism and the chemical bond* (Interscience Publishers).
Gorodetsky, G. and B. Lüthi (1970), Phys. Rev. B **2**, 3688.
Gorodetsky, G., B. Lüthi and T. J. Moran (1971), Int. J. Magnetism **1**, 295.
Gorodetsky, G., A. Shaulov, V. Volterra and J. Makovsky (1976a), Phys. Rev. B **13**, 1205.
Gorodetsky, G., B. Lüthi, M. Eibschütz and H. J. Guggenheim (1976b), Phys. Lett. **56A**, 479.
Gorodetsky, G., S. Shaft and B. M. Wanklyn (1976c), Phys. Rev. B. **14**, 2051.
Greiner, J. D., R. J. Schiltz, Jr., J. J. Tonnies, F. H. Spedding and J. F. Smith (1973), J. Appl. Phys. **44**, 3862.
Grishmanovskii, A. N., V. V. Lemanov, G. A. Smolenskii and P. P. Syrnikov (1973), Sov. Phys. – Solid State **14**, 2050.
Guermeur, R., J. Joffrin, A. Levelut and J. Penné (1968), Solid State Commun. **6**, 519.
Hausch, G. (1973), Phys. Stat. Sol. (a) **15**, 501.
Hirotsu, S. (1977), J. Phys. C **10**, 967.
Horner, H. and C. M. Varma (1968), Phys. Rev. Lett. **20**, 845.
Hornreich, R. M. and S. Shtrikman (1976), J. Phys. C **9**, L683.
Huber, D. L. (1971), Phys. Rev. B **3**, 836.
Hutchings, M. T. (1964), in *Solid state physics*, Ed. by F. Seitz, and D. Turnbull (Academic Press New York). Vol. 16.
Ikushima, A. and R. Feigelson (1971), J. Phys. Chem. Solids **32**, 417.
Itoh, Y. (1975), J. Phys. Soc. Japan **38**, 336.
Jimbo, T. and C. Elbaum (1974), Phys. Rev. B **10**, 2131.
Kaganov, M. I. and Ya. M. Chikvashvili (1961), Sov. Phys. – Solid State **3**, 200.
Kamilov, I. K. and Kh. K. Aliev (1972), Zh. ETF-Fis. Red. **15**, 506.
Kamilov, I. K., Kh. K. Aliev and G. M. Shakhshaev (1973), Sov. Phys. – Solid State **15**, 632.
Kataoka, M. and J. Kanamori (1972), J. Phys. Soc. Japan **32**, 113.
Kawasaki, K. (1976), in *Phase transitions and critical phenomena*, Ed. by C. Domb and M. S. Green (Academic Press, New York), Vol. 5a.
Kino, Y. and S. Miyahara (1966), J. Phys. Soc. Japan **21**, 2732.
Kino, Y., M. E. Mullen and B. Lüthi (1972), J. Phys. Soc. Japan **33**, 687.
Kino, Y., M. E. Mullen and B. Lüthi (1973), Solid State Commun. **12**, 275.
Kittel, C. (1958), Phys. Rev. **110**, 836.
Kollewe, D. and W. M. Gibson (1978), Phys. Lett. **65A**, 253.
Kroese, C. J. and W. J. A. Maaskant (1974), Chem Phys. **5**, 224.

Le Craw, R. C. and R. L. Comstock (1967), in *Physical acoustics*, Ed. by W. P. Mason (Academic Press, New York), Vol. IIIB.

Lemanov, V. V., A. V. Pawlenko and A. N. Grishmanovskii (1971), JETP **32**, 389.

Levinson, L. M., M. Luban and S. Shtrikman (1969), Phys. Rev. **187**, 715.

Levy, P. (1973), J. Phys. C **6**, 3545.

Lovesay, S. W. (1976), Comments Solid State Phys. **7**, 117.

Ludwig, W. (1970), *Festkörperphysik I* (Akademische Verlagsgesellschaft, Frankfurt a.M.).

Lüthi, B. (1965), Appl. Phys. Lett. **6**, 234.

Lüthi, B. (1966), J. Appl. Phys. **37**, 990.

Lüthi, B. and F. Oertle (1964), Phys. Kondens. Materie **2**, 99.

Lüthi, B. (1976), AIP Conf. Proc. **34**, 7.

Lüthi, B., T. J. Moran and R. J. Pollina (1970), J. Phys. Chem. Solids **31**, 1741.

Lüthi, B., M. E. Mullen, K. Andres, E. Bucher and J. P. Maita (1973), Phys. Rev. B **8**, 2639.

Lüthi, B., P. S. Wang, Y. H. Wong, H. R. Ott and E. Bucher (1977), in *2nd international conference on CEF effects in metals and alloys*, Ed. by A. Furrer (Plenum Press, New York).

Maekawa, S., R. A. Treder, M. Tachiki, M. C. Lee and M. Levy (1976), Phys. Rev. B **13**, 1284.

Mason, W. P. (1954), Phys. Rev. **96**, 302.

Matthews, H. and R. C. Le Craw (1962), Phys. Rev. Lett. **8**, 397.

Melcher, R. L. (1970a), Phys. Rev. B **1**, 4493.

Melcher, R. L. (1970b), Phys. Rev. B **2**, 733.

Melcher, R. L. (1970c), Phys. Rev. Lett. **25**, 1201.

Melcher, R. L. (1972), in *Proceedings of the international school of physics "Enrico Fermi", Varenna*, Ed. by E. Burstein (Academic Press, New York), Course LII.

Melcher, R. L. (1976), in *Physical acoustics*, Ed. by W. P. Mason and R. N. Thurston (Academic Press, New York), Vol. XII.

Melcher, R. L. and D. I. Bolef (1969a), Phys. Rev. **178**, 864.

Melcher, R. L. and D. I. Bolef (1969b), Phys. Rev. **186**, 491.

Melcher, R. L. and B. A. Scott (1972), Phys. Rev. Lett. **28**, 607.

Melcher, R. L., E. Pytte and B. A. Scott (1973), Phys. Rev. Lett. **31**, 307.

Merry, J. B. and D. I. Bolef (1969), Phys. Rev. Lett. **23**, 126.

Moran, T. J. and B. Lüthi (1969), Phys. Rev. **187**, 710.

Moran, T. J. and B. Lüthi (1971), Phys. Rev. B **4**, 122.

Moran, T. J. and B. Lüthi (1970), J. Phys. Chem. Solids **31**, 1735.

Morin, P., J. Rouchy and D. Schmitt (1978), Phys. Rev. B **17**, 3684.

Mullen, M. E., B. Lüthi, P. S. Wang, E. Bucher, L. D. Longinotti, J. P. Maita and H. R. Ott (1974), Phys. Rev. B **10**, 186.

Nickolson, K., U. Häfner, E. Müller-Hartmann and D. Wohlleben (1978), Phys. Rev. Lett. **41**, 1325.

Ott, H. R. and B. Lüthi (1977), Z. Phys. B **28**, 141.

Penney, T., R. L. Melcher, F. Holtzberg and G. Güntherodt (1975), AIP Conf. Proc. **29**, 392.

Phillips, T. G. and R. L. White (1967a), Phys. Rev. **153**, 616.

Phillips, T. G. and R. L. White (1967b), J. Appl. Phys. **38**, 1222.

Pytte, E. (1965), Ann. Phys. (N. Y.) **32**, 377.

Rado, G. T. and H. Suhl, editors (1963–73), *Magnetism* (Academic Press, New York), Vol. I–[illegible]

Re[illegible]. A., C. A. Wert and M. Metzger (1974), *Methods of experimental physics*, (Academic Press, New York), Vol. 6a.

Rimai, D. S. (1977), Phys. Rev. B **16**, 4069.

Sandercock, J. R. (1975), in *Festkörperprobleme XV*, Ed. by H. J. Queisser (Pergamon-Vieweg, Braunschweig), p. 183.

Sandercock, J. R., S. B. Palmer, R. J. Elliott, W. Hayes, S. R. P. Smith and A. P. Young (1972), J. Phys. C **5**, 3126.
Schaack, G. (1977), Z. Physik B **26**, 49; 1977, Physica **89B**, 195.
Schlömann, E. (1960), J. Appl. Phys. **31**, 1647.
Schlömann, E. (1961), in *Adv. quantum electronics*, Ed. by J. R. Singer (Columbia University Press), p. 437.
Shamir, N., M. Melamud, H. Shaked and M. Weger (1978), Physica **94B**, 225.
Shapira, Y. (1971), J. Appl. Phys. **42**, 1588.
Shapira, Y. and C. C. Becerra (1978), Phys. Rev. B **16**, 4920.
Sheard, F. W., T. F. Smith, G. K. White and J. A. Birch (1977), J. Phys. C **10**, 645.
Silverstein, S. D. (1963), Phys. Rev. **132**, 997.
Smith, A. B. and R. V. Jones (1963), J. Appl. Phys. **34**, 1283.
Smith, A. W. (1968), Phys. Rev. Lett. **20**, 334.
Stevens, K. W. H. (1952), Proc. Phys. Soc. **A65**, 209.
Stevens, K. W. H. and E. Pytte (1973), Solid State Comun. **13**, 101.
Strauss, W. (1968), in *Physical acoustics*, Ed. by W. P. Mason (Academic Press, New York), Vol. IVB.
Sturge, M. (1967), in *Solid state physics*, Ed. by F. Seitz and D. Turnball (Academic Press, New York), Vol. 19.
Tachiki, M., S. Maekawa, R. Toeder and M. Levy (1975), Phys. Rev. Lett. **34**, 1579.
Taylor, K. N. R. (1972), *Physics of rare earth solids* (Chapman Hall, London).
Taylor, R. H. (1975), Advan. Phys. **24**, 681.
Terauchi, H., M. Mori and Y. Yamada (1972), J. Phys. Soc. Japan **32**, 1049.
Thalmeier, P. and P. Fulde (1975), Z. Phys. B **22**, 359.
Thalmeier, P. and P. Fulde (1977), Z. Phys. B **26**, 323.
Thomas, H. (1977), in *Electron–phonon interactions and phase transitions*, Ed. by T. Ristel (Plenum Press, New York).
Tiersten, H. F. (1964), J. Math. Phys. **5**, 1298.
Tinkham, M. (1964), *Group theory and quantum mechanics* (McGraw-Hill, New York).
Truell, R., C. Elbaum and B. B. Chick (1969), *Ultrasonic methods in solid state physics* (Academic Press, New York).
Tucker, E. B. (1967), in *Physical acoustics*, Ed. by W. P. Mason and R. N. Thurston (Academic Press, New York), Vol. IVA.
Turner, E. H. (1960), Phys. Rev. Lett. **5**, 100.
Van Vleck, J. H. (1930), *Electric and magnetic susceptibilities*, (Oxford University Press).
Van Vleck, J. H. (1939), J. Chem. Phys. **7**, 72.
Wallace, D. C. (1972), *Thermodynamics of crystals* (Wiley, New York).
Waller, I. (1932), Z. Phys. **79**, 370.
Wang, P. S. and B. Lüthi (1977), Phys Rev. B **15**, 2718.
Walther, K. (1971), Phys. Rev. B **4**, 3873.
Walther, K. (1973), Phys. Stat. Sol. (b) **58**, K1.
White, G. K. (1961), Cryogenics **1**, 151.

Author Index

Subject Index